Acta Physica Austriaca

Supplementum XXIV

Proceedings of the
XXI. Internationale Universitätswochen für Kernphysik 1982
der Karl-Franzens-Universität Graz
at Schladming (Steiermark, Austria)
February 25th—March 6th, 1982

Sponsored by
Bundesministerium für Wissenschaft und Forschung
Steiermärkische Landesregierung
International Centre for Theoretical Physics, Trieste
Sektion Industrie der Kammer der
Gewerblichen Wirtschaft für Steiermark

1982

Springer-Verlag
Wien GmbH

Electroweak Interactions

Edited by
Heinrich Mitter, Graz

With 88 Figures

1982

Springer-Verlag
Wien GmbH

Organizing Committee

Chairman

Prof. Dr. H. Mitter
Institut für Theoretische Physik
der Universität Graz

Committee Members

F. Gesztesy
L. Pittner
K. Schwarz
F. Widder
H. Zankel

Secretary

Mrs. E. Neuhold
Miss E. Tandl

Originally published by Springer-Verlag/Wien in 1982.
Softcover reprint of the hardcover 1st edition 1982

Library of Congress Cataloging in Publication Data

Internationale Universitätswochen für Kernphysik der Karl-Franzens-Universität Graz (21st : 1982 ; Schladming, Austria)
Electroweak interactions.

(Acta physica Austriaca. Supplementum ; 24)
Includes bibliographical references.
1. Electromagnetic interactions—Congresses.
2. Weak interactions (Nuclear physics)--Congresses.
I. Mitter, Heinrich. II. Austria. Bundesministerium für Wissenschaft und Forschung. III. Title. IV. Series.
QC794.8.E4I63 1982 539.7'54 82-19205

ISSN 0065-1559
ISBN 978-3-7091-4033-8 ISBN 978-3-7091-4031-4 (eBook)
DOI 10.1007/978-3-7091-4031-4

CONTENTS

FOREWORD

The papers contained in this volume are invited lectures presented at the 21st "Universitätswochen für Kernphysik" in Schladming in February 1982. To consider electromagnetic and weak interactions as manifestations of a single theory is a standpoint, which is generally accepted by now. The goal of the school was to outline the present state of this unified theory and to discuss possible future developments. Thanks to the generous support provided by the Austrian Ministry of Science and Research, the Styrian Government and other sponsors, it was again possible to invite experts in the field as lecturers. The lecture notes have been reexamined by the authors and are now published in their final form to enable a larger number of physicists to profit from them. Since the lectures are already quite voluminous, we have decided to restrict the publication to the lectures themselves and omit all seminars, interesting as they were, as well as all details connected with the meeting. It is a pleasure to thank all the lecturers for their efforts, making it possible to speed up publication. Thanks are also due to L.Pittner for organisation and proof-reading as well as to Mrs. Krenn and Mrs. Neuhold for the careful typing of the papers.

H.Mitter

Acta Physica Austriaca, Suppl. XXIV, 3–62 (1982)

INTRODUCTION TO GAUGE THEORIES OF ELECTRO-WEAK INTERACTIONS[+]

by

G. ECKER
Institut für Theoretische Physik
Universität Wien, Austria

TABLE OF CONTENTS

I. INTRODUCTION

The last decade has witnessed an enormous progress in our understanding of the fundamental interactions. Fifteen years ago it would have been very bold if not

---

[+]Lectures given at the XXI. Internationale Universitätswochen für Kernphysik, Schladming, Austria, February 25 - March 6, 1982.

presumptuous to assert that all of particle physics can be described in terms of gauge theories. Today we speak of the standard gauge model of electromagnetic, weak and strong interactions. In this development, the way was undoubtedly led by the unification of weak and electromagnetic forces into a gauge theory of electroweak interactions [1].

These lectures are meant to serve a twofold purpose. First, they contain an introduction to electroweak gauge theories for those participants who are not (or not yet) specialized in particle theory. Hopefully, this introduction will also be useful as a reminder for others. For pedagogical reasons I shall not follow strictly the historical development of the subject but rather try to emphasize the physical requirements which almost inevitably lead to the concept of gauge theories.

The second aim of these introductory lectures is to provide a background for the more advanced courses of this School, in particular on grand unified theories and on dynamical symmetry breaking [2]. Therefore, the emphasis will mostly be on the properties of a general gauge theory, and the standard model [3] will then be discussed as the simplest example to illustrate those properties. The lectures will only be concerned with the perturbative aspects of gauge theories,and the phenomenological implications will be treated in more detail by other lecturers [4].

In Section II the unitarity problem of the V-A Fermi theory [5] of charged weak currents is discussed as the motive for the introduction of charged intermediate vector bosons. The requirement of smooth high energy behaviour for the Born terms of a general theory of massive vector bosons interacting with fermions leads to a Lie algebra structure for the various couplings. For longitudinally polarized vector bosons there are however residual divergences in the Born terms at high energies. It is argued that the introduction of scalar fields with suitable couplings to fermions and vector fields may solve that problem.

Starting from the Schrödinger equation in the presence of an electromagnetic field the general gauge field theory is set up in Section III. Since gauge invariance forbids an explicit mass term for vector fields in the Lagrangian, the gauge symmetry must be spontaneously broken. Spontaneous symmetry breaking (SSB) in quantum mechanics is discussed by studying a linear chain of oscillators. It is shown that, strictly speaking, SSB can only occur in an infinite system. We investigate the simplest example of a spontaneously broken gauge field theory (SBGT), the abelian Higgs model [6], and discuss the limit in which the gauge field decouples. The similarities to solid state physics are stressed. Generalizing the Higgs model to an arbitrary non-abelian gauge theory it is argued that SSB does not change the high energy structure of the symmetric theory leading to a renormalizable quantum field theory. The quantization of gauge theories is reviewed only qualitatively emphasizing the dual roles of unitary and renormalizable gauges.

Using the principle of simplicity we arrive at the minimal gauge group of electroweak interactions in Section IV. The weak neutral current of the SU(2) × U(1) gauge theory is determined. Calculating the vector boson masses for a general Higgs potential we find that the strength of neutral currents is accounted for by a single scalar SU(2) doublet. In addition, the standard model [3] has all left-handed fermions in doublets and right-handed fermions as singlets. A few comments are made concerning the alternative left-right symmetric gauge theory [7] based on SU(2) × SU(2) × U(1).

In Section V fermion mass generation is treated for a general gauge theory with scalar fields. Diagonalization of the mass matrix leads to a set of neutral Majorana fermions and a number of charged Dirac fermions. The diagonalizing unitary matrix mixes fermions of the same charge leading in general to flavour changing neutral currents. The experimentally indicated requirement of natural flavour conservation imposes stringent constraints which are discussed for SU(2) ×

× U(1). Properties of the mixing matrix in charged currents are noted including a comment about CP violation in the standard model with three generations [8]. Finally, we make some remarks about neutrino masses and the generation structure of quarks and leptons.

Section VI contains a rather extensive discussion of anomalies in quantum field theory starting from two-dimensional QED. The necessary regularization of quantum field theory is responsible for anomalous divergences of currents which are conserved on the classical level. However, such anomalies are shown to clash with the requirement of perturbative unitarity in gauge theories. The general condition for the absence of anomalies for currents associated with generators of the gauge group is derived. Application to SU(2) × U(1) produces another strong indication for the generation structure of quarks and leptons. Although the anomalies must cancel for the gauge currents it is emphasized that there are in general other currents in the theory associated with global symmetries which possess anomalous divergences.

In the last Section VII radiative corrections to the vector boson masses M_W, M_Z and to the Weinberg angle Θ_W are considered. The importance of distinguishing between different renormalization conditions in the literature is stressed. The dependence of the mass shifts on the Higgs mass and on the quark masses is examined. The mass shifts due to one-loop radiative corrections can be determined rather accurately and turn out to be sizeable. Through the calculations of radiative corrections the standard model is now in good shape to be tested at the one-loop level by accurate measurements of M_W, M_Z and $\sin^2\Theta_W$.

II. MOTIVATION FOR GAUGE THEORIES OF ELECTROWEAK INTERACTIONS

The motivation for a gauge theory of weak interactions has its origin in an inherent short-coming of the V-A Fermi theory [5]$^+$

$$L_{V-A} = -\frac{G_F}{\sqrt{2}} j_{CC\mu} j_{CC}^{\mu\dagger} \tag{2.1}$$

with the charged current

$$j_{CC}^{\mu} = \sum_{\text{leptons}} \bar{\psi}_{\nu_\ell} \gamma^{\mu}(1 + \gamma_5)\psi_\ell + j_{hadr}^{\mu} \tag{2.2}$$

and the Fermi coupling constant $G_F = 1.1663 . 10^{-5}$ GeV^{-2}. This shortcoming is known as the unitarity problem of the Fermi theory and it first [9] shows up in the reaction

$$\nu_\mu + e^- \rightarrow \nu_e + \mu^- \quad . \tag{2.3}$$

To lowest order in (2.1)

$$\sigma_{tot}^{V-A}(\nu_\mu e^- \rightarrow \nu_e \mu^-) = 2G_F^2 \, s/\pi \tag{2.4}$$

for large enough center of mass energy $E_{CM} = \sqrt{s}/2$ such that lepton masses can be neglected. Because of the point coupling (2.1) only the lowest partial wave ($\ell = 0$) can contribute to the scattering amplitude. But from the conservation of probability (unitarity) in quantum mechanics we know that

$$\sigma_{inelastic}^{\ell=0} \leq 4\pi/s \tag{2.5}$$

for any scattering process. Since

$^+$The summation convention applies to all indices appearing more than once in an expression unless stated otherwise or if the index appears only once on the other side of an equation.

$$\sigma^{V-A}_{tot}(\nu_\mu e^- \to \nu_e \mu^-) \leq \sigma^{\ell=0}_{inelastic}(\nu_\mu e^-)$$

we find that (2.4) is consistent with unitarity only for

$$E_{CM} \leq \left(\frac{\pi\sqrt{2}}{4G_F}\right)^{1/2} = 309 \text{ GeV} . \qquad (2.6)$$

On the other hand, why should we believe the first-order result for such high energies? Unfortunately, it is not a matter of belief but an unpleasant matter of fact that we cannot calculate higher-order contributions. The quantum field theory based on the Lagrangian (2.1) is nonrenormalizable, it does not allow for a well-defined perturbation expansion.

At this point we fall back upon the most successful quantum field theroy we have at our disposal: QED. As far back as 1935 [10] it was suggested to consider (2.1) only as a low energy approximation to the more fundamental interaction

$$L = g_{CC}\, j^\mu_{CC}\, W_\mu + \text{h.c.} \qquad (2.7)$$

with a charged vector boson W of mass M_W. Consistency with (2.1) at low energies requires $g^2_{CC}/M^2_W = G_F/\sqrt{2}$. Although W is a massive charged field the resemblance of (2.7) to QED is obvious.

Working out the scattering amplitude for (2.3) one gets

$$A(s, \cos\Theta_{CM}) \underset{s>>m^2_\mu}{\sim} \frac{g^2_{CC}}{1 + 2M^2_W/s - \cos\Theta_{CM}} \qquad (2.8)$$

instead of $A \sim G_F s$ for the V-A theory. The limit $s \to \infty$ exists for all scattering angles Θ_{CM} except in the forward direction and the unitarity problem is postponed to ultra-high energies.

Having been rescued from the unitarity abyss by the introduction of the <u>intermediate vector boson</u> the question arises whether we should already be satisfied. As a charged particle the W will of course interact with the electromagnetic field and so we have to look a little further into the possible couplings of the W.

Due to its spin a massive vector field differs in two important aspects from a scalar field as far as calculating S-matrix elements in perturbation theory is concerned. For one, the propagator of a vector particle in momentum space is

$$\Delta_{\mu\nu}(k) = \frac{-g_{\mu\nu} + k_\mu k_\nu/M_W^2}{k^2 - M_W^2 + i\varepsilon} \tag{2.9}$$

instead of $(k^2-m^2+i\varepsilon)^{-1}$ for a scalar field and therefore does not fall off for large momenta k. Secondly, S-matrix elements contain the wave functions characterizing the spin state of external vector particles. If the spin projection on the direction of momentum vanishes (helicity zero) the corresponding wave function is called the longitudinal polarization vector $\varepsilon_\mu^L(k)$. Whereas the other two (transverse) polarization vectors approach finite limits for $k \to \infty$ one obtains

$$\varepsilon_\mu^L(k) \xrightarrow[k\to\infty]{} k_\mu/M_W \quad . \tag{2.10}$$

For the high energy behaviour of an S-matrix element each longitudinally polarized vector boson will therefore produce one extra power of momentum.

Let us now investigate the possible self-couplings of a set of massive hermitian vector fields W_μ^α $(\alpha = 1,\dots,N)$. Because each W_μ^α has canonical dimension one we can have at most quartic couplings if we wish to avoid coupling constants with negative mass dimensions in order not to jeopardize renormalizability from the very beginning. The most general interaction has the form [11]

$$L_{int}(W) = \frac{1}{2} a_{\alpha\beta\gamma} W_\mu^\gamma (W^{\beta,\nu} \partial^\mu W_\nu^\alpha - (\alpha \leftrightarrow \beta)) +$$

$$+ \frac{1}{2} b_{\alpha\beta\gamma} W_\mu^\gamma (W^{\beta,\nu} \partial^\mu W_\nu^\alpha + (\alpha \leftrightarrow \beta)) +$$

$$+ \frac{1}{2} c_{\alpha\beta\gamma} \varepsilon^{\mu\nu\lambda\rho} W_\mu^\alpha W_\nu^\beta \partial_\lambda W_\rho^\gamma + \frac{1}{4} d_{\alpha\beta\gamma\delta} W_\mu^\alpha W^{\beta,\mu} W_\nu^\gamma W^{\delta,\nu} +$$

$$+ \frac{1}{4} e_{\alpha\beta\gamma\delta} \varepsilon^{\mu\nu\lambda\rho} W_\mu^\alpha W_\nu^\beta W_\lambda^\gamma W_\rho^\delta \tag{2.11}$$

with obvious symmetry relations for the real coupling matrices a, b, c, d, e. Taking into account the derivative cubic couplings and remembering (2.9) and (2.10) one finds that the Born terms for WW → WW shown in Fig. 1 may diverge as badly as k^6 for large k. However, for certain values of the coupling constants miraculous cancellations take place. It turns out [11] that the most gentle high energy behaviour can be obtained if the following relations hold:

$$b = c = e = 0 \tag{2.12a}$$

$a_{\alpha\beta\gamma}$ totally antisymmetric (2.12b)

$$a_{\alpha\beta\gamma} a_{\alpha\delta\varepsilon} + a_{\alpha\beta\delta} a_{\alpha\varepsilon\gamma} + a_{\alpha\beta\varepsilon} a_{\alpha\gamma\delta} = 0 \tag{2.12c}$$

$$2d_{\alpha\beta\gamma\delta} = a_{\varepsilon\alpha\delta} a_{\varepsilon\gamma\beta} + a_{\varepsilon\alpha\gamma} a_{\varepsilon\delta\beta} \ . \tag{2.12d}$$

Note that even with the relations (2.12) the scattering amplitude for WW → WW still diverges like k^2 if all external particles are longitudinally polarized.

What about the couplings to fermions as in (2.7)? The most general interaction of dimension ≤ 4 is

$$L_{int}(\psi,W) = \bar{\psi}_a \gamma^\mu (L_{\alpha,ab} \frac{1+\gamma_5}{2} + R_{\alpha,ab} \frac{1-\gamma_5}{2}) \psi_b W_\mu^\alpha \tag{2.13}$$

where again $\alpha = 1,\dots,N$; a, b label the different fermions in the theory and L_α, R_α are hermitian coupling matrices. Calculating the amplitudes for the scattering process $F\bar{F} \to WW$

from the diagrams in Fig. 2 one finds that the amplitude behaves like k^2 for large k if both vector mesons are longitudinally polarized unless

$$[L_\alpha, L_\beta] = i\, a_{\alpha\beta\gamma}\, L_\gamma \tag{2.14a}$$

$$[R_\alpha, R_\beta] = i\, a_{\alpha\beta\gamma}\, R_\gamma \ . \tag{2.14b}$$

The requirement of smooth high energy behaviour has led to a remarkable algebraic structure: the relations (2.12b) and (2.12c) define a Lie algebra and the coupling matrices for both left-handed $(1+\gamma_5)$ and right-handed $(1-\gamma_5)$ fermions must be representations of this Lie algebra. Note that the left- and right-handed fermions may define completely independent representations L_α, R_α.

As before, the relations (2.14) are in general not sufficient to guarantee a satisfactory high energy behaviour for longitudinally polarized vector bosons, but instead the amplitudes increase as $m_F k$ for large k with a generic fermion mass m_F. With only elementary spin 1/2 or 1 fields there is nothing more that can be done. But what if we allow scalar fields in the theory shying away from the complications of field theories with spin > 1? For a qualitative discussion we write the possible couplings of a scalar field Φ to fermions and vector fields as

$$L_S = g_\Phi\, W_\mu\, W^\mu\, \Phi + h_\Phi\, \bar{\psi}\, \psi\, \Phi + \ldots \quad . \tag{2.15}$$

The corresponding diagrams fro WW → WW and $F\bar{F}$ → WW are given in Fig. 3. For longitudinally polarized vector bosons the diagrams a) contribute $g_\Phi^2 s/M_W^4$ to the WW → WW amplitude in the high energy limit compared to $g_W^2\, s/M_W^2$ for the diagrams of Fig. 1. Similarly, diagram b) gives $h_\Phi g_\Phi \sqrt{s}/M_W^2$ for $F\bar{F}$ → WW in comparison with $g_W^2\, m_F \sqrt{s}/M_W^2$. Therefore, the remaining high energy divergences may indeed cancel in the presence of scalar fields and we find

$$g_\Phi \sim g_W M_W$$
$$h_\Phi \sim g_W m_F/M_W \qquad (2.16)$$

as necessary conditions.

It should be kept in mind that the residual divergences only appear if at least some of the external vector bosons are longitudinally polarized. For transverse polarizations the algebraic structure expressed in (2.12) and (2.14) is already sufficient to guarantee an acceptable high energy behaviour of Born terms.

III. SPONTANEOUSLY BROKEN GAUGE THEORIES (SBGT)

The Schrödinger equation for a non-relativistic particle of charge q in an external, purely electric potential $A_o(x)$ is

$$i \frac{\partial\psi}{\partial t} = (-\frac{\Delta}{2m} + q\, A_o)\psi \; . \qquad (3.1)$$

Under a gauge transformation

$$A_\mu(x) \rightarrow A_\mu(x) + \partial_\mu \Lambda(x) \qquad (3.2)$$

the observable electromagnetic fields $F_{\mu\nu} = \partial_\mu A_\nu - \partial_\nu A_\mu$ are unchanged but (3.1) is obviously not invariant. Every student of quantum mechanics knows that the correct Schrödinger equation is obtained by applying the substitution rule

$$\partial_\mu \rightarrow D_\mu = \partial_\mu + i\, q\, A_\mu \qquad (3.3)$$

to the free equation yielding

$$(i\frac{\partial}{\partial t} - qA_o)\psi = -\frac{1}{2m}(\partial_j + iqA_j)(\partial_j + iqA_j)\psi, \; j = 1,2,3. \qquad (3.4)$$

Given a solution $\psi(x)$ of (3.4)

$$\psi'(x) = e^{-iq\Lambda(x)}\psi(x) \tag{3.5}$$

is a solution for the gauge transformed potentials (3.2). Since only states with the same charge can be superposed (charge superselection rule) the gauge transformed $\psi'(x)$ is indeed physically equivalent to the original $\psi(x)$ as it must be.

Turning now to field theory we postulate that the Lagrangian $L(\Phi,\partial_\mu\Phi)$, where Φ stands for a collection of fields Φ_i of either spin 0 or 1/2, is invariant under the transformations of a compact Lie group+ G with N real parameters $\Lambda_1,\Lambda_2,\ldots,\Lambda_N$. More precisely, L is invariant under the unitary transformations

$$\Phi(x) \rightarrow U(\Lambda)\Phi(x) = \exp\{-i\Lambda_\alpha T_\alpha\}\Phi(x) \tag{3.6}$$

where the hermitian matrices T_α define a representation of the corresponding Lie algebra of G, i.e.

$$[T_\alpha,T_\beta] = i\, c_{\alpha\beta\gamma}\, T_\gamma \tag{3.7}$$

with totally antisymmetric real structure constants $c_{\alpha\beta\gamma}$.

What happens if we promote the global symmetry G (constant Λ_α) to a local symmetry by allowing space-time dependent $\Lambda_\alpha(x)$ in analogy to (3.5)? Obviously, $L(\Phi,\partial_\mu\Phi)$ will no longer be invariant because the derivative ∂_μ does not commute with $U(\Lambda(x))$.

From our experience with the Schrödinger equation we are led to introduce [12] N hermitian vector fields W_μ^α into the theory by defining a covariant derivative

$$D_\mu\Phi = (\partial_\mu + i\, g_\alpha\, T_\alpha\, W_\mu^\alpha)\Phi \tag{3.8}$$

+A Lie group is compact if its parameters Λ_α only vary in a bounded domain of R_N. Compactness guarantees that all finite-dimensional representations are equivalent to unitary representations.

with real coupling constants g_α.

If we can ensure that under a local symmetry transformation $\Phi(x) \to U(\Lambda(x))\Phi(x)$ we also have

$$D_\mu \Phi \to U(\Lambda(x)) D_\mu \Phi \quad , \tag{3.9}$$

the Lagrangian $L(\Phi, D_\mu\Phi)$ will be locally invariant if $L(\Phi, \partial_\mu\Phi)$ possesses the corresponding global symmetry. It is easily checked that (3.9) requires the gauge fields W_μ^α to transform as

$$W_\mu \to U(W_\mu + i\, U^{-1}\partial_\mu U)U^{-1} \tag{3.10}$$

with

$$W_\mu = g_\alpha\, T_\alpha\, W_\mu^\alpha \quad . \tag{3.11}$$

Contrary to appearance the transformation law (3.10) does not depend on the representation T_α as can be seen from the infinitesimal form (no sum over α)

$$g_\alpha\, W_\mu^\alpha \to g_\alpha\, W_\mu^\alpha + \partial_\mu\, \Lambda_\alpha - g_\beta\, c_{\alpha\beta\gamma}\, W_\mu^\beta\, \Lambda_\gamma \quad . \tag{3.12}$$

For a complete gauge field Lagrangian we still need a kinetic term for the vector fields. It is at this point that we run into difficulties with arbitrary coupling constants g_α. A gauge invariant kinetic term can only be constructed if

$$g_\alpha = g_\beta = g_\gamma \qquad \text{for} \qquad c_{\alpha\beta\gamma} \neq 0 \tag{3.13}$$

and it is given by

$$L_{kin}(W) = -\frac{1}{4}\, G_{\mu\nu}^\alpha\, G^{\alpha,\mu\nu} \tag{3.14}$$

with

$$G_{\mu\nu}^\alpha = \partial_\mu\, W_\nu^\alpha - \partial_\nu\, W_\mu^\alpha - g_\alpha\, c_{\alpha\beta\gamma}\, W_\mu^\beta\, W_\nu^\gamma \quad . \tag{3.15}$$

Any compact Lie group can be written as a direct product

$$G = G_1 \times G_2 \times \ldots \times G_k \tag{3.16}$$

where the G_i cannot be decomposed any further (G_i is either the abelian one-parameter Lie group U(1) or a so-called simple group). The conditions (3.13) imply that there is one independent coupling constant for each factor in the decomposition (3.16).

For a non-abelian group ($c_{\alpha\beta\gamma} \neq 0$) the Lagrangian (3.14) contains cubic and quartic terms. Comparing (3.14) with the general interaction Lagrangian (2.11) we find immediately that the relations (2.12) are fulfilled (for $a_{\alpha\beta\gamma} = - g_\alpha c_{\alpha\beta\gamma}$) which guarantee the maximal possible damping of Born terms at high energies.

What about the gauge boson couplings to fermions? According to the prescription (3.8) they are contained in the gauge invariant kinetic fermion Lagrangian

$$\bar{\psi}\, i\gamma^\mu D_\mu\, \psi = \bar{\psi}\, i\gamma^\mu \partial_\mu\, \psi - g_\alpha\, \bar{\psi}\, \gamma^\mu T_\alpha\, \psi\, W_\mu^\alpha \quad . \tag{3.17}$$

Setting

$$- g_\alpha T_\alpha = L_\alpha \frac{1+\gamma_5}{2} + R_\alpha \frac{1-\gamma_5}{2} \tag{3.18}$$

the Lie algebra commutation relations (3.7) are seen to imply the relations (2.14) among the coupling matrices L_α, R_α.

For phenomenological reasons some of the gauge vector bosons must be massive. However, a general mass term

$$L_M(W) = \frac{1}{2} M_{\alpha\beta}^2\, W_\mu^\alpha\, W^{\beta,\mu} \tag{3.19}$$

necessarily breaks the local symmetry: invariance under (3.10) or (3.12) implies $M_{\alpha\beta}^2 = 0$. We recall from Sect. II that the residual high energy divergences of Born terms were due to the longitudinal polarization vectors $\varepsilon_\mu^L(k)$. But a massless vector field has only two degrees of freedom corresponding to the two transverse polarization vectors. Thus one finds that the local gauge symmetry which requires massless vector fields is indeed sufficient for the Born terms to behave as they should in a renormalizable field theory: they approach finite limits for high energies.

On the other hand, we must break the symmetry in some way because the intermediate vector bosons of the weak interactions are certainly massive. The way out of this dilemma is to break the local symmetry in a more subtle way which allows for mass differences between the vector bosons but which must not ruin the good high energy behaviour of the massless theory. The solution is called spontaneous symmetry breaking (SSB) [13].

In quantum theory we usually encounter the following situation: given a Hamiltonian H and a symmetry operator S_Λ with $[S_\Lambda, H] = 0$ one finds that states which are related by S_Λ are degenerate:

$$H|n\rangle = E_n|n\rangle \qquad \text{and} \qquad |n_\Lambda\rangle = S_\Lambda|n\rangle$$

imply

$$H|n_\Lambda\rangle = HS_\Lambda|n\rangle = S_\Lambda H|n\rangle = E_n|n_\Lambda\rangle \quad . \tag{3.20}$$

Let us specialize to a Poincaré invariant theory with a continuous symmetry parametrized by Λ and generated by a current $j^\mu(x)$ such that

$$S_\Lambda = e^{i\Lambda Q} \qquad \text{with} \qquad Q = \int d^3x\, j^o(x) \quad . \tag{3.21}$$

Calculating the norm of the state $Q|0\rangle$ with $|0\rangle$ the ground state one finds

$$\| Q|0\rangle \|^2 = \langle 0|Q^2|0\rangle = \int d^3x \langle 0|Qj^o(x)|0\rangle = \int d^3x \langle 0|Qj^o(0)|0\rangle \tag{3.22}$$

because the ground state is assumed to be invariant under translations. The last equality in (3.22) shows that

either $Q|0\rangle = 0$
or $Q|0\rangle$ is not in the Hilbert space under consideration.

In the first case the ground state is invariant under S_Λ and (3.20) holds. In the second case, S_Λ leads out of Hilbert space: the ground state does not exhibit the symmetry and the eigenstates of H show no degeneracy. The symmetry is said to be spontaneously broken although formally $[S_\Lambda, H] = 0$.

Although the above arguments are tailored to a Poincaré invariant field theory the phenomenon of SSB is well-known in solid state physics. The ground state of a crystal does not exhibit the full translation symmetry of the Hamiltonian, the symmetry is spontaneously broken to a discrete lattice group. Analogously, a ferromagnet is not rotationally invariant. These two examples suggest that we should be able to understand the phenomenon of SSB already in non-relativistic quantum mechanics.

For this purpose let us study the Hamiltonian for a linear chain of oscillators with nearest neighbour couplings

$$H = \sum_{i=1}^{n} \{p_i^2/2 + \omega^2(q_i - q_{i+1})^2/2 + V(q_i)\} \tag{3.23}$$

where each oscillator feels a potential of the form

$$V(q_i) = \frac{\ell}{4}(q_i^2 - q_0^2)^2 , \qquad \ell > 0 , \tag{3.24}$$

shown in Fig. 4. H is obviously invariant under the parity transformation P: $q_i \to -q_i$, $p_i \to -p_i$.

The standard argument runs as follows: classically, for the system to be in the ground state all q_i must have the same value, but because of (3.24) this common value can be either

q_o or $-q_o$. One is tempted to infer that quantum mechanically the oscillators should be concentrated either all around q_o or all around $-q_o$ and these two configurations should be expected to be degenerate in energy. Neither of these two states $|q_o>$ and $|-q_o>$ exhibits the parity symmetry, so here we would have an example of SSB.

However, the argument is wrong in this simplified form: each oscillator can tunnel through to the other minimum with a tunnelling probability (calculated in WKB approximation)

$$P_{+-} \sim \exp\,(-\sqrt{\ell}\; q_o^3)\;, \tag{3.25}$$

so the asymmetric states $|\pm q_o>$ cannot be eigenstates of H. Diagonalizing H for a single oscillator one finds the true eigenstates $(|q_o> \pm |-q_o>)/\sqrt{2}$ which clearly do not break P spontaneously and which are separated by an energy difference proportional to $\exp\,(-\sqrt{\ell}\; q_o^3)$.

How can we then understand the phenomenon of SSB in a quantum mechanical system? The answer lies in the large number of degrees of freedom. As a matter of fact, the total tunnelling probability for n oscillators is

$$P_{+-}^{total} \sim \exp\,(-n\sqrt{\ell}q_o^3) \tag{3.26}$$

and this probability becomes extremely small for large n. We may also consider the standard continuum limit for (3.23) in a 3-dimensional volume V [14]. In this limit the q_i turn into a scalar field $\Phi(x)$ with the Lagrangian density

$$L = \frac{1}{2}\,\partial_\mu\,\Phi\,\partial^\mu\,\Phi - \frac{\lambda}{4}\,(\Phi^2 - v^2)^2 \tag{3.27}$$

and the barrier penetration factor becomes

$$P_{+-}^{cont} \sim \exp\,(-\sqrt{\lambda}\; v^3\; V)\;. \tag{3.28}$$

Even if the asymmetric state is not the ground state it will be an effectively stable (metastable) state because for macroscopic volumes $\sqrt{\lambda}\, v^3\, V >> 1$ and correspondingly the lifetime of such a state will be enormous. In practice, the crystal and the ferromagnet are stable configurations.

From the preceding qualitative arguments we expect that SSB will in any case occur for a field theory (3.27) in infinite space since for $V \to \infty$ the tunnelling probability (3.28) vanishes. At the same time, the two asymmetric states $|0_\pm\rangle$ which are characterized by $\langle 0_\pm|\Phi(x)|0_\pm\rangle = \pm\, v$ at the tree level will become degenerate and there will be two disjoint Hilbert spaces associated to those ground states in accordance with the conclusions after (3.22).

Another line of reasoning observes that in three dimensions the Hamiltonian contains a centrifugal term $\vec{L}^2/2I$ where $\vec{L}$ is the total angular momentum and I is the moment of inertia. Although for a finite system there will normally be a non-degenerate ground state with L = 0 states with different L may become degenerate as $I \to \infty$ (infinite system).

In the above example we have considered the spontaneous breaking of a discrete symmetry. Let us now turn to the simplest example of a SBGT [6]:

$$L_{Higgs} = (D_\mu\Phi)^\dagger D^\mu\Phi - \lambda(\Phi^\dagger\Phi - \frac{v^2}{2})^2 - \frac{1}{4} F_{\mu\nu} F^{\mu\nu} \qquad (3.29)$$

in analogy to (3.27) where $\Phi(x)$ is a non-hermitian scalar field and

$$\begin{aligned} D_\mu\Phi &= (\partial_\mu + ig\, W_\mu)\Phi \\ F_{\mu\nu} &= \partial_\mu W_\nu - \partial_\nu W_\mu \end{aligned} \qquad (3.30)$$

define a U(1) gauge theory. Through a gauge transformation

$$W_\mu(x) \to W_\mu(x) + \partial_\mu \Lambda(x)$$

$$\Phi(x) \rightarrow \exp\{-ig\,\Lambda(x)\}\Phi(x) = R(x)/\sqrt{2} \qquad (3.31)$$

we can always rotate $\Phi(x)$ into a hermitian field $R(x)$. Using W_μ again for the transformed gauge field, (3.29) can now be written

$$L_{Higgs} = \frac{1}{2}(\partial_\mu R \partial^\mu R + g^2 R^2 W_\mu W^\mu) - \frac{\lambda}{4}(R^2 - v^2)^2 - \frac{1}{4} F_{\mu\nu} F^{\mu\nu} . \qquad (3.32)$$

Assuming SSB to occur and $\langle 0|R(x)|0\rangle = v$ in Born approximation one defines a new field

$$H(x) = R(x) - v \qquad (3.33)$$

with vanishing vacuum expectation value in lowest order. In terms of the Higgs field $H(x)$

$$\begin{aligned} L_{Higgs} &= \frac{1}{2}\partial_\mu H \partial^\mu H + \frac{g^2}{2} W_\mu W^\mu (v^2 + 2vH + H^2) - \\ &- \frac{\lambda}{4}(H^2 + 2vH)^2 - \frac{1}{4}F_{\mu\nu}F^{\mu\nu} = \frac{1}{2}\partial_\mu H \partial^\mu H - \lambda v^2 H^2 - \qquad (3.34) \\ &- \frac{1}{4}F_{\mu\nu}F^{\mu\nu} + \frac{g^2 v^2}{2} W_\mu W^\mu + \frac{g^2}{2} W_\mu W^\mu (H^2 + 2vH) - \frac{\lambda H^2}{4}(H^2 + 4vH) \end{aligned}$$

we finally arrive at a field theory of a scalar particle of mass $\sqrt{2\lambda}\,v$ and of a massive vector boson with mass gv. This is known as the Higgs-Kibble mechanism [6,15] in particle physics: via SSB the originally massless gauge field has "eaten up" one scalar degree of freedom to become a massive vector field. However, it was already recognized earlier [16] that such a mechanism is at work in superconductors where the ground state contains weakly bound pairs of electrons (Cooper pairs) leading to a spontaneous breaking of electromagnetic gauge invariance and an effective photon mass (Meissner effect).

It is instructive to investigate (3.29) in the limit $g \rightarrow 0$ when the gauge field W_μ decouples. The resulting scalar Lagrangian [17]

$$L_{\text{Goldstone}} = \partial_\mu \Phi^\dagger \partial^\mu \Phi - \lambda(\Phi^\dagger \Phi - \frac{v^2}{2})^2 \qquad (3.35)$$

is invariant under the global U(1) transformation

$$\Phi(x) \rightarrow e^{-i\Lambda} \Phi(x) \qquad (3.36)$$

and this symmetry enables us to fix the phase of $\Phi(x)$ such that $\langle 0|\Phi(x)|0\rangle$ is positive.

Decomposing

$$\Phi(x) = [R(x) + iG(x)]/\sqrt{2}$$

$$\langle 0|R(x)|0\rangle = v, \quad \langle 0|G(x)|0\rangle = 0 \qquad (3.37)$$

one can write

$$L_{\text{Goldstone}} = \frac{1}{2}\partial_\mu R \partial^\mu R + \frac{1}{2}\partial_\mu G \partial^\mu G - \frac{\lambda}{4}(R^2 + G^2 - v^2)^2 . \qquad (3.38)$$

Introducing $H(x)$ as in (3.33) one obtains

$$\begin{aligned} L_{\text{Goldstone}} &= \frac{1}{2}\partial_\mu H \partial^\mu H + \frac{1}{2}\partial_\mu G \partial^\mu G - \frac{\lambda}{4}(H^2 + G^2 + 2vH)^2 = \\ &= \frac{1}{2}\partial_\mu H \partial^\mu H - \lambda v^2 H^2 + \frac{1}{2}\partial_\mu G \partial^\mu G - \frac{\lambda}{4}(H^2 + G^2)(H^2+G^2+4vH) . \end{aligned} \qquad (3.39)$$

We conclude that there is again a massive scalar field $H(m_H = \sqrt{2\lambda}\, v)$ in the theory, but in addition we now find a massless scalar field $G(x)$, the Goldstone boson of the model (3.35).

It must be emphasized that both the existence of massless quanta in theories with a spontaneously broken continuous global symmetry (Goldstone theorem [17,18]) and the transmutation of those quanta into the longitudinal degrees of freedom of massive vector fields in gauge theories (Higgs-Kibble mechanism [6,15]) are very general phenomena which are not restricted to the simple models discussed so far. In solid state physics the

Goldstone theorem predicts the existence of excitations with vanishing energy in the limit of infinite wavelength (longitudinal phonons in a crystal, spin waves in a ferromagnet) whereas in a superconductor all excitations have finite energy in the same limit (mass gap). Superconductivity is also an instructive example for SSB in a gauge theory without fundamental scalar degrees of freedom. Gauge theories of electroweak interactions with only fermions and gauge fields run under the heading "dynamical symmetry breaking" and are reviewed at this School by P. Sikivie [2].

For an arbitrary N-parameter Lie group G [15] the ground state will in general break the symmetry spontaneously to an L-parameter subgroup G_o of G. This means that the operators $\hat{T}_\alpha$ representing the generators of G_o in Hilbert space annihilate the ground state, but the remaining symmetries in G cannot be unitarily represented (see the discussion after (3.22)).

In terms of the vacuum expectation values

$$\langle 0|\Phi_i(x)|0\rangle = v_i \qquad (i = 1,\dots,n_s)$$

for n_s hermitian scalar fields Φ_i we have

$$\exp(-i\Lambda_\beta T_\beta)v = v \qquad \beta = 1,\dots,L \qquad (3.40)$$

$$\text{or} \quad T_\beta v = 0$$

where the T_β are n_s-dimensional matrix representations of the Lie algebra of G_o. It can then be shown [1] that in the so-called unitary gauge which is the generalization of (3.31) for arbitrary G the Lagrangian contains N-L massive and L massless vector fields and n_s-N+L hermitian scalar fields generalizing the model (3.29). The remaining N-L scalar degrees of freedom have been transformed into the longitudinal parts of the massive vector fields. For electroweak gauge theories $G_o = U(1)_{em}$ and therefore L = 1 (photon), whereas for grand unified theories $G_o = U(1)_{em} \times SU(3)_{colour}$ and $L = 9$ (1 photon, 8 gluons).

We have succeeded in making vector bosons massive without destroying the symmetry of the Lagrangian

$$L = -\frac{1}{4} G^{\alpha}_{\mu\nu} G^{\alpha,\mu\nu} + L(\psi, D_\mu\psi, \Phi, D_\mu\Phi) \tag{3.41}$$

by requiring the gauge symmetry to be spontaneously broken. By rewriting (3.41) in terms of the shifted scalar fields (3.33) one does not change the Lagrangian at all. Instead, such a shift only makes the particle content of the theory transparent in the unitary gauge. It is therefore very plausible that the high energy behaviour of Born terms is the same as for the symmetric theory with massless vector fields. As a simple consistency check let us look once more at (3.34). The cubic coupling between W_μ and H is $g^2v = gM_W$ in accordance with the qualitative result (2.16).

That the spontaneous breaking of a continuous symmetry is a "soft" breaking which does not modify the high energy structure of the symmetric theory is also the fundamental reason for the renormalizability of such theories [1,19]. It is beyond the scope of these lectures to treat the quantization of gauge theories [1,20] in any more than very qualitative terms. In order to exhibit the dynamical degrees of freedom which are needed for the canonical quantization procedure one gauge condition must be chosen for each gauge field. Similarly to the classical problem of solving the Euler-Lagrange equations for a system with constraints two approaches are in principle possible. One can either try to solve the gauge conditions explicitly and insert the solutions into the equations of motion or one includes the constraints in the Lagrangian using the method of Lagrange multipliers. For the quantization of gauge theories Faddeev and Popov [21] developed this "Lagrange multiplier method" into an extremely useful framework for perturbation theory. In general, the naive Feynman rules must be modified by the introduction of fictitious scalar fields with Fermi statistics (Faddeev-Popov ghosts) as internal lines in Feynman diagrams to guarantee perturbative unitarity.

Although the unitary gauge is ideally suited to display the particle spectrum of the theory it is not convenient for perturbation theory. The renormalizability of SBGT is more transparent in the so-called renormalizable gauges [1,20] of which the covariant R_ξ gauges [19,22] are an especially popular subset. In that gauge the gauge field propagator is given by

$$\Delta^{\alpha\beta}_{\mu\nu} = \frac{-\delta_{\alpha\beta}}{k^2-M_\alpha^2+i\varepsilon} \{g_{\mu\nu} - (1 - 1/\xi) \frac{k_\mu k_\nu}{k^2-M_\alpha^2/\xi}\} \quad (3.42)$$

which shows the high momentum behaviour of a scalar propagator for all values of the gauge parameter $\xi \neq 0$ and thus guarantees renormalizability by the usual power counting arguments. In order to cancel the spurious singularities of (3.42) at $k^2 = = M_\alpha^2/\xi$ the Feynman rules contain the Faddeev-Popov ghosts and in addition N-L scalar fields with propagators

$$(k^2 - M_\alpha^2/\xi)^{-1} \quad . \quad (3.43)$$

These unphysical scalar fields are sometimes called would-be Goldstone bosons: in the limit $\xi \to 0$ they decouple completely whereas the gauge field propagator (3.42) approaches the standard propagator (2.9) of a massive vector boson. Thus, for $\xi \to 0$ we approach the unitary gauge where the N-L would-be Goldstone bosons are swallowed by the N-L massive vector bosons.

The unitarity of the perturbation expansion requires that S-matrix elements are independent of the gauge parameter ξ. The consequences of gauge invariance for Green's functions can be formulated in terms of the so-called Slavnov-Taylor identities [23] which generalize the Ward-Takahashi identities [24] in QED. Using those identities one can indeed prove that an arbitrary S-matrix element is independent of ξ (or gauge independent in general) up to the subtle question of anomalies which will be discussed at length in Sect. VI. With this proviso a unitary and renormalizable perturbation expansion can be set up for any SBGT.

IV. CHOOSING THE GAUGE GROUP FOR ELECTROWEAK INTERACTIONS: THE STANDARD MODEL

With all the ingredients of SBGT at our disposal we must address the question how to find the "correct" gauge group G for electroweak interactions. The number of parameters N of G equals the number of gauge bosons or the number of currents in the theory so that we must have $N \geq 3$ to account for the electromagnetic and charged weak interactions. Particle states are eigenstates of charge which must therefore be among the simultaneously diagonalizable generators of the Lie algebra of G (the maximal number of commuting generators is called the rank of the Lie algebra).

In Table 1 all compact Lie groups with $3 \leq N \leq 7$ (up to local isomorphisms) are exhibited except for the groups $U(1)^N = U(1) \times U(1) \times \ldots \times U(1)$ which cannot account for charged currents. In the spirit of simplicity which has scored so many successes in physics we shall start with the unique 3-parameter group of rank 1, SU(2). The charge operator must be proportional to one of the three generators, say T_3. In order to get any interactions at all we must put the fermions into a non-trivial representation of SU(2). The smallest irreducible representation is the doublet which seems just right for the leptons ν_e, e. However, the eigenvalues of the charge operator aT_3 are $\pm\, a/2$ so that there is no room for a neutral particle and we must move on to the triplet with charges a,0,-a. Because of the vector-, axial-vector structure of weak and electromagnetic currents we cannot put e^-, ν_e,

μ^+ (or τ^+) into a triplet. Unless we accept an unknown and therefore very heavy lepton [25] we must have $N \geq 4$. Table 1 then implies that for $N < 8$ we need at least two independent coupling constants for electroweak interactions.

Nature seems to prefer the second simplest possibility $SU(2) \times U(1)$ which has rank 2 and therefore two neutral vector bosons and two neutral currents.

The interaction between fermions and the gauge fields W_μ^α ($\alpha = 1,2,3$), B_μ is determined by

$$\bar{\psi}\, i\gamma^\mu D_\mu \psi = \bar{\psi}\, i\gamma^\mu (\partial_\mu + ig\, T_\alpha W_\mu^\alpha + ig' \tfrac{Y}{2} B_\mu)\psi \tag{4.1}$$

and the charge (in units of the positron charge) is a linear combination

$$Q = T_3 + Y/2 \tag{4.2}$$

of weak isospin T_3 and weak hypercharge Y. The factor 1 in front of T_3 is dictated by $|\Delta Q| = 1$ in each lepton or quark multiplet whereas $Y/2$ is a nostalgic convention. Because of (4.2) neither W_μ^3 nor B_μ can be the electromagnetic field A_μ, but A_μ and the additional massive neutral vector field Z_μ must be related to them by an orthogonal transformation

$$\begin{bmatrix} Z_\mu \\ A_\mu \end{bmatrix} = \begin{bmatrix} \cos\Theta_W & -\sin\Theta_W \\ \sin\Theta_W & \cos\Theta_W \end{bmatrix} \begin{bmatrix} W_\mu^3 \\ B_\mu \end{bmatrix} \tag{4.3}$$

where Θ_W is the famous Weinberg angle.

Thus,

$$gT_3 W_\mu^3 + g' \frac{Y}{2} B_\mu = g \sin\Theta_W A_\mu (T_3 + \frac{g'}{g}\frac{\cos\Theta_W}{\sin\Theta_W}\frac{Y}{2}) +$$

$$+ \frac{g}{\cos\Theta_W} Z_\mu (\cos^2\Theta_W T_3 - \frac{g' \sin\Theta_W \cos\Theta_W}{g}\frac{Y}{2}) \tag{4.4}$$

and we obtain both the desired electromagnetic interaction for

$$\begin{aligned} \mathrm{tg}\,\Theta_W &= g'/g \\ e &= g \sin\Theta_W \end{aligned} \tag{4.5}$$

and a new neutral current interaction with

$$L_{NC} = - \frac{g}{\cos\Theta_W} Z_\mu \bar{\psi} \gamma^\mu (T_3 - \sin^2\Theta_W Q)\psi \quad . \tag{4.6}$$

The overwhelming evidence from charged and neutral current phenomenology [26,27] is that all left-handed fermions ν_e, e^-, u, d, ... are in SU(2) doublets and their right-handed counterparts are singlets. Since a neutral SU(2) singlet field does not couple to any gauge fields the right-handed neutrinos are usually eliminated from the theory altogether. We shall come back to these assignments in Sect. V when discussing fermion masses.

The charged current interaction is then

$$\begin{aligned} L_{CC} &= - \frac{g}{2\sqrt{2}} j^\mu_{CC} W^+_\mu + \text{h.c.} \\ W^+_\mu &= (W^1_\mu - i W^2_\mu)/\sqrt{2} \end{aligned} \tag{4.7}$$

with the usual normalization of j^μ_{CC} as in (2.2). For momenta small compared to M_W we recover the V-A Fermi theory (2.1) with

$$G_F/\sqrt{2} = \frac{g^2}{8M_W^2} = \frac{e^2}{8M_W^2 \sin^2\Theta_W} = \frac{\pi\alpha}{2M_W^2 \sin^2\Theta_W} \tag{4.8}$$

and therefore

$$M_W = 37.28 \text{ GeV}/\sin\Theta_W \quad . \tag{4.9}$$

In the same way one obtains the effective low energy neutral current Lagrangian

$$L_{NC}^{eff} = - \frac{4G_F}{\sqrt{2}} \rho \, j_{NC}^{\mu} \, j_{NC\mu} \qquad (4.10)$$

with

$$j_{NC}^{\mu} = \bar{\psi}\gamma^{\mu}(T_3 - \sin^2\Theta_W \, Q)\psi, \qquad T_3 = \frac{\tau_3}{2} \frac{1+\gamma_5}{2} \qquad (4.11)$$

and

$$\rho = \frac{M_W^2}{M_Z^2 \cos^2\Theta_W} \quad . \qquad (4.12)$$

The Lagrangian (4.10) accounts successfully for all available neutral current data with [27]

$$\sin^2\Theta_W = 0.234 \pm 0.013 \; (\pm 0.009)$$
$$\rho = 1.002 \pm 0.015 \; (\pm 0.011) \qquad (4.13)$$

where the errors in brackets estimate theoretical uncertainties of the analysis.

The vector boson masses M_W, M_Z are generated by SSB. Let us consider a general gauge invariant Lagrangian for an n-component scalar field Φ

$$L_S = (D_\mu \Phi)^\dagger D^\mu \Phi - V(\Phi) \qquad (4.14)$$

where $V(\Phi)$ is a general potential inducing non-vanishing vacuum expectation values

$$\langle 0|\Phi_i(x)|0\rangle = v_i/\sqrt{2} \, , \qquad i = 1,\dots,n \qquad (4.15)$$

at the tree level. Since electromagnetic gauge invariance must not be broken we have

$$Qv = 0 \rightarrow \frac{Y}{2} v = - T_3 v \qquad (4.16)$$

where Q, Y, T_3 are now n-dimensional matrix representations of the corresponding generators in the space of scalar fields Φ. Performing the shift $\Phi = \Phi' + v/\sqrt{2}$ and using (4.16) one arrives at the following vector boson mass terms from (4.14):

$$L_M = \frac{1}{2}v^\dagger(-ig\,T_\alpha\,W^\alpha_\mu + ig'\,T_3\,B_\mu)(ig\,T_\beta\,W^{\beta,\mu} - ig'\,T_3\,B^\mu)v \tag{4.17}$$

or

$$L_M = \frac{g^2}{2}W^+_\mu W^{-\mu}v^\dagger(T_+T_-+T_-T_+)v+\frac{1}{2}(gW^3_\mu-g'B_\mu)(gW^{3,\mu}-g'B^\mu)v^\dagger T_3^2 v \tag{4.18}$$

with $T_\pm = (T_1 \pm iT_2)/\sqrt{2}$.

Recalling (4.3) and (4.5) together with

$$T_+\,T_- + T_-\,T_+ = \vec{T}^2 - T_3^2 \tag{4.19}$$

we find

$$L_M = \frac{g^2}{2}W^+_\mu W^{-\mu}\sum_{i=1}^{n}[t_i(t_i+1)-t_{3i}^2]|v_i|^2 + \frac{g^2}{2\cos^2\Theta_W}Z_\mu Z^\mu \sum_{i=1}^{n} t_{3i}^2|v_i|^2 \tag{4.20}$$

with

$$\vec{T}^2 v_i = t_i(t_i+1)v_i, \quad T_3 v_i = t_{3i}v_i \quad . \tag{4.21}$$

From (4.20) we read off the vector boson masses

$$M_W^2 = \frac{g^2}{2}\sum_{i=1}^{n}\;[t_i(t_i+1) - t_{3i}^2]|v_i|^2 \tag{4.22a}$$

$$M_Z^2 = \frac{g^2}{\cos^2\Theta_W}\sum_{i=1}^{n}\;t_{3i}^2\;|v_i|^2 \tag{4.22b}$$

and therefore

$$\rho = \sum_{i=1}^{n}\;[t_i(t_i+1)-t_{3i}^2]|v_i|^2/(2\sum_{i=1}^{n}\;t_{3i}^2|v_i|^2) \quad . \tag{4.23}$$

How can we understand the experimental result $\rho \simeq 1$? It is amazing that nature seems to have chosen once more the simplest possibility$^+$ that all Φ_i are in SU(2) doublets so that

$$t_i(t_i+1) - t_{3i}^2 = 2t_{3i}^2 = \frac{1}{2} \rightarrow \rho = 1 \ . \tag{4.24}$$

As a matter of fact, one scalar doublet

$$\Phi = \binom{\Phi_+}{\Phi_o} \quad \text{with} \quad \langle O|\Phi_o|O\rangle = v/\sqrt{2}$$

and

$$|v|^2 = 4M_W^2/g^2 = (\sqrt{2}G_F)^{-1} = (246 \text{ GeV})^2 \tag{4.25}$$

is enough. This simplest choice of SSB together with the above-mentioned fermion assignments define the standard model of electroweak interactions [3].

There is at the moment absolutely no necessity for phenomenological reasons to go beyond the standard model. From an aesthetic point of view one may however find it unnatural that a theory unifying electromagnetism with the weak interactions is intrinsically parity violating. Some theoreticians consider it more satisfactory for the basic Lagrangian to conserve parity so that parity violation of the weak interactions would be a low energy phenomenon caused by SSB. Of course, the scale where parity is spontaneously broken must be larger than $|v| = 246$ GeV to account for the observed structure of charged and neutral currents. In particular, the parity transformed charged currents which are right-handed must couple predominantly to additional charged vector bosons which are heavier than (4.9) and thus $N \geq N_{\text{standard}} + 2 = 6$. Looking again at Table 1 we find that SU(2) × SU(2) is not acceptable for the same reason as SU(2) and SU(2) × $U(1)^3$ only contains two charged vector bosons. The simplest possibility

$^+$Of course, SU(2) singlet scalar fields do not contribute to M_W, M_Z.

is therefore SU(2) × SU(2) × U(1) which has three coupling constants at first sight. However, in order to have parity invariance one has to demand that the Lagrangian is invariant under the automorphism that exchanges the two SU(2) algebras leading to the left-right symmetric $SU(2)_L \times SU(2)_R \times U(1)$ gauge theory [7] with only two coupling constants. I cannot discuss the attractive features of this model here, but let me only mention that neutral current phenomenology requires the ratio of the two massive neutral vector bosons to obey [28]

$$M_{Z_2}/M_{Z_1} > 2.1 \qquad (2\sigma \text{ limit}) \ . \tag{4.26}$$

V. FERMION MASSES

A general gauge theory of electroweak interactions contains left-handed fermion fields ϕ_{aL} $(a = 1,\ldots,n_L)$ and right-handed fermion fields χ_{bR} $(b = 1,\ldots,n_R)$ with given transformation properties under the gauge group. For each charged field ϕ_L there must be a right-handed counterpart χ_R of the same charge because otherwise the electromagnetic current would violate parity. However, the number of neutral fields (neutrinos) may be different so that in general $n_L \neq n_R$ as is the case in the standard model. It is important to realize that the fields ϕ, χ do not correspond to definite particles yet because they are all massless for the time being. In order to make this more explicit I have chosen different letters for the left- and right-handed fermion fields.

The fermion kinetic Lagrangian is

$$L_F^{kin} = \bar{\phi}_L \, i\gamma^\mu(\partial_\mu + ig_\alpha T_{\alpha L} W_\mu^\alpha)\phi_L + \bar{\chi}_R \, i\gamma^\mu(\partial_\mu + ig_\alpha T_{\alpha R} W_\mu^\alpha)\chi_R \tag{5.1}$$

where $T_{\alpha L}$, $T_{\alpha R}$ are hermitian n_L, n_R-dimensional matrix representations of the generators of the gauge group. We may now choose to write (5.1) in terms of right-handed fields only be using charge conjugate fields. Given a fermion field ψ one defines the charge conjugate field ψ^C by

$$\psi^C = C\,\gamma_o^T\,\psi^* \tag{5.2}$$

where C is the Dirac charge conjugation matrix which obeys

$$C^{-1}\,\gamma_\mu\,C = -\,\gamma_\mu^T$$

$$C^T = -\,C\,, \qquad C^\dagger = C^{-1} \tag{5.3}$$

in an arbitrary representation of the γ-matrices. Because of

$$C^{-1}\,\gamma_5\,C = \gamma_5^T = \gamma_5^* \tag{5.4}$$

one finds

$$\psi^C_{\substack{L\\R}} = C\,\gamma_o^T\,\psi^*_{\substack{R\\L}} \tag{5.5}$$

and

$$\overline{\psi^C_{\substack{L\\R}}} = -\,\psi^T_{\substack{R\\L}}\,C^{-1}\quad . \tag{5.6}$$

Allowing for a partial integration in the action for (5.1) and taking the anticommutativity of fermion fields into account we may write the first part of (5.1) as

$$\phi_L^T\,i\gamma^{\mu T}(\partial_\mu - ig_\alpha T^T_{\alpha L}W^\alpha_\mu)\,\gamma_o^T\phi_L^* \quad . \tag{5.7}$$

Using (5.3) - (5.6) we get

$$L_F^{kin} = \overline{\phi_R^C}i\gamma^\mu(\partial_\mu - ig_\alpha T^*_{\alpha L}W^\alpha_\mu)\,\phi_R^C + \overline{\chi_R}i\gamma^\mu(\partial_\mu + ig_\alpha T_{\alpha R}W^\alpha_\mu)\,\chi_R \tag{5.8}$$

because $T^T_{\alpha L} = T^*_{\alpha L}$ due to the hermiticity of $T_{\alpha L}$.

Defining an $n_L + n_R = n_F$-component fermion field

$$\omega_R = \begin{bmatrix} \phi_R^C \\ \\ \chi_R \end{bmatrix} \tag{5.9}$$

and an n_F-dimensional representation

$$T_\alpha = \begin{bmatrix} -T_{\alpha L}^* & 0 \\ \\ 0 & T_{\alpha R} \end{bmatrix} \tag{5.10}$$

one obtains finally

$$L_F^{kin} = \overline{\omega_R}\, i\gamma^\mu (\partial_\mu + ig_\alpha T_\alpha W_\mu^\alpha)\, \omega_R \quad . \tag{5.11}$$

It was assumed in (5.1) that the fields ϕ_L and χ_R transform separately among themselves under the gauge group so that the representation (5.10) is obviously reducible. Although this is the case for the electroweak gauge theories based on SU(2) × × U(1) or $SU(2)_L \times SU(2)_R \times U(1)$ it is not necessary in a general gauge theory. In grand unified theories fields of opposite charge are often put into irreducible representations so that (5.11) is the general kinetic fermion Lagrangian but with T_α an arbitrary hermitian representation of the Lie algebra of G.

In addition to (5.11) the fermions ω_R may appear in bilinear mass terms and in Yukawa couplings to the scalar fields Φ_i (i = 1,...,n). Of course, all couplings must respect gauge invariance. The only way to make a Lorentz scalar out of right-handed fields ω_R is $\omega_R^T\, C^{-1}\, \omega_R$. Thus, the fermion Lagrangian L_F has the general form

$$L_F - L_F^{kin} = \frac{1}{2}\omega_R^T\, C^{-1}\, M_o\, \omega_R + \frac{1}{2}\, \omega_R^T\, C^{-1}\, \Gamma^i\, \omega_R \Phi_i + \text{h.c.} =$$

$$= -\frac{1}{2}\, \overline{\omega_L^C}\, M_o\, \omega_R - \frac{1}{2}\, \overline{\omega_L^C}\, \Gamma^i\, \omega_R\, \Phi_i + \text{h.c.} \tag{5.12}$$

where the n_F-dimensional matrices M_o, Γ^i are symmetric because of $C^T = -C$ and the anticommutativity of fermion fields.

After SSB (5.12) yields a lowest-order mass term

$$L_M = \frac{1}{2} \omega_R^T C^{-1} M \omega_R + \text{h.c.} \tag{5.13}$$

with

$$M = M_o + \Gamma^i v_i$$

$$v_i = \langle 0 | \Phi_i(x) | 0 \rangle \quad . \tag{5.14}$$

Because of charge conservation the fermion mass matrix M can only connect fields with opposite charge. Grouping together fields with the same absolute value of Q, M is block-diagonal in this basis and we can treat the different submatrices separately.

Only for neutral fields M can have non-vanishing diagonal elements which are called Majorana mass terms. Such terms violate all conservation laws related to U(1) symmetries. In particular, lepton number is violated by such a mass term for neutrinos leading to the phenomenon of neutrino oscillations.

For neutral fields ω_R we can always perform a unitary transformation

$$\omega_R = U_o \Omega_R \tag{5.15}$$

such that $U_o^T M U_o$ is a diagonal, positive definite mass matrix M^d and

$$L_M = \frac{1}{2}\Omega_R^T C^{-1} U_o^T M U_o \Omega_R + \text{h.c.} = -\frac{1}{2}\overline{\Omega_L^C} M^d \Omega_R + \text{h.c.} = -\frac{1}{2}\bar{N} M^d N \tag{5.16}$$

with

$$N = \Omega_R + \Omega_L^C \quad . \tag{5.17}$$

Since

$$(\Omega_{\substack{L\\R}})^C = \Omega^C_{\substack{R\\L}}$$

we find that the general mass matrix for neutral fermion fields leads to a number of massive so-called Majorana fields N which obey $N^C = N$. In case of degeneracy two massive Majorana fields may be grouped together to form a Dirac field.

For charged fields the same procedure could be followed but it is a little more convenient to write

$$\omega_R = \begin{bmatrix} \phi^C_R \\ \chi_R \end{bmatrix}$$

again where ϕ and χ each contain n_Q fields with the same charge Q. In this case,

$$L_M = \frac{1}{2} (\phi^{C^T}_R \ \chi^T_R) C^{-1} \begin{bmatrix} 0 & M_Q \\ M^T_Q & 0 \end{bmatrix} \begin{bmatrix} \phi^C_R \\ \chi_R \end{bmatrix} + \text{h.c.} \tag{5.18}$$

is the general form for the mass terms with an arbitrary n_Q-dimensional matrix M_Q. Using again (5.3) - (5.6) this can also be written

$$L_M = \phi^{C^T}_R C^{-1} M_Q \chi_R + \text{h.c.} = - \overline{\phi_L} M_Q \chi_R + \text{h.c.} \tag{5.19}$$

which is the conventional expression for a Dirac mass term. Defining a Dirac field $F = F_L + F_R$ with

$$\phi_L = U_{QL} F_L$$

$$\chi_R = U_{QR} F_R \tag{5.20}$$

such that $U^\dagger_{QL} M_Q U_{QR}$ is a diagonal, positive definite matrix M^d_Q we arrive at the mass Lagrangian

$$L_M = - \bar{F} M^d_Q F \tag{5.21}$$

of n_Q massive charged fermions.

Altogether, diagonalization of the mass matrix (5.14) leads to fermion fields N, F corresponding to massive particles and related to the original fields by a unitary transformation U which mixes fields of the same charge. In the original basis all generators commuting with Q are represented by diagonal matrices. In the basis of mass eigenfields N, F this will in general only be true for Q itself because $[Q,U] = 0$ and thus the electromagnetic current is again diagonal. Since experimentally non-diagonal (flavour changing) neutral currents are heavily suppressed [26] one usually imposes the requirement of <u>natural flavour conservation</u> [29]: neutral currents should be diagonal[+] independent of the details of the mass matrix and therefore for every U. This requirement imposes severe constraints on any gauge theory.

After this general discussion let us specialize to the gauge group SU(2) × U(1) which in addition to the electromagnetic current has one neutral current

$$j^{\mu}_{NC} = \bar{\psi}\gamma^{\mu}(T_3 - \sin^2\Theta_W Q)\psi \qquad (5.22)$$

in terms of the original weak eigenfields ψ.

For (5.22) to conserve flavour naturally T_3 must commute with U for every possible U and therefore T_3 must be proportional to Q. Actually, strangeness and charm changing neutral currents are experimentally suppressed even to $O(G_F\alpha)$ in the effective neutral current Lagrangian so that they must not be generated by lowest-order radiative corrections. This leads to the following condition [29]:

All fields of the same charge must have the same eigenvalues of T_3 and $\vec{T}^2$ (for left- and right-handed components separately).

[+]Unless the masses of the mediating neutral vector bosons are large enough to sufficiently suppress flavour changing neutral currents.

This is a very strong indication for the generation structure of fundamental fermions as embodied in the standard model where all left-handed fermions are SU(2) doublets and all right-handed fermions are singlets.

Let us now investigate the charged current for quarks ($Q_p = 2/3$, $Q_n = -1/3$)

$$\begin{bmatrix} p_a \\ n_a \end{bmatrix}_L , \; p_{aR}, \; n_{aR}, \qquad a = 1,2,\ldots,n_G \tag{5.23}$$

where n_G denotes the number of generations:

$$j^\mu_{CC} = \overline{p_a}\, \gamma^\mu (1+\gamma_5) n_a = 2\, \overline{p_{aL}}\, \gamma^\mu \; n_{aL} \quad . \tag{5.24}$$

Diagonalization of the mass matrix in the Q = 2/3, -1/3 subspaces leads to massive quarks u, c, t,...,d,s,b,... . Recalling (5.20), j^μ_{CC} is seen to take the form

$$j^\mu_{CC} = \overline{2\,(u,c,t,\ldots)_L}\;\; \gamma^\mu\, U_C \begin{bmatrix} d \\ s \\ b \\ \cdot \\ \cdot \\ \cdot \end{bmatrix}_L \tag{5.25}$$

where $U_C = U^\dagger_{pL}\, U_{nL}$ is an n_G-dimensional unitary matrix called mixing matrix or generalized Cabibbo matrix.

An n_G-dimensional unitary matrix is characterized by n_G^2 real parameters of which $n_G(n_G-1)/2$ can be regarded as generalized Euler angles which parametrize an n_G-dimensional orthogonal matrix. The matrices U_{pL}, U_{nL} are determined by the requirement that the mass matrix in (5.21) be diagonal and positive definite. Multiplying each mass eigenfield by a phase factor does not change (5.21) but modifies U_C in (5.25) except that one overall phase drops out. Physically relevant are therefore only

$$N_{ph} = n_G^2 - n_G(n_G-1)/2 - (2n_G-1) = (n_G-1)(n_G-2)/2 \qquad (5.26)$$

phases which imply in general CP violation from charged current interactions for $n_G \geq 3$ [8].

For $n_G = 3$ we have $N_{ph} = 1$ and $n_G(n_G-1)/2 = 3$ angles. The mixing matrix is usually written in the form

$$U_{KM} = \begin{bmatrix} c_1 & s_1c_3 & s_1s_3 \\ -s_1c_2 & c_1c_2c_3-s_2s_3e^{i\delta} & c_1c_2s_3+s_2c_3e^{i\delta} \\ s_1s_2 & -c_1s_2c_3-c_2s_3e^{i\delta} & -c_1s_2s_3+c_2c_3e^{i\delta} \end{bmatrix} \qquad (5.27)$$

with $s_i = \sin\Theta_i$, $c_i = \cos\Theta_i$ $(i=1,2,3)$ and the CP violating phase δ.

It is not my task to discuss the phenomenological implications [30] of (5.27), but let me note an interesting property of this mixing matrix which is characteristic for three generations. It can easily be checked that one can turn U_{KM} into an orthogonal matrix by redefining quark phases if either one of the angles $\Theta_i = 0$ or $\pi/2$. If charged currents are the only source of CP violation any CP violating amplitude must therefore be proportional to

$$c_1c_2c_3s_1s_2s_3 \sin\delta \quad . \qquad (5.28)$$

Thus, the observed smallness of CP violation may either be due to the smallness of $\sin\delta$ or of s_2s_3 $(s_1 \simeq 0.23)$.

What about leptonic charged currents? In the standard model with a Higgs doublet and only left-handed neutrinos there can neither be Majorana nor Dirac mass terms for the neutrinos which must therefore be massless. In the present context it is important that they are degenerate because in that case I can define arbitrary unitary combinations as mass eigenstates without changing any physics. But this freedom can be used to transform away the leptonic analogoue of U_C in (5.25) so that the effective mixing matrix is the unit matrix. In SU(2) × U(1) gauge models with Higgs triplets or singlets (Majorana mass terms) and/or right-handed neutrinos (Dirac mass terms) [31] or in theories with larger gauge groups the neutrinos will in general be non-degenerate so that a non-trivial mixing matrix emerges with interesting phenomenological consequences.

The generation structure of fundamental fermions (quarks and leptons) is one of the big puzzles of present-day particle physics. The standard model or any other gauge theory of the sequential type cannot by itself shed light on this question. It has therefore been conjectured that there is some additional symmetry which relates the different generations to each other and which must of course be spontaneously broken since quarks and leptons show no degeneracy. Where such a horizontal symmetry comes from is anybody's guess: grand unified theories [2], dynamical symmetry breaking [2], composite models [32],...? What I find intriguing is the following rather surprising result: For a given gauge group with a certain particle content, including scalar or effective scalar fields, there are usually only very few types of mass matrices which are consistent with any horizontal symmetry at all [33]. So the situation may not be quite as hopeless as looking for a needle in a stack of hay.

VI. THE CASE FOR AND AGAINST ANOMALIES

In spite of their discriminating name anomalies are rather the normal thing in a quantum field theory with vector- and axial-vector currents. Their physical relevance is especially transparent in 1+1-dimensional QED.

Consider first the Dirac equation for a positively charged fermion of mass m in an external potential A_o:

$$i\,\frac{\partial\psi}{\partial t} = (-i\alpha\,\frac{\partial}{\partial x} + \beta m + eA_o)\,\psi \quad . \tag{6.1}$$

In two dimensions the Dirac matrices are the Pauli matrices and a possible choice is

$$\alpha = \sigma_1, \quad \gamma^o = \beta = \sigma_3, \quad \gamma^1 = \beta\alpha = i\sigma_2, \quad \gamma_5 = \alpha \ . \tag{6.2}$$

We can define a vector and an axial-vector current in the usual way

$$j^\mu = \bar{\psi}\,\gamma^\mu\,\psi \quad , \qquad j_5^\mu = \bar{\psi}\,\gamma^\mu\,\gamma_5\,\psi \tag{6.3}$$

with

$$\partial_\mu\, j^\mu = 0 \ , \qquad \partial_\mu\, j_5^\mu = 2im\,\bar{\psi}\,\gamma_5\,\psi \tag{6.4}$$

using the Dirac equation (6.1). In the limit $m \to 0$ which will be assumed from now on we get two conserved charges

$$Q = \int dx\ \psi^\dagger\,\psi \quad , \qquad Q_5 = \int dx\ \psi^\dagger\,\gamma_5\,\psi \ . \tag{6.5}$$

For the second-quantized theory we Fourier decompose $\psi(t,x)$ as usual into particle annihilation operators $b(p)$ and anti-particle creation operators $d^\dagger(p)$. Inserting into (6.5) one easily obtains

$$Q = \int dp\ [b^\dagger(p)b(p) - d^\dagger(p)d(p)] = N - \bar{N}$$

$$Q_5 = \int dp\ \mathrm{sgn}\,p[b^\dagger(p)b(p) - d^\dagger(p)d(p)] = N_R - N_L + \bar{N}_L - \bar{N}_R \tag{6.6}$$

$$N = N_L + N_R \ , \qquad \bar{N} = \bar{N}_L + \bar{N}_R$$

where $N_{L,R}$ ($\bar{N}_{L,R}$) are the number operators for left-, right-moving (anti-)particles.

Let us now look [34] at the specific potential $eA_o(x)$ shown in Fig. 5. The electric field near the origin will produce particle-antiparticle pairs (the general condition

$2m < V$ is of course always fulfilled for $m = 0$) with the positively charged particles moving to the left and the antiparticles to the right. But for such an emission process $\Delta Q_5 < 0$, so Q_5 cannot possibly be time-independent in contrast to Q since $\Delta Q = 0$.

To elucidate this puzzle consider the current correlation functions [35]

$$\Pi^V_{\mu\nu}(k) = \int d^2x\, e^{ikx} \langle 0|T\{j_\mu(x), j_\nu(0)\}|0\rangle$$

$$\Pi^A_{\mu\nu}(k) = \int d^2x\, e^{ikx} \langle 0|T\{j_{5\mu}(x), j_{5\nu}(0)\}|0\rangle \tag{6.7}$$

with the Lorentz covariant decomposition

$$\Pi^{V,A}_{\mu\nu}(k) = (g_{\mu\nu}k^2 - 2k_\mu k_\nu) B^{V,A}(k^2) + g_{\mu\nu} C^{V,A}(k^2) \; . \tag{6.8}$$

From (6.2) it can immediately be checked that

$$\gamma_\mu \gamma_5 = \varepsilon_{\mu\nu} \gamma^\nu \qquad (\varepsilon_{\mu\nu} = -\varepsilon_{\nu\mu},\ \varepsilon_{01} = 1) \tag{6.9}$$

and thus

$$j_{5\mu} = \varepsilon_{\mu\nu} j^\nu \quad . \tag{6.10}$$

Using (6.10) and the relations (remember that we are in two dimensions)

$$\varepsilon_{\mu\lambda} \varepsilon^{\nu\lambda} = -\delta^\nu_\mu$$

$$\varepsilon_{\mu\lambda} \varepsilon_{\nu\rho} k^\lambda k^\rho = -g_{\mu\nu} k^2 + k_\mu k_\nu \tag{6.11}$$

one gets

$$B^V(k^2) = B^A(k^2)$$

$$C^V(k^2) = -C^A(k^2) \quad . \tag{6.12}$$

Vector current conservation implies $k^\mu \Pi^V_{\mu\nu} = 0$ and therefore

$$C^V(k^2) = k^2 B^V(k^2) \quad . \tag{6.13}$$

On the other hand,

$$k^\mu \Pi^A_{\mu\nu} = (-k^2 B^A + C^A) k_\nu = (-k^2 B^V - C^V) k_\nu = -2C^V k_\nu \tag{6.14}$$

so that the axial-vector current is indeed not conserved unless $C^V(k^2) = 0$. But $C^V(k^2)$ must be different from zero because otherwise

$$\Pi^V_{\mu\nu} = \Pi^A_{\mu\nu} = 0 \text{ for } k^2 \neq 0$$

which is certainly not the case in perturbation theory.

The lesson to be learnt from this 2-dimensional example is also valid for the real world with four dimensions: applying equations of motion to products of field operators at the same space-time point is a dangerous undertaking! The correct procedure is to regularize the theory by introducing a cutoff, to establish relations between operators and to study these relations in the limit of infinite cutoff. Imagine e.g. a Pauli-Villars regularization with heavy fermions of mass M. By introducing heavy fermions one breaks of course the axial symmetry, but vector current conservation can be maintained. The general result (6.14) shows that there must be a non-vanishing divergence $\partial_\mu j^\mu_5$ in the limit $M \to \infty$. But the above discussion implies the much stronger conclusion that there cannot possibly exist any regularization procedure that respects both vector and axial-vector conservation.

It is not difficult to show (again for two and four dimensions) that the problem only comes from one-loop diagrams because higher-order loops can always be regulated in a way consistent with vector- and axial-vector conservation [36]. Such a one-loop calculation shows that the correct divergence equation in two dimensions is [37]

$$\partial_\mu j^\mu_5 = -\frac{e}{2\pi} \varepsilon_{\mu\nu} F^{\mu\nu} \tag{6.15}$$

which can describe pair creation in the external field A_o in a completely satisfactory way [34].

The corresponding problem in four dimensions first[+] shows up in the 3-point function of two vector and one (or 3) axial currents and is known as the Adler-Bell-Jackiw anomaly [38]. It is also called the triangle anomaly and can be calculated from the diagrams in Fig. 6.

Let us work out the triangle anomaly for the general gauge theory based on (5.11) with couplings (no sum over α)

$$-ig_\alpha \, T_{\alpha,ab} \, \gamma^\mu \, \frac{1-\gamma_5}{2} \tag{6.16}$$

of two fermions a,b to the gauge field α. In view of the preceding discussion the anomaly is related to the (linear) divergence of the triangle diagrams. It is not difficult to convince oneself that the ultraviolet divergent terms containing γ_5 are the same [38] for the two diagrams of Fig. 6 except for the different order of the matrices T_β, T_γ. Summing over all fermions in the loop the anomaly which is independent of fermion masses is therefore proportional to

$$A_{\alpha\beta\gamma} = \mathrm{Tr}\ (T_\alpha\{T_\beta,T_\gamma\}) \ . \tag{6.17}$$

If T_α is of the form (5.10) we can also write

$$A_{\alpha\beta\gamma} = \mathrm{Tr}(T_{\alpha R}\{T_{\beta R},T_{\gamma R}\}) - \mathrm{Tr}(T^*_{\alpha L}\{T^*_{\beta L},T^*_{\gamma L}\}) =$$

$$= \mathrm{Tr}(T_{\alpha R}\{T_{\beta R},T_{\gamma R}\}) - \mathrm{Tr}(T_{\alpha L}\{T_{\beta L},T_{\gamma L}\}) \tag{6.18}$$

using the hermiticity of $T_{\alpha L}$ and the invariance of the trace with respect to transposition. Thus, in the usual electroweak gauge theories the total anomaly is given by the difference

---

[+] In general there are also 4-point and 5-point anomalies but they are related to the 3-point anomaly; in particular, all anomalies vanish if the triangle anomaly is zero [39].

between right-handed and left-handed contributions.

In order to simplify matters let us investigate a general abelian gauge theory in which scalar fields do not couple to fermions and which has no gauge invariant fermion mass terms so that (5.12) vanishes and all fermions are massless. Denoting the total amplitude in Fig. 6 by $S^{\lambda\mu\nu}_{\alpha\beta\gamma}(p_1,q_1)$, a non-vanishing anomaly $A_{\alpha\beta\gamma}$ implies that the normal Ward identity

$$k_\lambda\ S^{\lambda\mu\nu}_{\alpha\beta\gamma} = 0 \tag{6.19}$$

cannot be maintained for all α, β, γ in analogy to the 2-dimensional situation in (6.14). The current associated with the generator T_α has an anomalous divergence for $A_{\alpha\beta\gamma} \neq 0$ yielding an anomalous Ward identity $k_\lambda\ S^{\lambda\mu\nu}_{\alpha\beta\gamma} \neq 0$.

On the other hand, (6.19) is absolutely essential for perturbative unitarity in the gauge theory under consideration. The problem can be exemplified by studying the amplitude in Fig. 7 where inclusion of crossed diagrams as in Fig. 6 is understood. In the R_ξ gauge with gauge field propagator (3.42) we get

$$S^{\lambda\mu\nu}_{\alpha\beta\gamma}(p_1,q_1)\frac{-i\delta_{\alpha\delta}}{k^2-M_\alpha^2+i\varepsilon}[g_{\lambda\rho}-(1-1/\xi)\frac{k_\lambda k_\rho}{k^2-M_\alpha^2/\xi}]S^{\rho\sigma\tau}_{\delta\varepsilon\eta}(p_2,q_2)\,. \tag{6.20}$$

This amplitude has a spurious singularity at $k^2 = M_\alpha^2/\xi$ which is obviously unphysical because it depends on the parameter ξ. In our case (massless fermions, no Yukawa couplings) this singularity cannot be cancelled by another 2-loop diagram with the same external legs. In order to ensure perturbative unitarity the residue of this unphysical pole must therefore vanish. Since the residue is given by (no sum over α)

$$\frac{-i}{M_\alpha^2}\, k_\lambda\ S^{\lambda\mu\nu}_{\alpha\beta\gamma}\ k_\rho\ S^{\rho\sigma\tau}_{\alpha\varepsilon\eta} \tag{6.21}$$

the normal Ward identities (6.19) must hold and consequently $A_{\alpha\beta\gamma} = 0$.

In the general non-abelian case with massive fermions the Ward identities are more complicated and additional diagrams (scalar exchange) contribute to the residue at $k^2 = M_\alpha^2/\xi$. However, concerning anomalies the situation is exactly the same as before mainly because the anomalies are independent of fermion masses.

The condition

$$\mathrm{Tr}(T_\alpha\{T_\beta,T_\gamma\}) = 0 \tag{6.22}$$

imposes severe constraints on the fermion representations [40]. Let us apply (6.22) to the gauge group SU(2) × U(1) with the generators T_1, T_2, T_3, Y.

We first employ the fact that all representations of SU(2) are equivalent to their complex conjugate representations so that

$$T_\alpha^* = -U^\dagger\, T_\alpha\, U\,, \qquad \alpha = 1,2,3 \tag{6.23}$$

and therefore $A_{\alpha\beta\gamma} = 0$ automatically using the same arguments as for (6.18). Moreover,

$$\mathrm{Tr}(Y\{T_\alpha,T_\beta\}) \sim \delta_{\alpha\beta} \quad , \qquad \mathrm{Tr}(Y^2\, T_\alpha) = 0 \;, \tag{6.24}$$

and so we are left with the two conditions

$$\mathrm{Tr}\; Y\, T_3^2 = \mathrm{Tr}\; Y^3 = 0 \tag{6.25}$$

to ensure that all anomalies vanish. With $Y/2 = Q - T_3$ these conditions can also be written

$$\mathrm{Tr}\; Q\, T_3^2 = \mathrm{Tr}\; Q^3 = 0 \;. \tag{6.26}$$

But $\mathrm{Tr}\; Q^3 = 0$ is always fulfilled because for each charged fermion field its oppositely charged counterpart must also be in the representation vector (5.9). So, finally, the only non-trivial constraint for the general SU(2) × U(1) gauge theory is

$$\mathrm{Tr}\; Q\, T_3^2 = 0 \; . \tag{6.27}$$

In the standard model all fermions are either in doublets ($T_3^2 = 1/4$) or singlets ($T_3 = 0$) of SU(2), so (6.27) simplifies to

$$\mathrm{Tr}_d\; Q = 0 \tag{6.28}$$

where the trace is only to be taken over the doublet fields (Tr Q = 0 is again trivial).

For a lepton doublet $\binom{\nu_\ell}{\ell^-}_L$ Tr Q = -1, and for a quark doublet $\binom{p}{n}_L$ Tr Q = 1/3. There are two important conclusions [41]:

1) The anomalies do not cancel in the lepton sector.
2) The colour factor 3 is needed to fulfil (6.28).

This is another strong theoretical argument in favour of the observed generation structure of fermions: leptons and quarks of each family separately obey the anomaly condition. Is this maybe an indication that leptons and quarks are related to each other on a more fundamental level [32]?

After having argued that there must be no anomalies in gauge theories let me emphasize strongly that this applies only to currents associated to generators of the gauge group. In general there will be and according to our present understanding even must be other currents associated to global symmetries with anomalous divergences. In QED the anomaly of the axial isospin current is needed to understand the $\pi^0 \to 2\gamma$ decay [38], and in QCD the anomaly of the axial U(1) current allows for an understanding of the so-called η-problem [42]. More recently, the analysis of anomalies initiated by 't Hooft [43] has shed new light on chiral symmetry breaking in QCD and on possible chirality conservation in composite models of quarks and leptons [32].

Apart from currents associated with chiral transformations there is another class of currents with anomalous behaviour. Those currents are the Noether currents of scale and conformal transformations and they are anomalous for the same reason as the vector and axial-vector currents: the symmetries are broken by the necessary regularization of quantum field theory. The most familiar of those anomalies are known as anomalous dimensions and without them the renormalization group equations for Green's functions or coefficient functions of operator product expansions would be trivial.

VII. RADIATIVE CORRECTIONS IN THE STANDARD MODEL

An important incentive for a gauge theory of electroweak interactions was the need for a renormalizable quantum field theory with a well-defined perturbation expansion. The first calculations of radiative corrections mainly served the purpose of exhibiting the miraculous cancellations we have discussed for Born terms also for higher orders. Then the activity in that field decreased for some time for two reasons: first of all, the corrections were indeed $O(\alpha)$ compared to the tree level results and therefore too small in most cases to be of practical interest; secondly, one must usually calculate a considerable number of diagrams before one arrives at this sobering conclusion.

In the last two or three years the situation has changed: the experimental determination of $\sin^2\Theta_W$ has reached an accuracy of about 4% and radiative corrections may be important. Other observables of paramount interest are the vector boson masses M_W, M_Z. I shall concentrate in this review on the radiative corrections to those quantities in the standard model. In the pedagogical spirit of these lectures I shall not just write down the final numbers but also try to exhibit how renormalization theory is put to work in a specific problem.

Unlike QED there are a lot of parameters in the Lagrangian even for the standard model. In addition to the fermion masses and the Higgs boson mass there are the two gauge coupling constants g, g' and the vacuum expectation value v of the scalar field. All other quantities can be expressed in terms of these parameters: in particular, the Yukawa couplings of fermions to the Higgs boson are given by m_F/v, and they can therefore usually be neglected for first-order corrections (recall that v = 246 GeV) unless the top quark or any additional fermions are very heavy.

It may not be obvious to everybody how a consistent perturbation expansion is set up in a theory with different couplings which even appear with different powers in the Lagrangian. The guideline is gauge invariance: only if one expands S-matrix elements in the number of loops are the results guaranteed to be gauge invariant in every order. The loop expansion has this pleasant property because it can be viewed as an expansion in powers of $\hbar$ [44] and such an expansion must of course be gauge invariant order by order.

The basic parameters g, g', v may be replaced by e, $\sin^2\Theta_W$, G_F through the relations (cf. Sect. IV)

$$\mathrm{tg}\Theta_W = g'/g, \qquad e = g\,\sin\Theta_W, \qquad G_F^{-1} = \sqrt{2}\,v^2. \tag{7.1}$$

At the tree level the vector boson masses are given by

$$M_W^2 = \frac{\sqrt{2}\,e^2}{8G_F\sin^2\Theta_W}, \qquad M_Z^2 = M_W^2/\cos^2\Theta_W \quad . \tag{7.2}$$

At the one-loop level the poles of the vector boson propagators which define the physical masses are shifted from their values (7.2) by self-energy corrections shown in Fig. 8 for the charged W boson. Denoting the self-energies $\Pi_W(q^2)$, $\Pi_Z(q^2)$, the mass shifts are

$$\delta M_W^2 = \text{Re}\ \Pi_W(M_W^2)$$

$$\delta M_Z^2 = \text{Re}\ \Pi_Z(M_Z^2) \quad . \tag{7.3}$$

If one actually calculates the diagrams of Fig. 8 and similar ones for the Z boson in a suitable regularization scheme (practically everybody uses dimensional regularization [45]), $\Pi_W(M_W^2)$ and $\Pi_Z(M_Z^2)$ turn out to be ultraviolet divergent.

Depending on one's familiarity with renormalization theory one might be inclined to invoke mass counterterms in the Lagrangian to repair those infinities as one has learned in QED in connection with the electron self-energy. On second thought, this cannot be the right remedy. As we have argued in Sect. III a SBGT shows the same high energy behaviour as the unbroken theory and therefore the necessary counterterms can be chosen to be the same as in the symmetric case. But in the symmetric theory there can neither be bare masses nor mass counterterms for the vector bosons because of gauge invariance. Having to introduce such counterterms would contradict the assertion that gauge theories can be renormalized in a gauge invariant way.

Of course, the solution to this artificially created problem is that the parameters of the tree level masses (7.2) are the bare quantities in the Lagrangian which receive one-loop corrections themselves. At this point we have to make up our mind how the parameters e, G_F, $\sin^2\Theta_W$ are defined in terms of observable quantities. Unfortunately, there are almost as many different renormalization conventions as there are authors in this field.

Everybody agrees that e should be related to the physical charge of the positron. A possible procedure is to consider Coulomb scattering of muons on electrons and to require that the amplitude

$$A(\mu e \rightarrow \mu e) \xrightarrow[k^2 \rightarrow 0]{} \frac{e_r^2}{k^2} \bar{\mu} \gamma_\lambda \mu \quad \bar{e} \gamma^\lambda e \tag{7.4}$$

where k is the momentum transfer and e_r is the renormalized charge with $e_r^2/4\pi = \alpha = 1/137.036$. Calculating $A(\mu e \rightarrow \mu e)$ to one-loop accuracy one finds

$$e^2 = e_r^2 (1 + \delta_1) \tag{7.5}$$

in terms of the bare charge and an ultraviolet divergent contribution δ_1.

There is almost unanimous agreement that G_F should be normalized to μ-decay. In fact, the muon life-time is usually written

$$\tau_\mu^{-1} = \frac{G_\mu^2 m_\mu^5}{192\pi^3} (1-8m_e^2/m_\mu^2) (1 + \frac{\alpha}{2\pi} (\frac{25}{4} - \pi^2)) \tag{7.6}$$

where the $O(\alpha)$ terms are the purely photonic one-loop corrections of the V-A Fermi theory which are finite and gauge invariant by themselves [46]. The remaining radiative corrections to one-loop order in the standard model amount to

$$G_F = G_\mu (1 + \delta_2) \tag{7.7}$$

with δ_2 again ultraviolet divergent.

It is the renormalization convention for $\sin^2\Theta_W$ which differs from author to author. This renormalization dependence must not be forgotten when comparing results from different papers. Let us follow, at least for the time being, the suggestion of Veltman and others [47,48] to define $\sin^2\Theta_W$

by the ratio of the total cross-sections for $\bar{\nu}_\mu e \to \bar{\nu}_\mu e$ and $\nu_\mu e \to \nu_\mu e$ at some convenient energy $E_o \ll M_Z$. This seems to be a suitable convention in order to minimize the effect of strongly interacting particles. In practice, one defines

$$\left.\frac{\sigma(\bar{\nu}_\mu e \to \bar{\nu}_\mu e)}{\sigma(\nu_\mu e \to \nu_\mu e)}\right|_{E_o \ll M_Z} = \frac{1-4\sin^2\Theta_V + 16\sin^4\Theta_V}{3-12\sin^2\Theta_V + 16\sin^4\Theta_V} \tag{7.8}$$

which is the tree level result for $\sin^2\Theta_V = \sin^2\Theta_W$ (V stands for Veltman). In the one-loop approximation one then obtains

$$\sin^2\Theta_W = \sin^2\Theta_V\ (1+\delta_3) \tag{7.9}$$

and the one-loop correction δ_3 is again divergent.

Putting (7.2), (7.3), (7.5), (7.7) and (7.9) together we find the physical vector boson masses to one-loop accuracy:

$$\begin{aligned} M_W^2 &= \frac{\pi\alpha}{\sqrt{2}\, G_\mu \sin^2\Theta_V}(1+\delta_1-\delta_2-\delta_3) + \mathrm{Re}\,\Pi_W(M_W^2) \equiv \\ &\equiv \frac{\pi\alpha}{\sqrt{2}\, G_\mu \sin^2\Theta_V}(1+2\Delta_W)\ , \end{aligned} \tag{7.10}$$

$$\begin{aligned} M_Z^2 &= \frac{\pi\alpha}{\sqrt{2}\, G_\mu \sin^2\Theta_V\cos^2\Theta_V}(1+\delta_1-\delta_2-\delta_3+\delta_3 tg^2\Theta_V) + \mathrm{Re}\,\Pi_Z(M_Z^2) \equiv \\ &\equiv \frac{\pi\alpha}{\sqrt{2}\, G_\mu \sin^2\Theta_V\cos^2\Theta_V}(1+2\Delta_Z)\ . \end{aligned}$$

As it must be, Δ_W and Δ_Z are finite as the cutoff goes to infinity (or rather, as $n \to 4$ in dimensional regularization). Since they are one-loop quantities they are unambiguously defined in terms of the tree level parameters.

The formulae for Δ_W, Δ_Z are too involved to be reproduced here [48,49] so I shall only make two remarks. One possible uncertainty in evaluating Δ_W, Δ_Z is the Higgs mass m_H about which practically nothing is known. Especially for $m_H > M_W$ terms proportional to m_H^2 could produce a drastic dependence on m_H. However, it turns out that for large m_H, Δ_W and Δ_Z depend only logarithmically on m_H [48]:

$$\Delta_W^{\text{Higgs}} \simeq \frac{\alpha}{96\pi \sin^2\Theta_W} \ln m_H^2/M_W^2$$

$$\Delta_Z^{\text{Higgs}} \simeq \frac{\alpha(1+10\text{tg}^2\Theta_W)}{96\pi \sin^2\Theta_W} \ln m_H^2/M_Z^2 \quad . \qquad (7.11)$$

This weak dependence on m_H is an example of the so-called "screening theorem" [50].

More important in practice are the fermionic contributions to the self-energies (diagram c in Fig.8). Each lepton doublet contributes

$$\Delta_W^{\text{lepton}} = \frac{\alpha}{24\pi \sin^2\Theta_W} (\ln M_W^2/m_\ell^2 - 5/3)$$

$$\Delta_Z^{\text{lepton}} = \frac{\alpha(1-2\sin^2\Theta_W+4\sin^4\Theta_W)}{24\pi \sin^2\Theta_W\cos^2\Theta_W} (\ln M_Z^2/m_\ell^2 - 5/3) \quad . \qquad (7.12)$$

The analogous contributions from quark loops therefore depend crucially on what values are used for the masses of light quarks u, d, s. The small current quark masses are certainly not suitable for this purpose because for light quarks non-perturbative strong interaction effects are expected to dominate. It has been suggested by several authors [48,49,51] to relate the hadronic contributions to the mass shifts to the total cross-section for $e^+e^- \to$ hadrons through a dispersive analysis. By disentangling the non-strange and the strange part (and possibly even the charm part [49]) of the experimentally measured cross-section one obtains a rather reliable estimate for the hadronic contributions. For the heavy quarks with masses large compared to the QCD scale parameter lowest-order perturbation theory is expected to be valid and so the only remaining uncertainty lies in the value for m_t.

However, all those uncertainties related to m_H and to the hadronic contributions are at the moment negligible compared to the experimental error of $\sin^2\Theta_W$. The practical drawback of the renormalization scheme discussed so far is that $\sin^2\Theta_V$ is not very accurately known due to the small statistics of $\nu_\mu e$ scattering. It is therefore advisable to turn to either deep inelastic neutrino nucleon scattering or polarized electron deuteron scattering for a precise determination of the Weinberg angle.

The radiative corrections for neutrino nucleon scattering have been calculated in [52,53] and in the leading logarithm approximation in [54]. Again, one cannot directly compare the results because different renormalization frameworks are employed. We follow here the work of Marciano and Sirlin [51,52] who use the conventional definitions of e and G_F but normalize the Weinberg angle such that (S for Sirlin)

$$\cos^2\Theta_S = M_W^2/M_Z^2 \quad . \tag{7.13}$$

At first sight, this definition of the renormalized Weinberg angle may give rise to some confusion because there are papers [55] reporting radiative corrections to the tree level relation $\cos^2\Theta_W = M_W^2/M_Z^2$. This is a good example for the importance of keeping in mind the different definitions of $\sin^2\Theta_W$. With the renormalization convention (7.13) the above-mentioned radiative corrections are zero to all orders by definition.

Of course, differently normalized $\sin^2\Theta_W$ can always be related to each other. From (7.10) and (7.13) one immediately finds

$$M_W^2/M_Z^2 = \cos^2\Theta_V\ [1 + 2(\Delta_W - \Delta_Z)] = \cos^2\Theta_S \tag{7.14}$$

and therefore

$$\sin^2\Theta_V = \sin^2\Theta_S + 2\cos^2\Theta_S(\Delta_W - \Delta_Z) \quad . \tag{7.15}$$

Using the explicit results for Δ_W, Δ_Z it actually turns out [49] that for $\sin^2\Theta_W = 0.22$ and $m_H = 100$ GeV the radiative corrections in (7.15) practically cancel so that $\Theta_V = \Theta_S$ to a very good approximation.

Experimentally, the Weinberg angle is determined from the ratios of neutral current to charged current cross-sections and thus radiative corrections must be applied to both processes. Taking into account experimental cuts in the hadronic energy Marciano and Sirlin [52] obtain from five high-statistics measurements of the neutrino ratio R_ν the value+

$$\sin^2\Theta_S = 0.217 \pm 0.010\ (\pm\ 0.004) \tag{7.16}$$

to be compared with the tree level result

$$\sin^2\Theta_W = 0.227 \pm 0.010\ (\pm\ 0.003) \quad . \tag{7.17}$$

With the renormalization convention (7.13) the radiatively corrected W and Z masses are [52]

$$M_W = 38.64\ {}^{+\ 0.07}_{-\ 0.09}\ \text{GeV}/\sin\Theta_S = 83.0\ {}^{+\ 3.0}_{-\ 2.8}\ \text{GeV} \tag{7.18}$$

$$M_Z = M_W/\cos\Theta_S = 93.8\ {}^{+\ 2.5}_{-\ 2.4}\ \text{GeV} \quad . \tag{7.19}$$

+The second error estimates the theoretical uncertainties.

Comparing with the tree level prediction (4.9) the total mass shifts for the mean values of charged and neutral vector boson masses due to first-order radiative corrections are 4.7 and 4.8 GeV, respectively. Note that these shifts are larger than the errors in (7.18), (7.19) which are essentially determined by the experimental error of $\sin^2\Theta_W$. When the W and Z bosons will be found the predictions (7.18), (7.19) together with a more accurate measurement of $\sin^2\Theta_W$ should provide the first decisive test of the standard model beyond the tree level. In addition, the radiative corrections for $\sin^2\Theta_W$ allow for another interesting conclusion. After an initial discrepancy the prediction of the simplest grand unified theory based on SU(5) for $\sin^2\Theta_W$ is now in impressive agreement with experiment [56].

ACKNOWLEDGEMENTS

I wish to thank H. Mitter and F. Widder for the invitation to Schladming and H. Kühnelt and H. Rupertsberger for discussions.

N	Lie group
3	$SU(2)$
4	$SU(2) \times U(1)$
5	$SU(2) \times U(1)^2$
6	$SU(2) \times SU(2)$, $SU(2) \times U(1)^3$
7	$SU(2) \times SU(2) \times U(1)$, $SU(2) \times U(1)^4$

Table 1. Complete list of Lie groups (compact, up to local isomorphisms) with $3 \leq N \leq 7$ parameters; the abelian groups $U(1)^N$ are omitted.

REFERENCES

1. Among the standard reviews of the subject are:
   E.S. Abers and B.W. Lee, Phys. Reports 9C (1973) 1.
   J. Bernstein, Rev. Mod. Phys. 46 (1974) 7.
   J.C. Taylor, Gauge Theories of Weak Interactions, Cambridge Univ. Press, Cambridge, 1976.
2. R. Barbieri and P. Sikivie, these Proceedings.
3. S.L. Glashow, Nucl. Phys. 22 (1961) 579.
   S. Weinberg, Phys. Rev. Lett. 19 (1967) 1264.
   A. Salam, Proc. 8-th Nobel Symposium, N. Svartholm, ed., Almqvist and Wiksell, Stockholm, 1968.
   S.L. Glashow, J. Iliopoulos and L. Maiani, Phys. Rev. D2 (1970) 1285.
4. R. Marshall, G. Altarelli, R.H. Dalitz, F. Wagner and J.J. Sakurai, these Proceedings.
5. E. Fermi, Z. Physik 88 (1934) 161.
   R.P. Feynman and M. Gell-Mann, Phys. Rev. 109 (1958) 193.
   R.E. Marshak and E.C.G. Sudarshan, Phys. Rev. 109 (1958) 1860.
   J.J. Sakurai, Nuovo Cimento 7 (1958) 649.
6. P.W. Higgs, Phys. Lett. 12 (1964) 132; Phys. Rev. Lett. 13 (1964) 508; Phys. Rev. 145 (1966) 1156.
   F. Englert and R. Brout, Phys. Rev. Lett. 13 (1964) 321.
   G.S. Guralnik, C.R. Hagen and T.W.B. Kibble, Phys. Rev. Lett. 13 (1964) 585.
7. J.C. Pati and A. Salam, Phys. Rev. D10 (1975) 275.
   R.N. Mohapatra and J.C. Pati, Phys. Rev. D11 (1975) 566 and 2558.
   H. Fritzsch and P. Minkowski, Nucl. Phys. B103 (1976) 61.
8. M. Kobayashi and K. Maskawa, Progr. Theor. Phys. 49 (1973) 652.
9. B.L. Ioffe et al., Soviet Physics JETP 20 (1965) 1281.
10. H. Yukawa, Proc. Phys.-Math. Soc. Japan 17 (1935) 48.
11. C.H. Llewellyn Smith, Phys. Lett. 46B (1973) 233.
    J.M. Cornwall, D.N. Levin and G. Tiktopoulos, Phys. Rev. D10 (1974) 1145.

12. C.N. Yang and R.L. Mills, Phys. Rev. 96 (1954) 191.
R. Utiyama, Phys. Rev. 101 (1956) 1597.
13. The terminology is due to M. Baker and S.L. Glashow, Phys. Rev. 128 (1962) 2462.
14. E.M. Henley and W. Thirring, Elementary Quantum Field Theory, McGraw-Hill, New York, 1962.
15. T.W.B. Kibble, Phys. Rev. 155 (1967) 1554.
16. P.W. Anderson, Phys. Rev. 110 (1958) 827 and Phys.Rev. 130 (1963) 439.
17. J. Goldstone, Nuovo Cimento 19 (1961) 154.
18. Y. Nambu, Phys. Rev. Lett. 4 (1960) 380.
J. Goldstone, A. Salam and S. Weinberg, Phys. Rev. 127 (1962) 965.
W. Gilbert, Phys. Rev. Lett. 12 (1964) 713.
19. G.'t Hooft, Nucl. Phys. B35 (1971) 167.
20. A more recent book on the quantization of gauge theories is: L.D. Faddeev and A.A. Slavnov, Gauge Fields: Introduction to Quantum Theory, Benjamin, Reading, 1980.
21. L.D. Faddeev and V.N. Popov, Phys. Lett. 25B (1967) 29.
22. K. Fujikawa, B.W. Lee and A.I. Sanda, Phys. Rev. D6 (1972) 2923.
23. A.A. Slavnov, Theor. and Math. Phys. 10 (1972) 99.
J.C. Taylor, Nucl. Phys. B33 (1971) 436.
24. J.C. Ward, Phys. Rev. 77 (1950) 2931.
E.S. Fradkin, JETP 29 (1955) 288.
Y. Takahashi, Nuovo Cimento 6 (1957) 370.
25. H. Georgi and S.L. Glashow, Phys. Rev. Lett. 28 (1972) 1494.
26. H. Harari, Phys. Reports 42C (1978) 235.
H. Fritzsch and P. Minkowski, Phys. Reports 73C (1981) 67.
27. J.E. Kim et al., Rev. Mod. Phys. 53 (1981) 211.
I. Liede and M. Roos, Nucl. Phys. B167 (1980) 397.
P.Q. Hung and J.J. Sakurai, The Structure of Neutral Currents, preprint LBL-12364, Aug. 1981.
28. G. Ecker, Proc. Symposium on Lepton and Hadron Interactions, F. Csikor et al., eds., Budapest, 1980.
29. S.L. Glashow and S. Weinberg, Phys. Rev. D15 (1977) 1958.

E.A. Paschos, Phys. Rev. D15 (1977) 1966.
30. R.H. Dalitz and J.J. Sakurai, these Proceedings.
31. T.P. Cheng and L.-F. Li, Phys. Rev. D22 (1980) 2860.
32. M. Peskin, Compositeness of Quarks and Leptons, Cornell preprint CLNS 81/516 (Oct. 1981).
33. G. Ecker, W. Grimus and W. Konetschny, Nucl. Phys. B191 (1981) 465 and references therein.
34. A.S. Blaer, N.H. Christ and J.-F. Tang, Phys. Rev. Lett. 47 (1981) 1364.
35. Y. Frishman et al., Nucl. Phys. B177 (1981) 157.
36. S.L. Adler and W.A. Bardeen, Phys. Rev. 182 (1969) 1517.
37. H. Georgi and J.M. Rawls, Phys. Rev. D3 (1971) 874.
38. S.L. Adler, Phys. Rev. 177 (1969) 2426.
J.S. Bell and R. Jackiw, Nuovo Cimento 60A (1969) 47.
39. W.A. Bardeen, Phys. Rev. 184 (1969) 1848.
J. Wess and B. Zumino, Phys. Lett. 37B (1971) 95.
40. H. Georgi and S.L. Glashow, Phys. Rev. D6 (1972) 429.
41. C. Bouchiat, J. Iliopoulos and P. Meyer, Phys. Lett. 38B (1972) 519.
D.J. Gross and R. Jackiw, Phys. Rev. D6 (1972) 477.
42. G.'t Hooft, Phys. Rev. Lett. 37 (1976) 8 and Phys. Rev. D14 (1976) 3432.
R. Crewther, Riv. Nuovo Cimento 2 (1979) 63.
43. G.'t Hooft, Recent Developments in Gauge Theories, G.'t Hooft et al., eds., Plenum Press, New York, 1980.
44. Y. Nambu, Phys. Lett. 26B (1968) 626.
45. G.'t Hooft and M. Veltman, Nucl. Phys. B44 (1972) 189.
C.G. Bollini and J.J. Giambiagi, Phys. Lett. 40B (1972) 566.
J.F. Ashmore, Nuovo Cimento Lett. 4 (1972) 289.
G.M. Cicuta and E. Montaldi, Nuovo Cimento Lett. 4 (1972) 329.
46. S.M. Berman, Phys. Rev. 112 (1958) 267.
T. Kinoshita and A. Sirlin, Phys. Rev. 113 (1959) 1652.
47. M. Green and M. Veltman, Nucl. Phys. B169 (1980) 137 and Err. ibid. B175 (1980) 547.

48. M. Veltman, Phys. Lett. 91B (1980) 95.
F. Antonelli et al., Phys. Lett. 91B (1980) 90 and Nucl. Phys. B183 (1981) 195.
49. W. Wetzel, Z. Phys. C11 (1981) 117.
50. M. Veltman, Acta Phys. Pol. B8 (1977) 475 and Phys. Lett. 70B (1977) 253.
51. A. Sirlin, Phys. Rev. D22 (1980) 971.
52. W.J. Marciano and A. Sirlin, Phys. Rev. D22 (1980) 2695.
A. Sirlin and W.J. Marciano, Nucl. Phys. B189 (1981) 442.
53. S. Sakakibara, Phys. Rev. D24 (1981) 1149.
E.A. Paschos and M. Wirbel, Dortmund preprint DO-TH 81/4, Apr. 1981.
C.H. Llewellyn Smith and J.F. Wheater, Phys. Lett. 105B (1981) 486.
54. F. Antonelli and L. Maiani, Nucl. Phys. B186 (1981) 269.
S. Dawson, J.S. Hagelin and L. Hall, Phys. Rev. D23 (1981) 2666.
55. M. Veltman, Nucl. Phys. B123 (1977) 89.
M. Chanowitz, M. Furman and I. Hinchliffe, Phys. Lett. 78B (1978) 285.
56. W.J. Marciano and A. Sirlin, Phys. Rev. Lett. 46 (1981) 163.

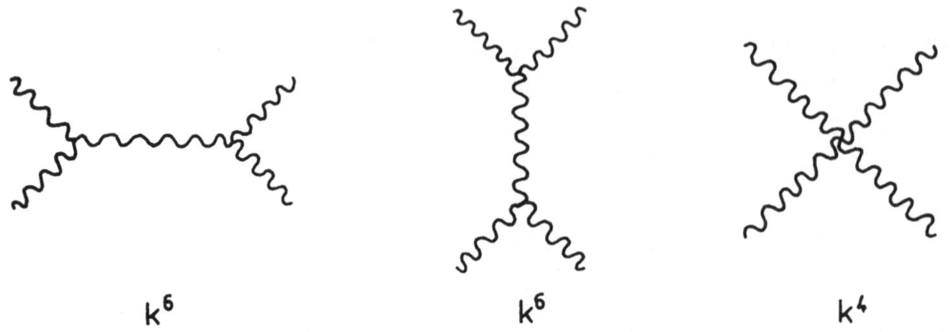

Fig. 1. Born terms for WW → WW; for each diagram the leading divergence for $k \to \infty$ is indicated if all external W are longitudinally polarized.

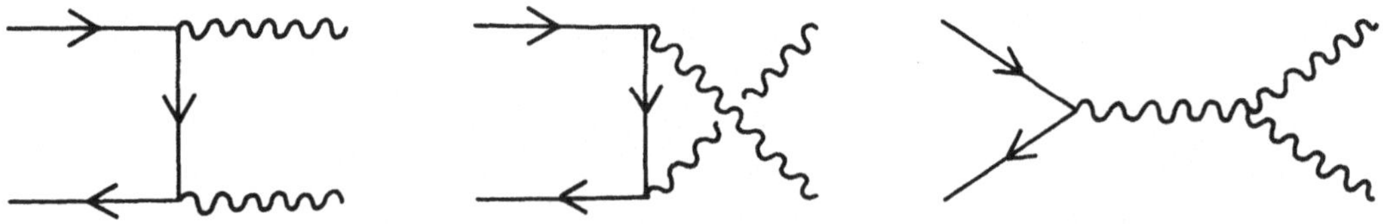

Fig. 2. Born terms for $F\bar{F} \to WW$; all diagrams increase as k^2 for large k if both W are longitudinally polarized and if the conditions (2.12) hold.

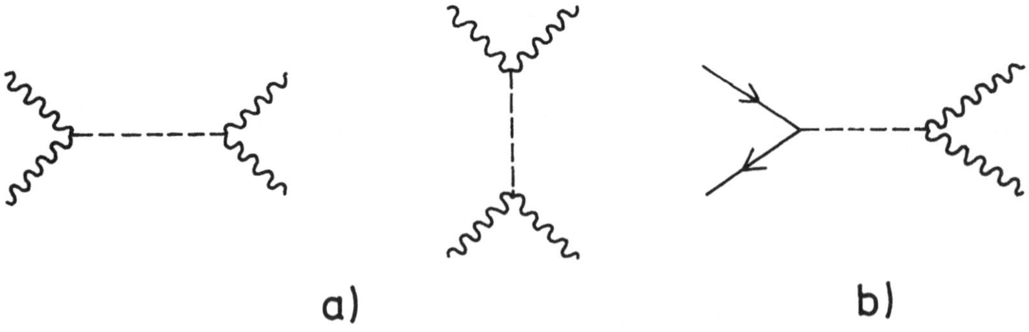

Fig. 3. Additional Born terms for WW → WW (diagrams a) and $F\bar{F} \to WW$ (diagram b) in the presence of scalar fields (dashed lines).

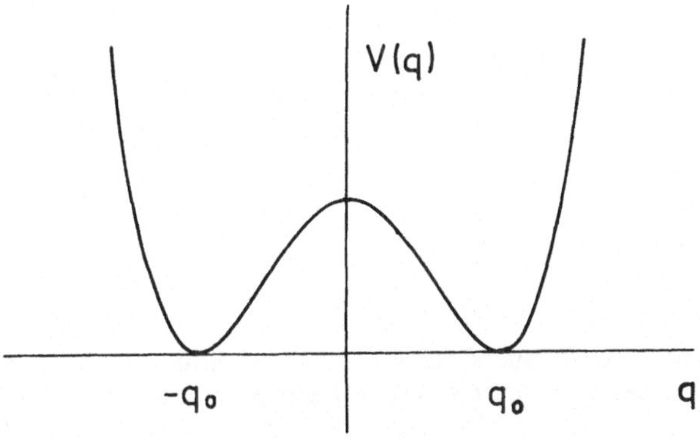

Fig. 4. Double-well potential (3.24) for each oscillator in the linear chain with Hamiltonian (3.23).

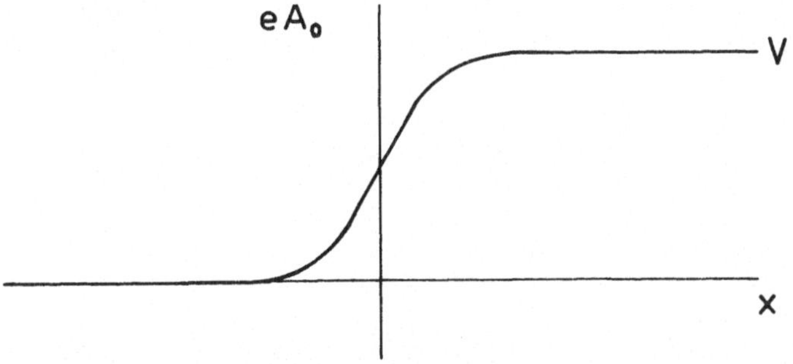

Fig. 5. One-dimensional potential $eA_o(x)$ for the Dirac equation (6.1).

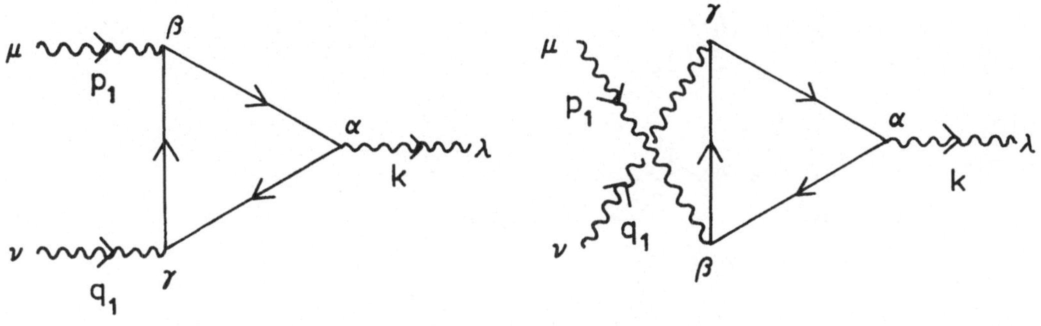

Fig. 6. Triangle diagrams.

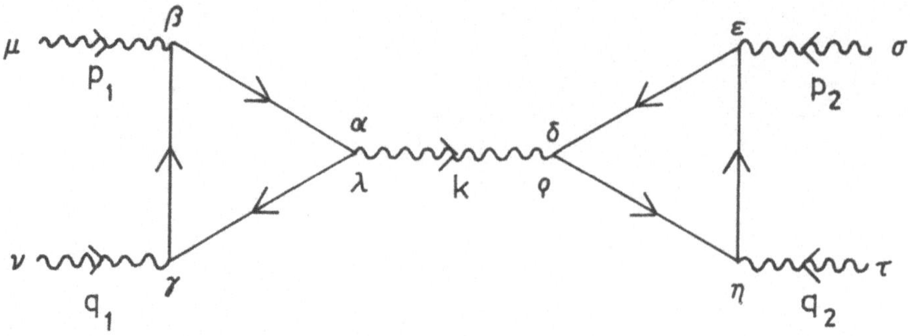

Fig. 7. 2-loop diagram for WW → WW; the total amplitude includes the crossed diagrams as in Fig. 6.

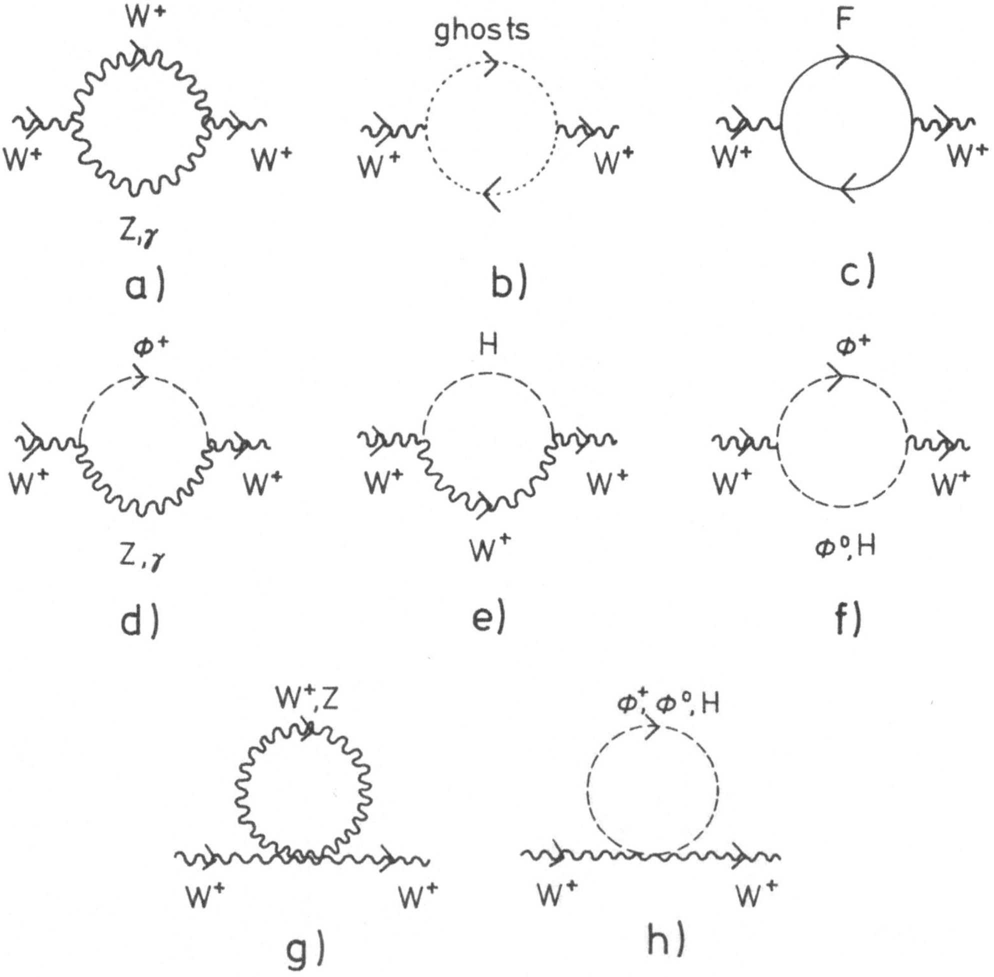

Fig. 8. Self-energy diagrams for the charged vector boson in the standard model. F denotes leptons and quarks, Φ^+ and Φ^0 are the would-be Goldstone bosons in a renormalizable gauge, and H stands for the physical Higgs field.

Acta Physica Austriaca, Suppl. XXIV, 63–124 (1982)

THE ELECTROWEAK INTERACTION IN e^+e^- ANNIHILATIONS+

by

R. MARSHALL
Rutherford Appleton Laboratory
Chilton, Didcot
Oxon, Great Britain

1. INTRODUCTION

The role of e^+e^- annihilation in studying the weak neutral current has taken on an increased significance during the last two or three years due to the operation of PETRA and PEP which provide values of s (= C.M. energy^2) in excess of 1000 GeV^2. Although it was clear from the outset that several years would be needed to collect enough data to make an accurate determination of the neutral current parameters, it was not long after PETRA had started operation that it was realised how even a modest amount of data could put tight limits on some of the lepton couplings [1]. The data also restricted the scope of possible extended gauge models.

The purpose of these lectures is to provide a general description of electroweak effects in e^+e^- annihilations guided perhaps in places by my own prejudices. The relationship with other areas of study, like neutrino polarised electron scattering will also be considered.

+Lectures given at the XXI. Internationale Universitätswochen für Kernphysik, Schladming, Austria, February 25 - March 6, 1982.

The amount of data is now quite substantial and it is possible to distinguish between problems caused by statistics or lack of energy, which time will solve, and problems caused by fundamental obstacles, which may survive until LEP experiments, or even longer.

The first theoretical discussion of possible weak effects in e^+e^- processes was made by Cabibbo and Gatto [2] in 1961. This was at the time when e^+e^- colliding beams were being proposed at Frascati and Stanford, proposals with led to the birth of the e^+e^- era. Now over 20 years later, we are about to see an e^+e^- machine take over the leading role in experimental particle physics research in Europe.

Although the work of Cabibbo and Gatto contains much that is still relevant and is well worth reading today, a more comprehensive treatment is contained in the paper by Ellis and Gaillard [3].

The main area of interest at present accelerators (PEP and PETRA) is in the production of a fermion-anti-fermion pair via s-channel e^+e^- annihilation:

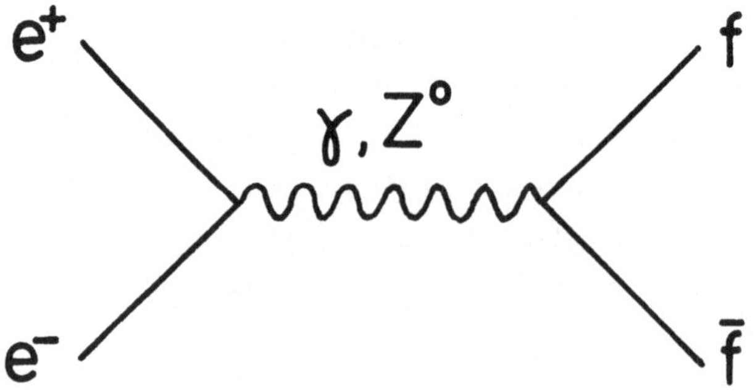

where the process is mediated by photon or neutral weak (Z^0) boson exchange. To first order (single photon or single weak gauge boson exchange) the relative strengths of the two contributions are obtained from the respective constants α and G_F and the $e^{\pm}$ energy. Single photon exchange has a $\frac{1}{s}$ dependence whereas Z^o exchange rises linearly with s (neglecting propagator effects). The γ-Z^o interference term is therefore constant with beam energy; it is however swamped at small s by the γ term and overwhelmed at large s by the pure weak term. In between, there is an energy region where the interference term alone provides information about the weak process. Moreover, some of the information (eg. the sign of the weak charges) can only be obtained in the interference region. This region starts at the present PETRA energies and ends at an energy determined by the weak gauge boson mass.

2. CLASSIFICATION OF FERMIONS AND THEIR ELECTROWEAK COUPLINGS

We start by assuming that the fermion neutral current coupling has a "Fermi type" vector component (V) and a "Gamow-Teller type" axial vector component (A).

The neutrino is believed to have V-A coupling since right-handed neutrinos decouple, ie. $(1-\gamma^5)$ only; the couplings of fermions with mass has to be established by experiment. Let us now consider the currently known "elementary" fermions, including the t quark which has not yet been observed but which is nevertheless suspected to exist. This gives a list of 12 which we can arrange according to electric charge and mass (or association in the case of ν) into three families or generations (I, II and III):

Q_F	I	II	III
0	ν_e	ν_μ	ν_τ
-1	e	μ	τ
$-\frac{1}{3}$	d	s	b
$\frac{2}{3}$	u	c	t

Table 1

We therefore have in general 24 possible different coupling constants (real numbers only if T invariance holds) if we admit a different V and A coupling for each fermion. I don't consider for the moment additional parameters like the mass of the gauge field or the mass of any associated scalar field (eg. Higgs). This maximal number of 24 can be successively reduced by a series of assumptions thus:

1. All the neutrinos are massless, eg. V-A defined for ν. This reduces the number to 21.
2. Generation universality. By assuming that each fermion in generation I has a counterpart in generation II and III with the same electric and weak charges. The number of independent numbers is now reduced by a factor 3 to 7.
3. If the gauge group is assumed to be SU(2) × U(1), spontaneously broken with the simplest Higgs structure, then the 7 independent parameters are reduced to only one, namely $\sin^2\Theta_W$, which defines all the couplings and at the same time the mass of the gauge boson. The mass of the Higgs field would still remain undetermined. The weak charges which arise from this maximal assumption in the standard SU(2) × U(1) model of Glashow, Weinberg and Salam can now be listed as follows:

	Electric charge	Weak axial charge	Weak vector charge
	Q_f	a	v
ν_e, ν_μ, ν_τ	0	1	1
e, μ, τ	-1	-1	$-1 + 4 \sin^2\Theta_W$
d, s, b	$-\frac{1}{3}$	-1	$-1 + \frac{4}{3} \sin^2\Theta_W$
u, c, t	$+\frac{2}{3}$	1	$1 - \frac{8}{3} \sin^2\Theta_W$

Table 2

At this stage, we can note that there are certain "special" values of $\sin^2\Theta_W$ which lead to a symmetric array of the electroweak charges (Q_f, v_f, a_f). In fact, the currently accepted value for $\sin^2\Theta_W$, namely 0.23, is rather close to one of these "special" values, eg. ($\sin^2\Theta_W = \frac{1}{4}$)

	Q_f	a_f	v_f
ν...	0	1	1
e, μ, τ	-1	-1	0
d, s, b	$-\frac{1}{3}$	-1	$-\frac{2}{3}$
u, c, t	$\frac{2}{3}$	1	$\frac{1}{3}$

Table 3

suggesting (perhaps) and underlying SU(3) symmetry or substructure of even more fundamental particles with EW charges 0 and $\pm\frac{1}{3}$.

Note that there is an alternative convention of defining the coupling constants, eg. for electrons, $g_V = \frac{1}{2}v$ and $g_a = \frac{1}{2}a$.

The range of possible numbers of independent parameters, from 24 down to one, represents a scale of options for the physicist. The experimentalist might wish to go to one end of the scale and measure as many of the 24 parameters as possible. This would among other things be a direct check of generation universality. At the other end of the scale, one would hope for an economic theory, and if all 24 parameters can be derived from the single number $\sin^2\Theta_W$, this would be a successful theoretical outcome.

However, e^+e^- processes at present accelerator energies are influenced by two additional factors, namely the value of Q^2 (= s) is higher than ever before and heavy quarks are produced democratically (no sea/valence suppression). This means that it could be more profitable to assume as little as possible, to measure as many fermion couplings as possible and thus maximise the possibility of learning something new.

Such an ambitious program might have to be curtailed in the face of reality. The neutrino couplings are hard to measure since it requires the detection of the process $e^+e^-\rightarrow\nu\nu$, not impossible but tricky since the method would be to rely on initial state radiation to provide a unique signature: $e^+e^- \rightarrow \nu\nu\gamma$. We shall see later that the measured spectrum of hard photons from initial state radiation agrees very well with the calculations of Berends and Kleiss [4] in the case of $e^+e^- \rightarrow \mu^+\mu^-\gamma$ when the muon pair is also observed. The γ spectrum for the neutrino process can therefore be reliably calculated. The measured yield of single photons therefore defines the number of possible neutrinos contributing to the process. Even if there is a series of heavier leptons with masses above the current accelerator energies, their associated neutrinos being massless (or at least light) can be produced. This reaction can therefore be used to measure, or put a limit on the number of neutrino flavours. This is a question of cosmological significance.

A further restriction on the program of measuring as many independent parameters as possible is imposed by the inability to separately measure the individual quark couplings. We shall see below that, at least at present, it is only possible to measure certain combinations of the various couplings.

Sakurai [5] has presented a complementary representation of the fundamental couplings in the form of his "tetragon":

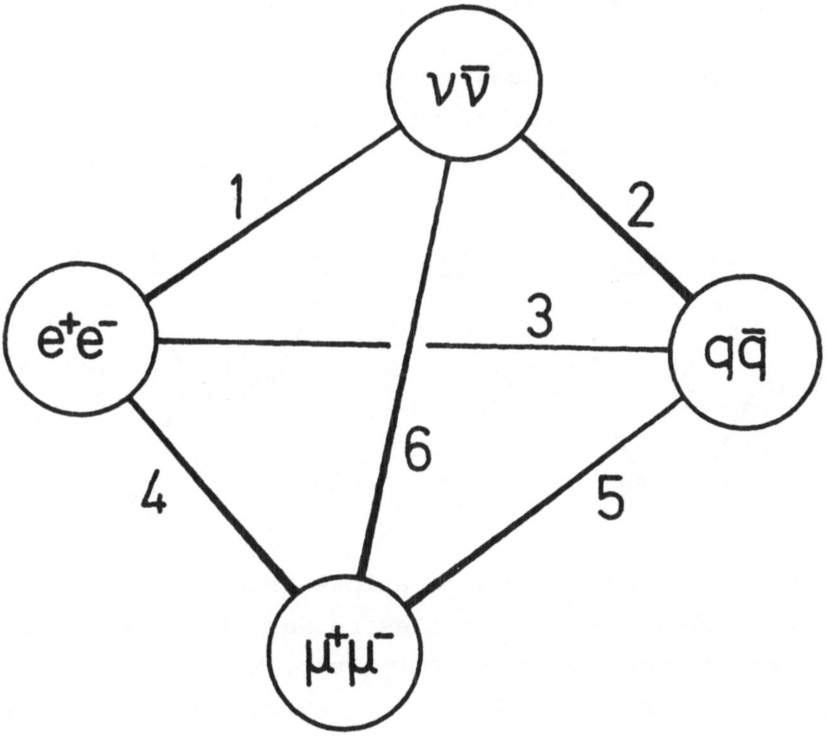

Each link between any two apexes in the figure corresponds to a particular reaction $f_1\bar{f}_1 \rightarrow f_2\bar{f}_2$ or its line-reversed equivalent. Sakurai thus thinks in terms of the product of any two fundamental fermion couplings. Moreover, if the neutral current is mediated by a series of gauge bosons rather than a single Z^o then the product of the two couplings becomes a sum over the products, ie. the couplings no longer factorise. The various processes represented by the above figure are as follows:

1. $\nu e \rightarrow \nu e$ and $e\bar{e} \rightarrow \nu\bar{\nu}$
2. $\nu q \rightarrow \nu q$
3. $eq \rightarrow eq$ and $e\bar{e} \rightarrow q\bar{q}$
4. $e\bar{e} \rightarrow \mu\bar{\mu}$
5. $q\bar{q} \rightarrow \mu\bar{\mu}$
6. $\mu\mu \rightarrow \nu\bar{\nu}$

With 6 links and V,A couplings, there are 12 different numbers that can be measured. If the neutral current is mediated by a single gauge boson, then one can apply factorisation as follows:

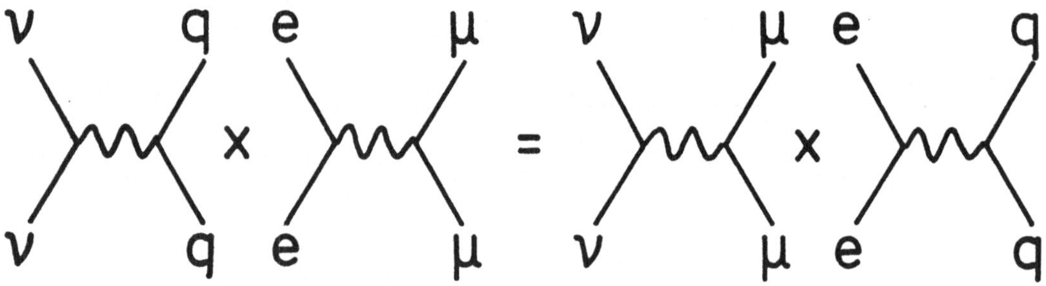

Taking all factorisation relations into account, there remain 7 idependent quantities to describe the couplings. The two schemes are thus equivalent, these 7 quantities also reduce to the single $\sin^2\Theta_W$ in the standard model.

To conclude this section we can remark that it is the aim of experiments at PETRA, PEP and eventually LEP to measure the processes:

$$
\begin{aligned}
e^+e^- \rightarrow\ & e^+e^- \\
& \mu^+\mu^- \\
& \tau^+\tau^- \\
& \nu\bar{\nu}(\gamma) \\
& q\bar{q} \rightarrow \text{hadrons}
\end{aligned}
$$

in order to determine as many of the neutral current fermion couplings as possible. The parameters thus determined can be compared with the prediction of the standard (or any other) model.

The way in which the vector and axial vector couplings appear in the total cross sections, the angular distributions and the polarisation dependences will be discussed in the next section.

3. ELECTROWEAK EFFECTS IN $e^+e^- \to f\bar{f}$

I shall only consider fermion antifermion production via the s channel processes: $e^+e^- \to \gamma, Z^o \to f\bar{f}$.

The case where f = e (Bhabha scattering) can also proceed via t channel γ and Z exchange and this makes the formalism more complicated [6].

a) As a first step consider this process with the exchange of a γ only. This demonstrates the symmetry properties of the electromagnetic interaction which are broken by the weak interaction.

From the interaction

$$L^{EM} = -\frac{e^2 Q_f}{s}(\bar{\psi}_e \gamma_\mu \psi_e)(\bar{\psi}_f \gamma^\mu \psi_f) \quad (1)$$

it is instructive to separate out the separate helicity components [7] (in the limit of massless leptons):

$$e_L^- e_R^+ \to f_L \bar{f}_R \qquad \frac{4s}{\alpha^2} \cdot \frac{d}{d\Omega} = Q_f^2 (1 + \cos\theta)^2 \quad (2a)$$

$$e_L^- e_R^+ \to f_R \bar{f}_L \qquad = Q_f^2 (1 - \cos\theta)^2 \quad (2b)$$

$$e_R^- e_L^+ \to f_L \bar{f}_R \qquad = Q_f^2 (1 - \cos\theta)^2 \quad (2c)$$

$$e_R^- e_L^+ \to f_R \bar{f}_L \qquad = Q_f^2 (1 + \cos\theta)^2 \quad (2d)$$

where θ is the angle between the e^- and f. The subscripts R and L refer to right- and left-handed electrons or fermions. The single γ exchange (spin 1) means that the initial (and final) pair of fermions have opposite helicity. Moreover, as required by the fact that the electromagnetic interaction conserves parity, the substitution: $e^-_R \to e^-_L$ etc. ($\equiv$ parity transformation) leaves each separate helicity amplitude unaltered. Note also the form of the θ dependence. Due to angular momentum conservation, the substates $e_L \to f_R$ and $e_R \to f_L$ cannot occur at 0°, hence the $1-\cos\theta$. Similarly, the substates $e_L \to f_L$, which cannot take place at 180°, contain the factor $1 + \cos\theta$.

For unpolarised $e^\pm$ beams, the angular distribution is obtained by averaging over all four initial helicity states; the terms in $\cos\theta$ all cancel and one has the well known result

$$\frac{4s}{\alpha^2} \cdot \frac{d\sigma}{d\Omega} = Q_f^2(1+\cos^2\theta)$$

which for the case of the $\mu^+\mu^-$ final state ($Q_f = -1$) gives

$$\frac{4s}{\alpha^2} \cdot \frac{d\sigma}{d\Omega} = (1 + \cos^2\theta) \tag{3}$$

and integrating over all production angles, $\sigma = \frac{87}{s}$ nb.

b) Now add the weak neutral current and define the weak couplings of right- and left-handed electrons to be R and L respectively. For the final state fermion the notation is R_f and L_f. The vector and axial vector couplings are related to the R and L as follows:

$$a = L-R \qquad a_f = L_f-R_f \qquad v = L+R \qquad v_f = L_f+R_f. \tag{4}$$

The manifestation of parity violation in the weak interaction is that in general $L \neq R$ and $L_f \neq R_f$.

The four helicity states are now modified, thus

$$e^-_L e^+_R \to f_L \bar{f}_R \qquad \frac{4s}{\alpha^2}\frac{d\sigma}{d\Omega} = (1+\cos\theta)^2 |Q_f + LL_f g|^2 \tag{5a}$$

$$e^-_L e^+_R \to f_R \bar{f}_L \qquad (1-\cos\theta)^2 |Q_f + LR_f g|^2 \tag{5b}$$

$$e^-_R e^+_L \to f_L \bar{f}_R \qquad (1-\cos\theta)^2 |Q_f + RL_f g|^2 \tag{5c}$$

$$e^-_R e^+_L \to f_R \bar{f}_L \qquad (1+\cos\theta)^2 |Q_f + RR_f g|^2 \quad . \tag{5d}$$

The weak amplitude g is essentially the Fermi coupling constant times a few factors and an energy/propagator term,

$$g = \frac{\sqrt{2}G_F}{4\pi\alpha} \cdot \frac{s \cdot m_Z^2}{(s-m_Z^2+im_Z\Gamma_Z)} \quad .$$

In equations (5a-d) we have in fact explicitly assumed factorisation by writing the product LL_f etc.. To be more general, one can write instead $\varepsilon_{LL} = \Sigma LL_f$ etc. although the arguments that follow do not depend on this assumption.

Note that the individual helicity states are no longer invariant under parity transformation (unless L = R and $L_f = R_f$). However, as we shall now show it is necessary to observe the individual amplitudes (or asymmetric combinations of them) if the parity violating nature of the interaction is to be revealed.

3.1 Angular Distributions and Total Cross Section

If the beams are unpolarised and the final state polarisation is not measured, then the angular distribution is obtained by adding the four helicity components and dividing by 4.

Then

$$\frac{4s}{\alpha^2} \cdot \frac{d\sigma}{d\Omega} = \frac{1}{4}\ (1+\cos\theta)^2[(Q_f+LL_f g)^2 + (Q_f+RR_f g)^2]$$

$$+ \frac{1}{4}(1-\cos\theta)^2[(Q_f+LR_f g)^2 + (Q_f+RL_f g)^2] \quad . \tag{6}$$

Already it can be seen that the parity asymmetry has been removed by combining essentially states of opposite parity. A parity transformation L → R etc. leaves $\frac{d\sigma}{d\Omega}$ unaltered and so the angular distribution and anything derived from it by integrating over regions of θ (eg. total cross section, forward-backward angular symmetry) is intrinsically a parity conserving quantity.

Working out equation 6 and substituting a = L-R etc. leads to

$$\frac{4s}{\alpha^2}\frac{d\sigma}{d\Omega}(f\bar{f}) = (1+\cos^2\theta)\,(Q_f^2-\tfrac{1}{2}Q_f vv_f g + \tfrac{1}{16}(v^2+a^2)\,(v_f^2+a_f^2)g^2)$$

$$- 2\cos\theta\,(\tfrac{1}{2}Q_f aa_f g - \tfrac{1}{4}vav_f a_f g^2) \quad . \tag{7}$$

The interference terms (proportional to g) have a vv_f term associated with $(1+\cos^2\theta)$ and an aa_f term associated with cosθ. The vector and axial components can therefore be isolated by either integrating over cosθ (cosθ term drops out) or forming the forward-backward asymmetry $(1+\cos^2\theta$ term drops out).

Thus

$$\frac{3s}{4\pi\alpha^2} \cdot \sigma(f\bar{f}) = R_f = Q_f^2 - \tfrac{1}{2}Q_f vv_f g + \tfrac{1}{16}(v^2+a^2)\,(v_f^2+a_f^2)g^2, \tag{8}$$

whereas the forward-backward

$$\text{asymmetry} = A_f = \frac{\int_0^1 \frac{d\sigma}{d\Omega} - \int_{-1}^{0} \frac{d\sigma}{d\Omega}}{\int_{-1}^{1} \frac{d\sigma}{d\Omega}}$$

is given by

$$A_f = \frac{3}{4} \cdot \frac{-\frac{1}{2}Q_f a a_f g + \frac{1}{4} v a v_f a_f g^2}{Q_f^2 - \frac{1}{2}Q_f v v_f g + \frac{1}{16}(v^2+a^2)(v_f^2+a_f^2) g^2} \quad . \tag{9}$$

At PETRA energies, where the weak squared terms can be neglected, the following approximate but transparent formulae are obtained:

$$R_f = Q_f^2 - \frac{1}{2}Q_f v v_f g$$

$$A_f = -\frac{3}{8}Q_f a a_f g$$

ie. total cross sections measure vector couplings and asymmetries measure axial couplings.

At higher energies, above the interference region where the weak squared term dominates, both the total cross section and the F-B asymmetry measure quadratic combinations of the vector and axial couplings.

To illustrate the dependence of R_f and A_f on the variables of interest, s and $\sin^2\theta_w$, some examples are shown in figs. 1 and 2. Fig. 1 shows the variation of R_μ with $\sin^2\theta_w$ for various energies (s) in the PETRA energy range. If $\sin^2\theta_w$ is close to $\frac{1}{4}$ as indicated by νe, νq and eq experiments, then the change in R_μ due to weak effects is so small that it cannot be measured. On the other hand, measurable effects would occur if $\sin^2\theta_w$ differs appreciably from $\frac{1}{4}$, ie. even if v_e or v_μ are close to zero, they can still be observed. The muon asymmetry A_μ does not depend too strongly on $\sin^2\theta_w$ (at PETRA energies) since it enters only as a small perturbation to the denominator in eq.9, through the propagator term ($m_z^2 = 37.3^2/(\sin^2\theta_w\cos^2\theta_w)$) or through the weak squared term. Due to the aa_μ dependence, and with $a = a_\mu = -1$, the effect on the asymmetry can be substantial and observable at PETRA energies, as shown in fig. 2. The asymmetry A_μ will therefore measure

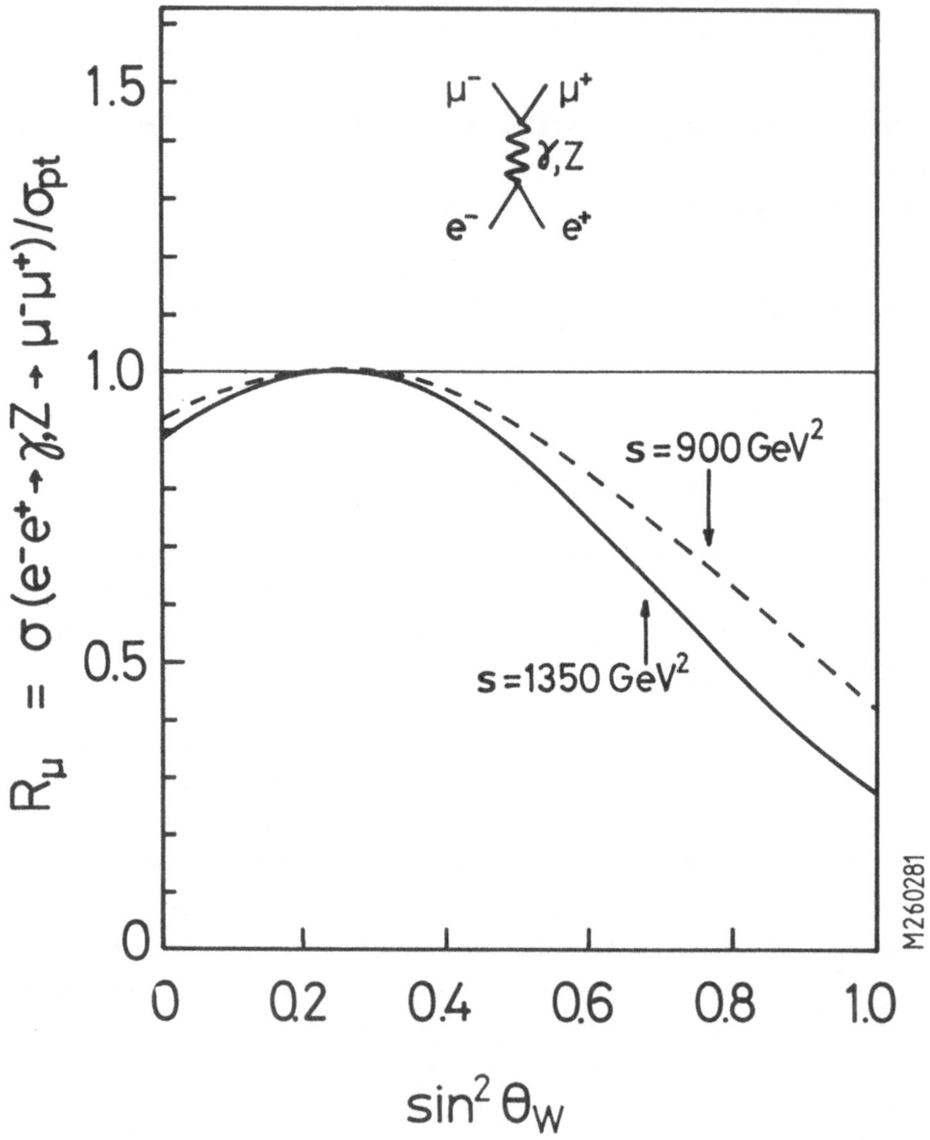

Fig. 1. The variation of R_μ with $\sin^2\theta_W$ for two values of s.

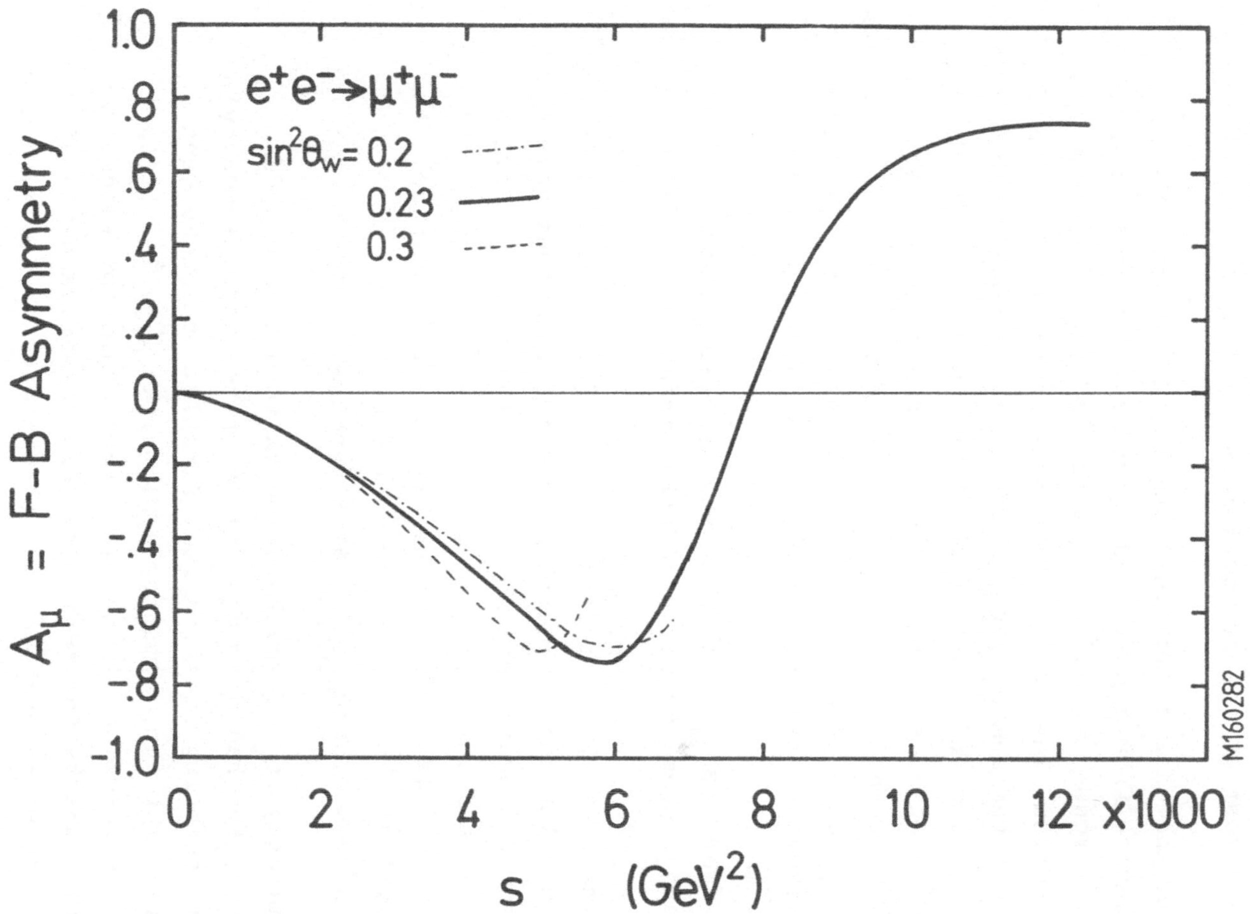

Fig. 2. The variation of A_μ with s for three values of $\sin^2\theta_w$.

the axial couplings via an observable non-zero effect, but due to its lack of dependence on $\sin^2\theta_w$ it does not offer a crucial test of the standard model.

In the case of $q\bar{q}$ final states, R_f corresponds to a contribution from a particular quark flavour to R, the total hadronic cross-section in units of the point like muon QED cross section. Then $R = 3 \sum_f R_f$ where the factor 3 arises from colour and the sum is carried out over all possible quarks, eg. d, u, s, c and b at present PETRA energies. The $\sin^2\theta_w$ dependence of R is shown in fig. 3 for several values of s. The ratio of the interference term to the γ term is proportional to vv_f/Q_f which is a factor 25 or 6 larger for d or u quarks than for muons (setting $\sin^2\theta_w = 0.23$) and therefore more readily measurable in hadron final states than $\mu^+\mu^-$. For values away from $\frac{1}{4}$, R changes substantially from 3.87 and can reach extreme values. A careful measurement of R can therefore measure $\sin^2\theta_w$ quite accurately. Note that a measurement of R at a single value of s would lead to two possible solutions for $\sin^2\theta_w$. The two solutions have different s dependences however (see the crossing regions in fig. 3) and a measurement of R over a wide range of s can lead to a single solution. The way in which the propagator term modifies the total cross sections is shown in fig. 4 which shows R_μ and $R_{hadrons}$ as a function of s.

Turning to the quark-antiquark forward-backward asymmetry (the hadronic equivalent of the muon asymmetry) a few interesting points arise. Firstly, it can be seen in equ. 9 that the asymmetry varies like $\frac{aa_f}{Q_f}$, ie. it is bigger for fractionally charged quarks than for muons. For d, s and b quarks the asymmetry could be as large as 30% at the highest PETRA energies. Unfortunately, we now encounter a fundamental obstacle; particles and antiparticles are distinguished by their charge. Thus a quark with positive charge would be either $Q_f = \frac{2}{3}$ quark or a $Q_f = \frac{1}{3}$ antiquark, and since the asymmetry depends on a_f/Q_f, which is always

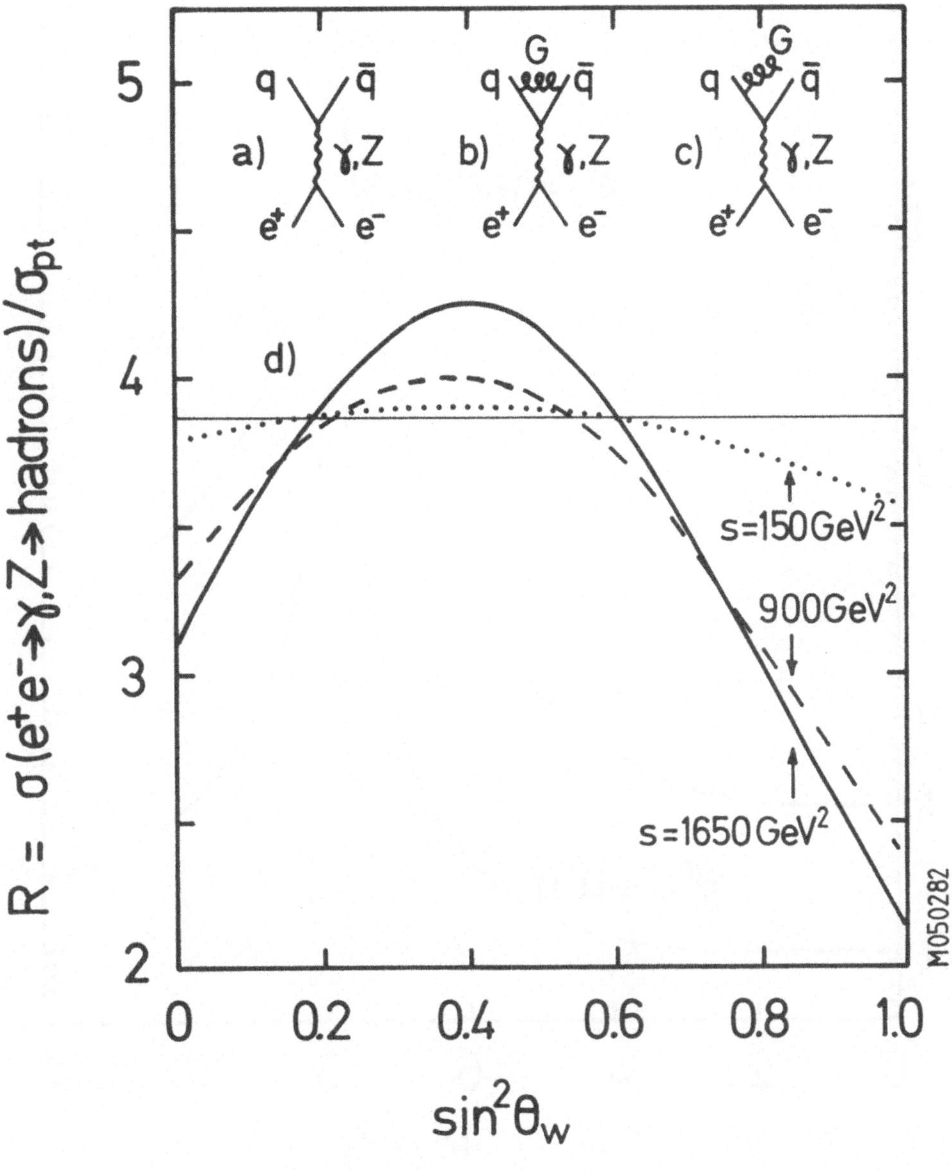

Fig. 3. The variation of $R_{hadrons}$ with $\sin^2\theta_W$ for three values of s.

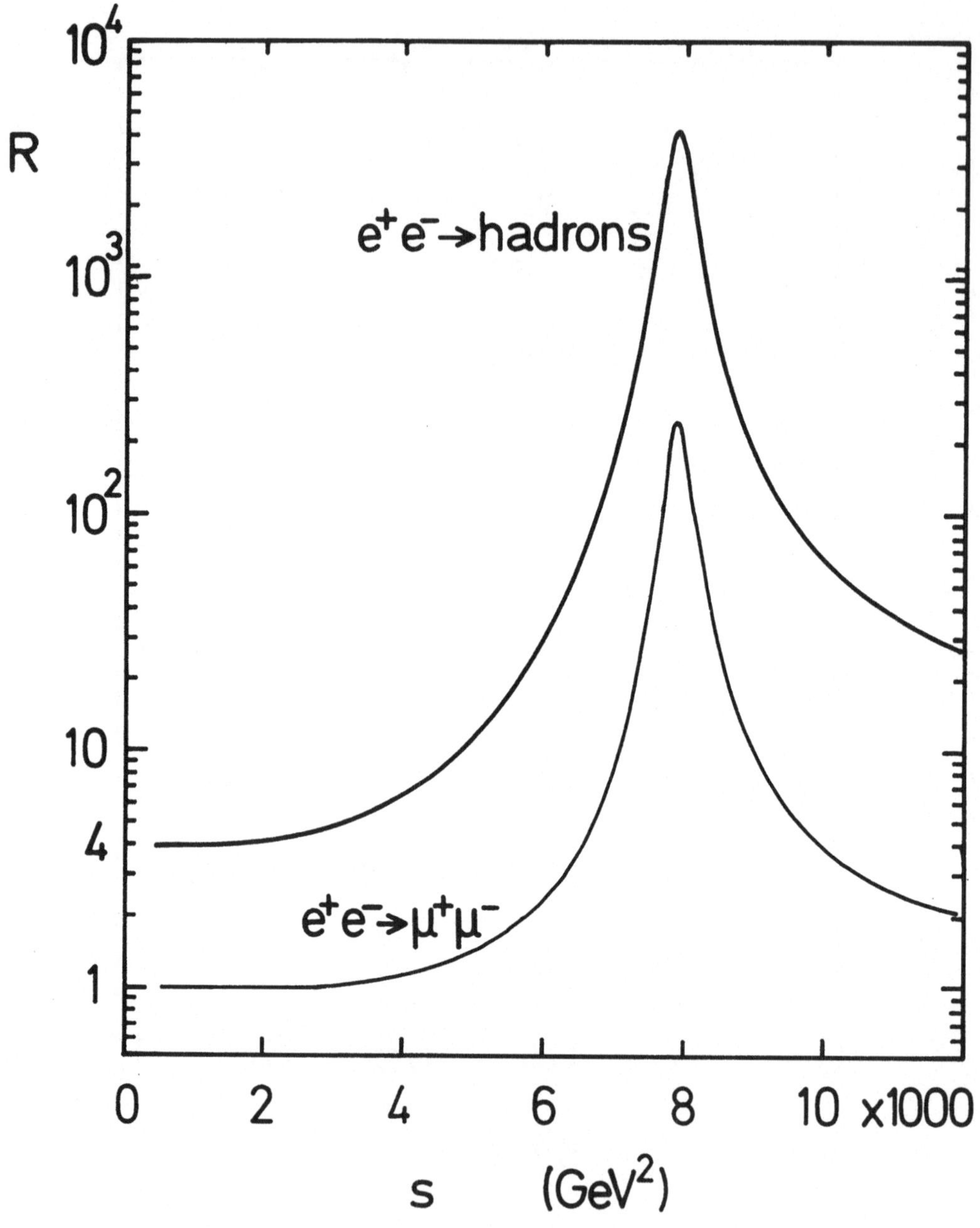

Fig. 4. The variation of $R_{hadrons}$ and R_{μ} as a function of s for $\sin^2\theta_W = 0.23$.

positive in the standard model, the asymmetries cancel in the summation over quark flavours. Quantitatively, this works as follows; summing over all $Q_f = \frac{2}{3}$ quarks and $Q_f = \frac{1}{3}$ antiquarks, the combined asymmetry becomes

$$A_{tot}(q\bar{q} \text{ or } \bar{q}q) = \frac{\sum |Q_f| a_f}{\sum Q_f^2} \cdot A_\mu$$

$$= \frac{\overset{d}{(-\frac{1}{3}} + \overset{u}{\frac{2}{3}} - \overset{s}{\frac{1}{3}} + \overset{c}{\frac{2}{3}} - \overset{b}{\frac{1}{3}})}{(\frac{1}{9} + \frac{4}{9} + \frac{1}{9} + \frac{4}{9} + \frac{1}{9})} A_\mu = \frac{3}{11} A_\mu \ .$$

So the resulting angular asymmetry would only be 3/11ths of the muon asymmetry and is not very promising at PETRA.

Possible methods of determining quark charges are as follows:

a) A determination of the momentum-weighted charges in a hadron jet can in principle be related to the parent quark charge - on a statistical basis [8]. For example,

$$\sum_i^N Q_i \, p_i^\gamma$$

has a distribution with a positive mean value for $Q_f = \frac{2}{3}$ quarks and a negative mean value for $Q_f = -\frac{1}{3}$ quarks. The summation is carried out over the N particles in a jet where the ith particle has charge Q_i and momentum p_i. The exponent γ is chosen empirically to maximise the separation between positively and negatively charged quarks. In practice, the measure is not a 100% reliable estimator since the widths of the two measured distributions are so large [9] due to experimental effects that they overlap almost completely. More work is necessary to show whether or not this technique will prove fruitful either at PETRA or LEP energies.

b) The semi-leptonic decays of quarks also offer a clue to the charge of the parent quark ,

eg.
$$b \to \mu^- \nu c$$
$$c \to \mu^+ \nu s \text{ etc.}$$

where we can now see that a μ^- is an indicator for a b or $\bar{c}$ etc.. So the asymmetry of hadron jets containing a μ^- (or μ^+ with $\theta_{jet} \to 180^o - \theta_{jet}$) will suffer from the same cancellation effects mentioned above. In this particular case however, one can do better. Firstly, the different lifetimes of successive species of quarks means that only those muons need be selected which originate close to the main hadron vertex (direct muons), the spatial resolution being of the order of a mm. This essentially excludes all decays except those from b and c.

Then

$$A_{c+\bar{b}} = \frac{\frac{2}{3} - \frac{1}{3}}{\frac{4}{9} + \frac{1}{9}} A_\mu = \frac{3}{5} A_\mu$$

still only a few percent but now possibly measurable and more important, providing a measurement of a_c and a_b. However, these measurements have to be made against a background from meson decays which reduce the statistical accuracy. A significant improvement can be made by measuring the transverse momentum of the muons relative to the jet axis. Muons from heavy b (or B meson) decay are likely to have a larger transverse momentum than these from c decay and this could be used to try and separate b and c jets. Even so, the $p_T(\mu)$ distributions from b and c decay overlap substantially and it is unlikely that the full 30% effect could be seen.

3.2 Parity

We saw that total cross-sections or forward-backward asymmetries for leptons and quarks are <u>all</u> parity conserving quantities - if the initial beams are unpolarised. This leads to certain advantages or disadvantages depending on one's point of view:

1) Since the electromagnetic and strong interactions conserve parity, they too can contribute to cross sections and angular asymmetries via single and multiple γ or gluon exchange or emission. Higher order QED processes otherwise known as radiative corrections must therefore be calculated carefully since they affect all the measured quantities by a few percent. QCD can also affect the value of R, eg. $R \rightarrow R(1+\alpha_s/\pi+...)$ and the quark asymmetry.

2) Because the parity conserving part of the neutral current is not measured in $e\!\uparrow\! q$ or ν scattering, e^+e^- can be viewed as providing fresh insight into the structure of the weak current - provided QED and QCD higher order contributions are understood. This can be an advantage.

3) But nevertheless, the classic, traditional signature for weak effects is parity violation. The observation of a parity violating quantity in e^+e^- could not be due to QED or QCD according to our present understanding and would be a clear unambiguous indication of a weak effect - just as a non-zero polarisation asymmetry in $e\!\uparrow\! q$ scattering means just that. The next section deals with possible ways of detecting parity violating effects in e^+e^-, now and in the future.

3.3 Polarisation Effects

Referring to equations 5a-d, the way in which the parity violating properties of the interaction can be revealed is to measure or define the spin state of the initial or final state fermions. For example, if the beams were polarised - say e_R^- only - then the cross section would be given by 5c + 5d which is no longer invariant under a parity operation ($L \rightarrow R$ etc.). We can now derive from equations 5a-d various helicity dependent cross sections.

a) Beam polarisation asymmetry

By analogy with the polarisation asymmetry in e↑q scattering, the difference between the cross sections for e^-_R and e^-_L scattering has the property that electromagnetic components disappear leaving a completely parity violating interference (or pure weak) term ;

ie.

$$\frac{4s}{\alpha^2}\left[\frac{d\sigma}{d\Omega}(e^-_R) - \frac{d\sigma}{d\Omega}(e^-_L)\right] = \frac{1}{2}(1+\cos\theta)^2[(Q_f+RR_f g)^2-(Q_f+LL_f g)^2]$$

$$+ \frac{1}{2}(1-\cos\theta)^2[(Q_f+RL_f g)^2-(Q_f+LR_f g)^2];$$

after some algebra, this leads to

$$\frac{4s}{\alpha^2}\left[\frac{d\sigma}{d\Omega}(e^-_R) - \frac{d}{d\Omega}(e^-_L)\right] = \frac{1}{2}(1+\cos^2\theta)[2Q_f av_f g-\frac{1}{2}va(v_f^2+a_f^2)g^2]$$

$$+ 2\cos\theta[Q_f va_f g-\frac{1}{4}(v^2+a^2)v_f a_f g^2]$$

or, integrating over cosθ (cosθ term vanishes) and summing over d, u, s, c, b quarks:

$$R(e^-_R) - R(e^-_L) = \sum_f[Q_f av_f g - \frac{1}{4}va(v_f^2+a_f^2)g^2].$$

It actually makes more sense to evaluate $(R(e^-_R) - R(e^-_L))/R$ because both QCD effects and experimental normalisation uncertainties cancel out. Note $2R = R(e_R) + R(e_L)$, and then

$$R_A = \frac{R(e^-_R)-R(e^-_L)}{R} = \frac{\sum[Q_f av_f g-\frac{1}{4}va(v_f^2+a_f^2)g^2]}{\sum[Q_f^2-\frac{1}{2}Q_f vv_f g+\frac{1}{16}(v^2+a^2)(v_f^2+a_f^2)g^2]} \qquad (10)$$

$$\approx \frac{\sum Q_f av_f}{\sum Q_f^2}\, g \text{ in the interference region.}$$

In this case, the interference term is proportional to v_f (not v_e) and therefore is substantial, even for $\sin^2\theta_W$ close to $\frac{1}{4}$. The behaviour of this quantity (equ.10) as a function of s ($\sin^2\theta_W = 0.23$) is shown in fig. 5. It would be quite easily measurable at PETRA energies although, alas, the beams are not polarised in this way, nor will they be at LEP at first. The prospects at the planned SLAC "single pass collider" are more promising because it will be possible to polarise the electron beam (but not positrons) and this is enough, for, as we saw above, a right-handed electron will only pick out left-handed positrons with which to interact. The flux of polarised positrons is essentially one half the total positron beam flux.

Equation 10 above also shows that the beam polarisation asymmetry for producing $\mu^+\mu^-$ will be close to zero due to the presence of v_f in the v interference term and v in the pure weak term.

b) Final state polarisation

The polarisation of final state fermions can be non-zero, even for unpolarised $e^{\pm}$ beams. Again from equations 5a-d, the polarisation is given by

$$P_f = \frac{\frac{d\sigma}{d\Omega}(f_R) - \frac{d\sigma}{d\Omega}(f_L)}{\frac{d\sigma}{d\Omega}(f_R) + \frac{d\sigma}{d\Omega}(f_L)}$$

where

$$\frac{4s}{\alpha^2}\frac{d\sigma}{d\Omega}(f_R)=\frac{1}{4}\{(1+\cos\theta)^2[Q_f+RR_f g]^2+(1-\cos\theta)^2[Q_f+LR_f g]^2\}$$

$$\frac{4s}{\alpha^2}\frac{d\sigma}{d\Omega}(f_L)=\frac{1}{4}\{(1+\cos\theta)^2[Q_f+LL_f g]^2+(1-\cos\theta)^2[Q_f+RL_f g]^2\}.$$

Then

$P_f(\cos\theta)$ =

$$\frac{(1+\cos^2\theta)(2Q_f va_f g-\frac{1}{2}(v^2+a^2)v_f a_f g^2)+2\cos\theta(2Q_f av_f g-\frac{1}{2}va(v_f^2+a_f^2)g^2)}{(1+\cos^2\theta)(4Q_f^2-2Q_f vv_f g+\frac{1}{4}(v^2+a^2)(v_f^2+a_f^2)g^2)-2\cos\theta(2Q_f aa_f g-vav_f a_f g^2)}. \quad (11)$$

Once again, QCD effects should cancel out when determining P_f.

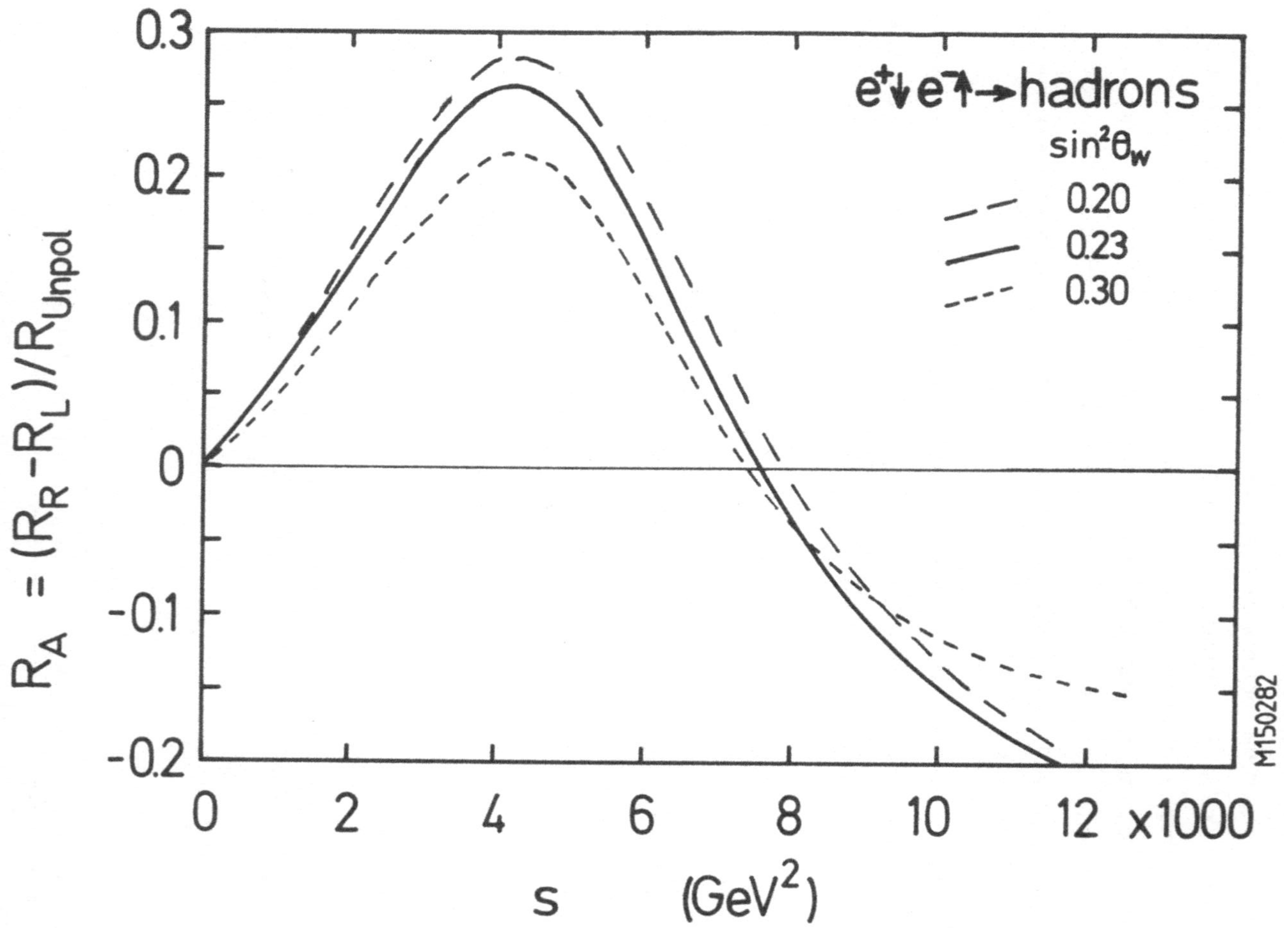

Fig. 5. The beam polarisation asymmetry R_A as a function of s for three values of $\sin^2\theta_w$.

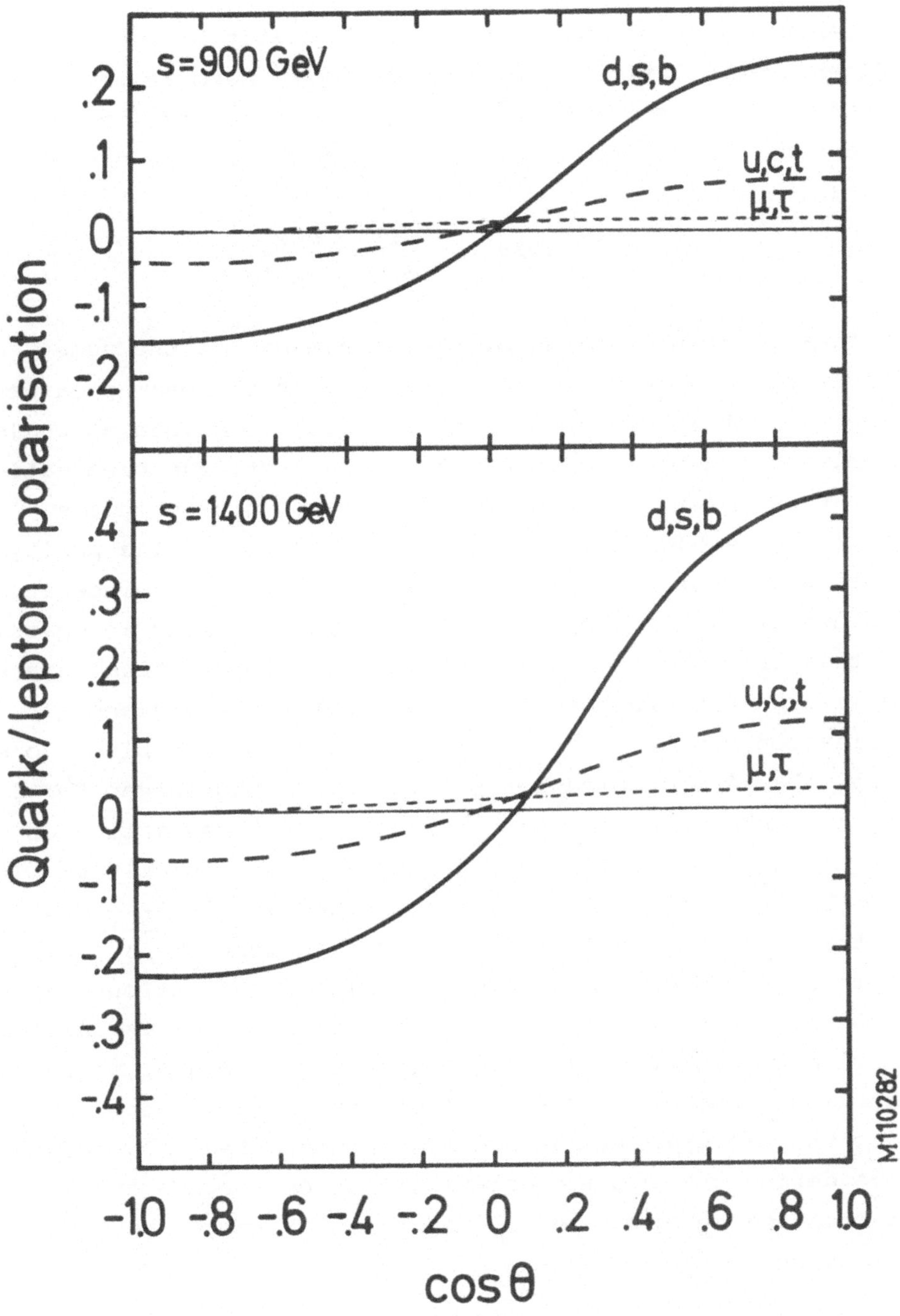

Fig. 6. The angular variation of the quark and lepton polarisation for two values of s. In this figure $\sin^2\theta_W = 0.23$.

Weak terms in the denominator can effectively be ignored at PETRA energies. Furthermore, the $(1+\cos^2\theta)$ term in the numerator contains $v(\approx 0)$, so this can be neglected in favour of the term proportional to $\cos\theta$. So we obtain the approximate formula

$$P_f(\cos\theta) = \frac{av_f g}{Q_f} \frac{\cos\theta}{1+\cos^2\theta} .$$

Note that the final state polarisation measures the same combination of weak couplings (ie. av_f) as the beam polarisation asymmetry. But if the beams can't be polarised then it would be the only source of this information. In fact, the observation of non-zero polarisation of final state particles in e^+e^- annihilation would be the first manifestation of the parity violating effects of the weak interaction in e^+e^- since it could not arise from the electromagnetic or strong interaction. Note that from equation (11) the polarisation of final state μ or τ is likely to be small due to v_f whereas the quark polarisation can be large. Using the numbers in table 2, one can see that the combination $\frac{av_f}{Q_f}$ is four times larger for d, s and b quarks than for u and c quarks. The observation of the s quark polarisation (via Λ decay) appears promising. It is worth noting that the angular dependence $\frac{\cos\theta}{(1+\cos^2\theta)}$ means that the average polarisation is close to zero! This can be seen in fig. 6 where the various fermion polarisations are plotted as a function of $\cos\theta$ for different s values. Although the polarisation for s quarks reaches large values, it changes sign and an experiment with limited statistics, averaging over production angle, would not see much of an effect. Therefore, let us define the average polarisation for forward and backward produced fermions, ie. P_F and P_B, and form two orthogonal combinations,

$\frac{1}{2}(P_F - P_B) := P_A$, the forward-backward polarisation asymmetry

$\frac{1}{2}(P_F + P_B) := \langle P \rangle$, average polarisation.

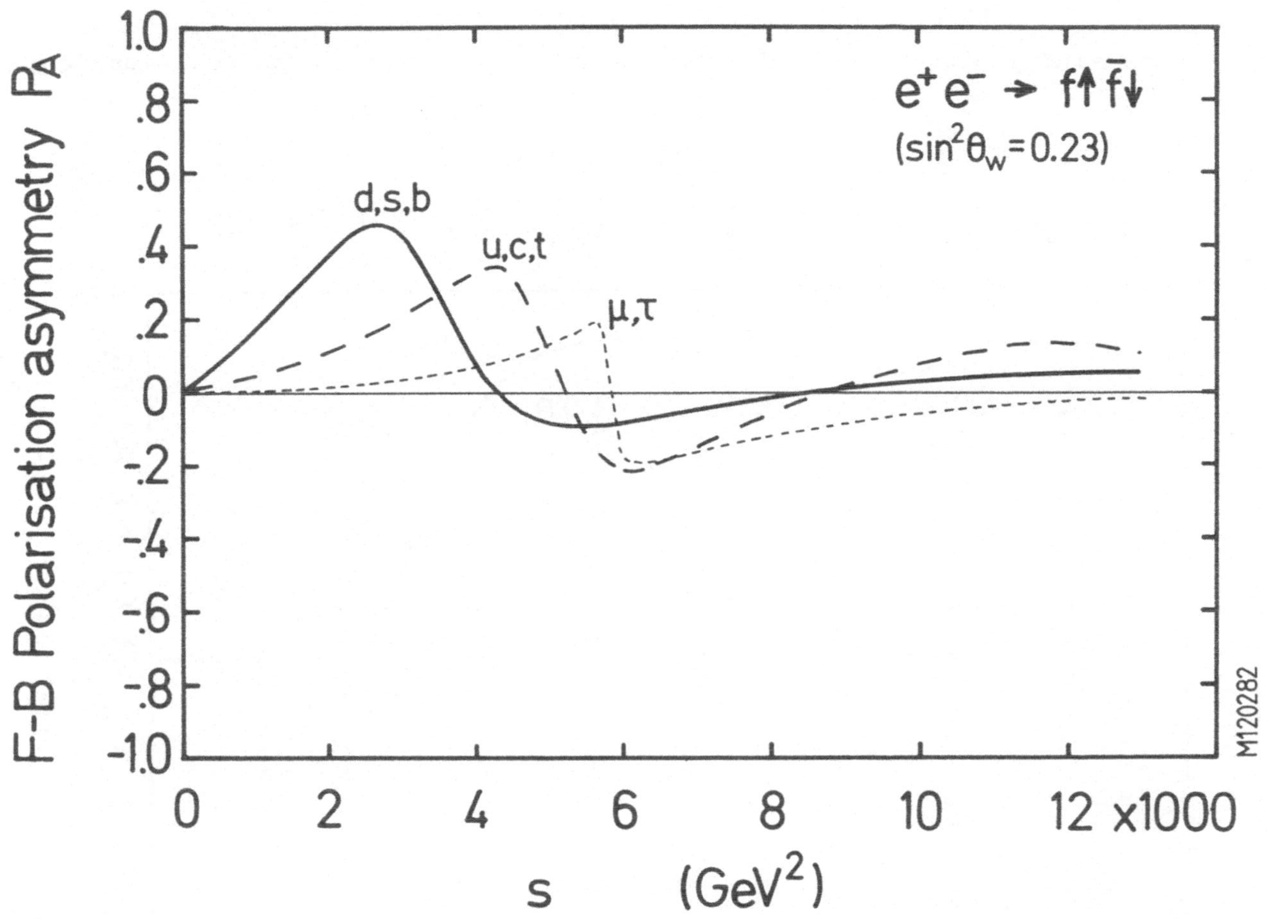

Fig. 7. The variation of P_A with s for quarks and leptons. In this figure $\sin^2\theta_w = 0.23$.

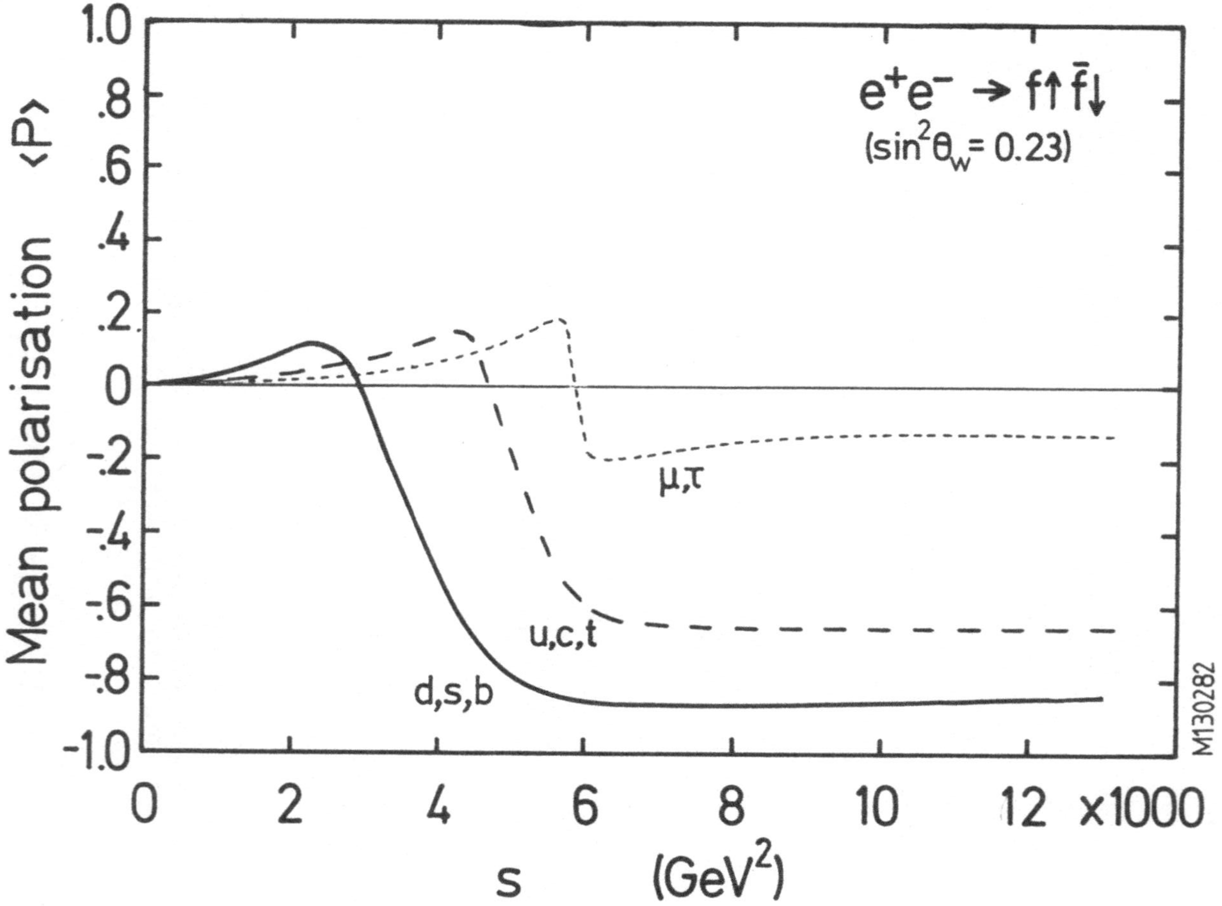

Fig. 8. The variation of <P> with s for quarks and leptons. In this figure $\sin^2\theta_w = 0.23$.

It then turns out that P_A is a good measure in the interference region (see fig.7), whereas <P> is close to zero in the interference region, but becomes large as the weak squared term takes over (fig. 8).

The classic methods of measuring the polarisation involve the decay angular distribution of the polarised particle, eg.

a)

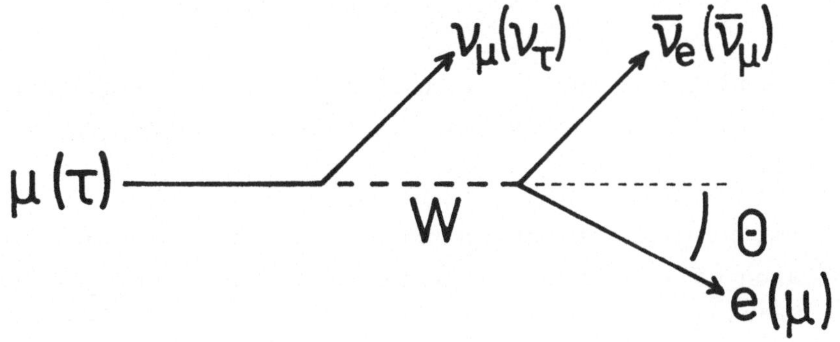

where $F(\theta) = 1 + P_\mu \cos\theta$,

or b)

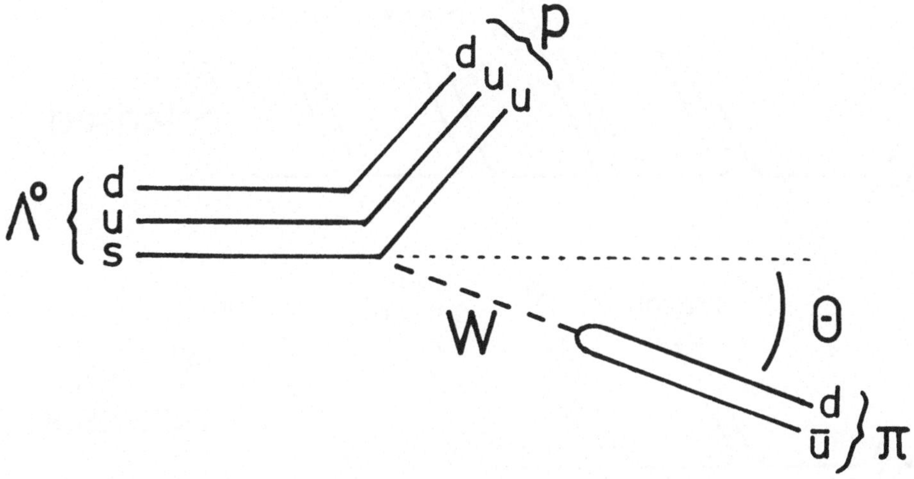

where now $F(\theta) = 1 + \alpha P_\Lambda \cos\theta$ with $\alpha = 0.64$.

Method a) would be suitable for τ decay, but not for μ decay due to the lifetime. Method b) can be used for Λ^{o} produced in $e^{+}e^{-}$ collisions, but the big question is whether or not the quark polarisation survives the fragmentation process and is transmitted to the Λ^{o}. This will depend crucially on the fragmentation process for baryon production. At the very least, one can measure the decay angular distribution of Λ^{o} and compare it with the expression

$$F(\theta) = 1 + 0.64\ \alpha_f P_{\Lambda} \cos\theta$$

where in addition to the traditional polarisation parameter $\alpha = 0.64$, there is an additional factor: the fragmentation depolarising parameter α_f.

The process of baryon production in connection with polarising effects has been considered by several authors [10]. Usually, to produce a baryon, a diquark is needed from the vacuum, eg.

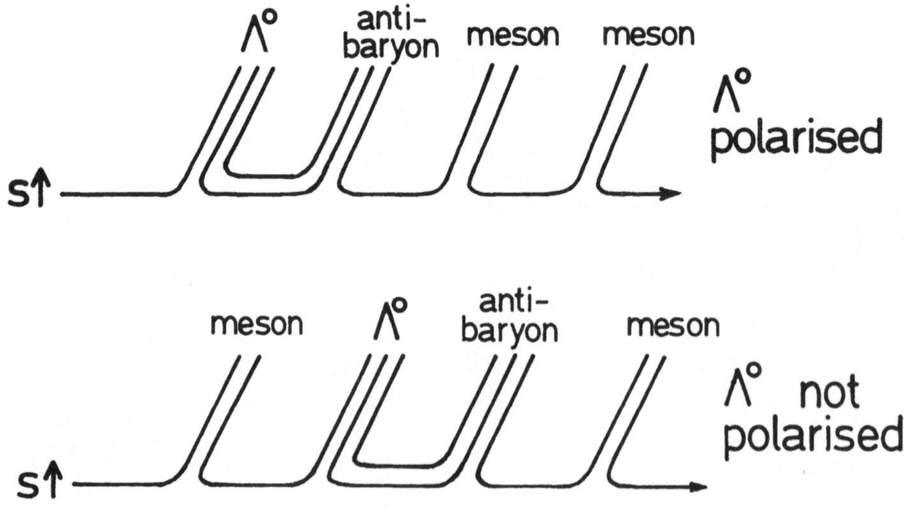

Ranft and Ranft for example [10] have calculated via a cascade Monte Carlo process that up to 50% of the initial polarisation will survive into so called "leading" Λ^o where "leading" means $x_\Lambda > 0.1$ (where $x_\Lambda = p_\Lambda/E_{beam}$).

In principle, another way of measuring the polarisation is via the semileptonic decays of heavy quarks. For example

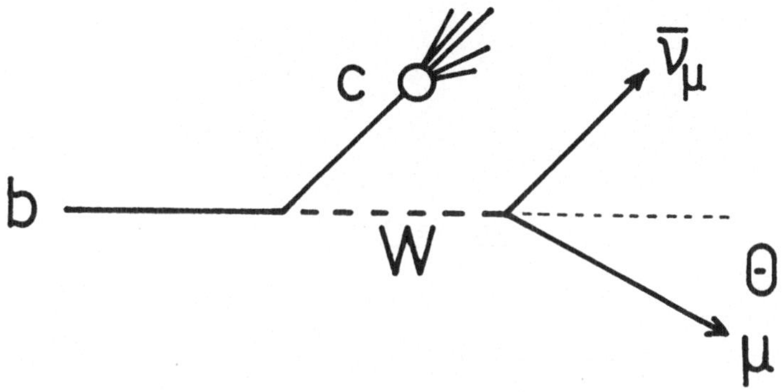

where $F(\theta) = 1 + \alpha P_b \cos\theta$,

but here, the probable mechanism is that the b quark forms a meson B^o which then decays weakly. The polarisation information is then lost in the spinless meson. One can make the nice observation that in general, the polarisation is induced by the weak neutral current and analysed by the weak charged current. In any case, it is likely that polarisation measurements will take on increasing importance in e^+e^- processes as both the energy and the statistics are increased.

To conclude this section, it is worth tabulating the various observables and listing the particular weak couplings or combination of couplings that each observable measures. This is shown in table 4.

	Interference region PEP, PETRA	← SLC →	"Resonance" region LEP
$\sigma(\mu^+\mu^-)$	vv_μ		$(v^2+a^2)(v_\mu^2+a_\mu^2)$
$A(\mu^+\mu^-)$	aa_μ		$vav_\mu a_\mu$
$R(e^+e^-\to hadrons)$	$v\sum_q v_q$		$(v^2+a^2)\sum_q(v_q^2+a_q^2)$
$A(q\bar{q})$	$a\sum_q Q_q a_q$		$va\sum_q v_q a_q$
$\sigma_R - \sigma_L$	$a\sum_q v_q$		$va\sum_q(v_q^2+a_q^2)$
Pol (Λ)	av_s		$va(v_s^2+a_s^2)$

Table 4

4. RADIATIVE CORRECTIONS

Since many of the observables used to determine the strength of electroweak effects in e^+e^- annihilations are intrinsically parity conserving quantities, QED and QCD processes (to all orders) can also contribute and they can cause effects similar to those produced by Z^o exchange. For example, higher order QED graphs change the total cross section for $\mu^+\mu^-$ and produce a forward-backward angular asymmetry. These effects are of the order of a few percent at PETRA energies $(1+\alpha)$. Similarly, QCD effects (gluon emission) change the total cross section for $q\bar{q}$ and produce a forward-backward quark angular asymmetry. These effects are also of the order of a few percent $(1+\alpha_s/\pi)$.

In order to extract the electroweak information as reliably as possible, it is necessary to understand these radiative corrections, to calculate them carefully or to measure them if possible.

Higher order QED processes are handled by using the

calculations of Berends and Kleiss [4], and all groups working at PETRA and PEP apply these (or similar) corrections to their data.

One of the PETRA experiments, namely JADE, has carried out a careful comparison of the Berends and Kleiss calculations with measurements of the processes $e^+e^- \to \gamma\, \mu^+\mu^-$ and $e^+e^- \to \gamma$ + + (hadrons = $q\bar{q}$) where the γ is radiated from one of the initial states $e^\pm$ and has an angle and energy which allows it to be detected by the apparatus. A comparison of the momentum spectrum of the γ and the $\mu^\pm$ angular distribution with the curves of Berends and Kleiss is shown in fig. 9 where good agreement can be seen. This provides confidence in using the calculations in regions where the γ is not detected (eg. small angles and low momentum). To quantify the situation, the QED contribution to the F-B asymmetry in $\mu^+\mu^-$ is 1.3% when integrated over the acceptance of the JADE detector. This can be compared with an expected asymmetry of electroweak origin of about 9% at present energies. Fig. 9 suggests that the error on the 1.3% correction is probably quite small. In fact, radiative corrections to lepton final states are less of a problem in present detectors with "4π" γ detection (eg. JADE Mark J) since radiative effects are largely observed and the final correction is small.

The problem of QCD corrections has not been investigated quite so well. Calculations have been made by Jersak et al.[11] who calculated R and the quark F-B asymmetry including graphs up to order α_s.

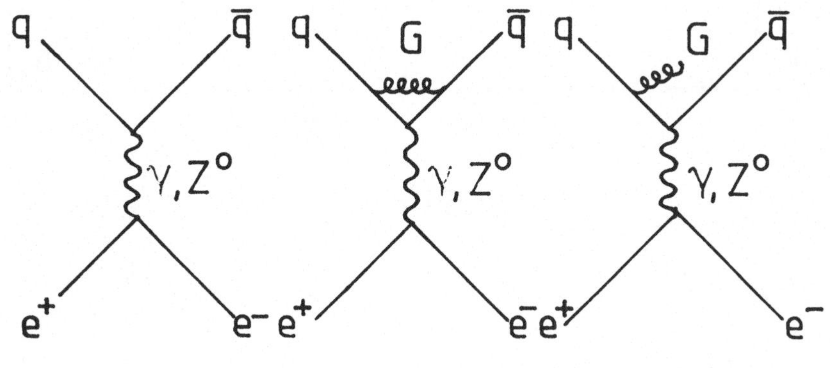

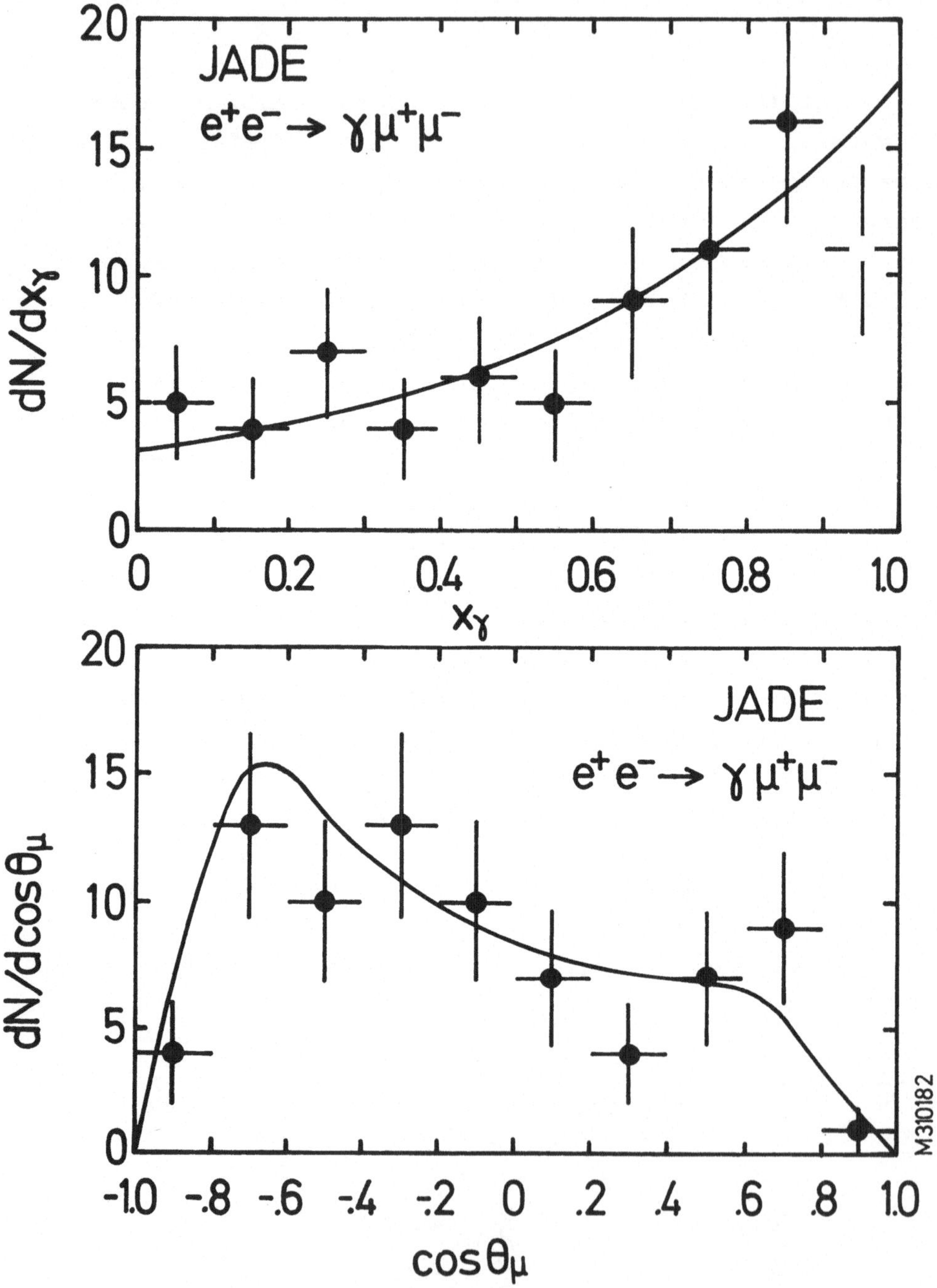

Fig. 9. The differential momentum and angular distribution for the radiative process $e^+e^- \to \mu^+\mu^-\gamma$.

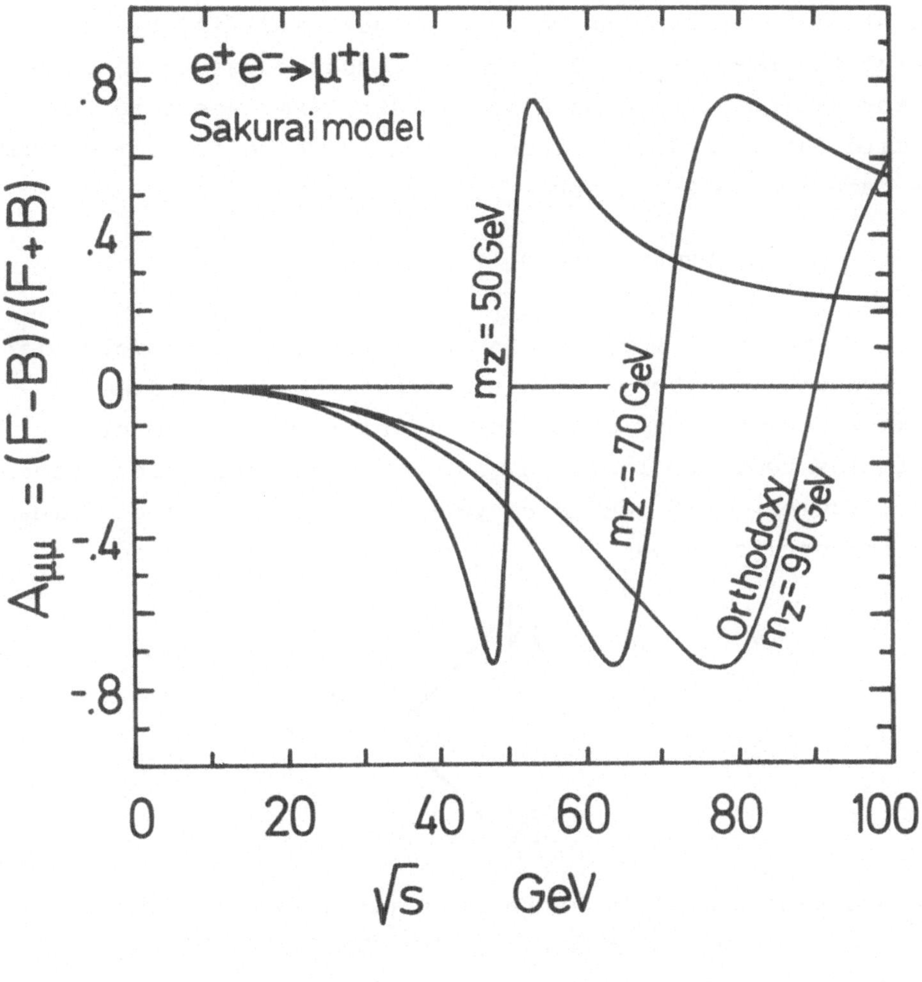

Fig. 10. The effect of a low mass Z^o on the $\mu^+\mu^-$ F-B asymmetry.

They also included finite quark mass effects which for example cause an axial suppression near threshold since a $q\bar{q}$ pair at rest cannot be in a $J^P = 1^+$ state. In general, the correction turns out to be <u>larger</u> than $(1 + \alpha_s/\pi)$. In the standard model, the value of R gets modified by up to 20% at the peak of the Z^o and the result suggests that more investigation, including higher order terms, needs to be carried out if reliable information is to be extracted from LEP data in the Z^o region.

5. EXTENSIONS TO THE STANDARD MODEL

There is no shortage in the literature of models with a "richer" gauge boson structure than the standard model. We know from history that a proliferation of "elementary" particles tends to imply an underlying substructure. Before the discovery of N^* and Δ baryons and the ρ, ω, ϕ etc. mesons, the nuclear force was thought to be mediated by pions.

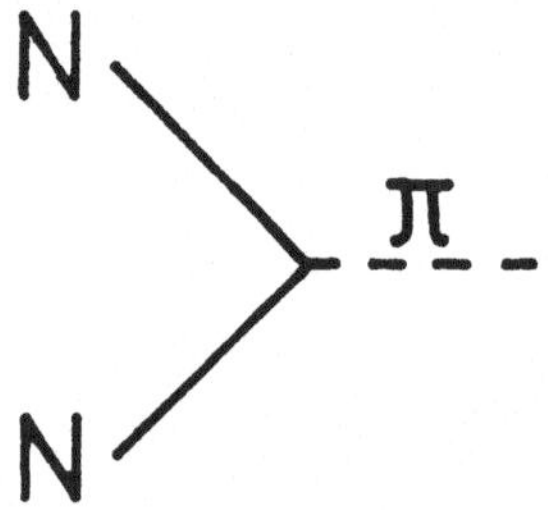

The eventual plurality of baryons and mesons and the subsequent experiments at large Q^2 at SLAC in the late 1960s put an end to this picture and we now have a substructure of quarks with their associated gauge particles-gluons. The neutral weak interaction of leptons and quarks can be drawn thus:

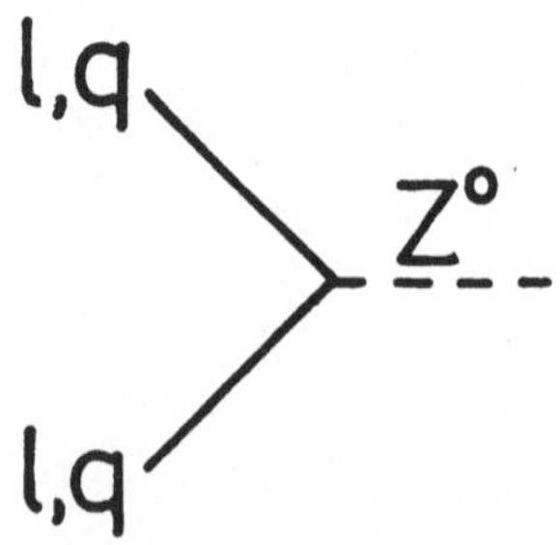

It is debatable whether there is a proliferation of leptons and quarks, the discovery of another heavy lepton would tend to suggest that there is! Similarly the existence of a complicated Z structure would also support this idea. The answer lies in spectroscopy and high Q^2 experiments.

In the meantime, the data from present experiments at PETRA and PEP can be used to exclude, restrict or eventually confirm these extended gauge models all of which incidentally have to coincide with the standard model at low Q^2 or to differ from the standard model by having extra parity conserving pieces only (since the parity violating part has been well tested already).

The various models achieve the richer gauge structure by adding an extra symmetry group to the standard model $SU(2) \times U(1)$,

eg.

Barger et al. [12]	$SU(2) \times U(1) \times SU(2)'$		
gauge particles	$W_1^{\pm}$		$W_2^{\pm}$
	Z_1	γ	Z_2
Elias, Pati & Salam [13]	$SU(2)_L \times SU(2)_R \times U(1)_L \times U(1)_R$		
	$W_L^{\pm}$	$W_R^{\pm}$	
	Z_L	Z_R	γ
de Groot et al.[14]	$SU(2) \times U(1) \times \tilde{U}(1)$		
	$W^{\pm}$		
	Z_1	γ	Z_2 .

In addition, a series of papers by Schildknecht et al.[15] has covered the properties of more general gauge group extensions to the standard model, namely $SU(2) \times U(1) \times G$.

The extra symmetry group corresponds to the insertion of an additional piece into the interaction Lagrangian, ie.

$$L = \frac{\sqrt{2}G_F}{4} \{[j_3 - \sin^2\theta_W j_{em}]^2 + Cj_{em}^2\} \qquad (12)$$

where C = 0 corresponds to the standard model. Since C is associated with j_{em} it is a parity conserving quantity, therefore not detectable in the electron deuterium polarisation asymmetry measurements at SLAC. Nor would it be observable in ν scattering experiments since the neutrino does not couple to the photon.

The extra symmetry group chosen together with the assumed structure of the Higgs sector determines the analytic form for C. For example, expressions for C for two of the above models are as follows:

Barger et al.[12] $$C = \sin^4\theta_W \frac{(m_2^2-m_Z^2)(m_Z^2-m_1^2)}{m_1^2 m_2^2}$$

and de Groot et al.[14] $$C = \cos^4\theta_W \frac{(m_2^2-m_Z^2)(m_Z^2-m_1^2)}{m_1^2 m_2^2} .$$

The strength of the extra piece of the Lagrangian is proportional to $\sin^4\theta_W$ ($\approx$ 0.04) in one case and $\cos^4\theta_W$ ($\approx$ 0.5) in the other; ie. there is a factor of 10 difference between the two and they can be used as extreme cases of these types of models. Moreover, since C is a positive definite quantity, the form of C above leads to the condition $m_1 < m_2$ (standard) $< m_2$ although more recent models [16] with a different Higgs structure and different m_1, m_2 dependence for C have $m_1, m_2 > m_Z$. C is also non-zero for models where the leptons and quarks are composite particles. Nevertheless, some of these models predict a gauge boson with a mass less than that of the standard model.

The masses are not predicted in these models, and in the limiting case m_1 or $m_2 = m_Z$ the other Z decouples and the model becomes identical to the standard model.

The quantity C also corresponds to a modification to the vector coupling (since C is associated with the vector electromagnetic current), thus $v^2 \rightarrow (-1+4\sin^2\theta_W)^2 + 16C$. So if v^2 varied away from its low Q^2 value of approximately zero as data at higher Q^2 (nearer the Z^o boson mass) is obtained, this would indicate the presence of the C term.

C can also be expressed in terms of the sum rule [17]

$$16C = \frac{\int\frac{ds}{s}\sigma(e^+e^- \rightarrow \gamma, Z_1 \ldots Z_N \rightarrow \text{all}) - \int\frac{ds}{s}\sigma(e^+e^- \rightarrow \gamma, Z(\text{standard}) \rightarrow \text{all})}{\int\frac{ds}{s}\,\sigma(e^+e^- \rightarrow \gamma, Z(\text{standard}) \rightarrow \text{all})}$$

ie. it represents the possible deviation from the standard model due to an arbitrary series of gauge bosons $Z_1 \ldots Z_N$.

Other deviations from the standard model have been proposed by Professor Sakurai [18] and will be discussed in more detail in his lectures. This γ-Z mixing (non-gauge invariant) model retains the v and a couplings with their values determined at low Q^2, but the relation between m_Z and $\sin^2\theta_W$, ie.

$$m_Z^2 = \frac{37.3^2}{\sin^2\theta_W \cos^2\theta_W}$$

is discarded and m_Z becomes a parameter to be determined by experiment. The effect of varying m_Z on the $\mu^+\mu^-$ asymmetry (eg. propagator effect) can be seen in fig. 10 which shows that a measurement of A_μ at PETRA energies can already say plenty about the value of m_Z in the context of this model.

6. INTERPRETATION OF e^+e^- DATA

This section is intended to review what has been learnt about the electroweak interaction from e^+e^- annihilation measurements. We shall be interested in the values of the lepton and quark vector and axial couplings v_f and a_f and a comparison with the predictions of the standard model. The value of $\sin^2\theta_W$ determined from lepton and hadron final states will also be considered. Finally the restrictions on the mass of the Z^o will be presented.

The data come mainly from five experiments which have been running on the e^+e^- storage ring PETRA. Experimental physicists these days tend to work under a cloak of acronymity; at PETRA for example the experiments are called CELLO, JADE, MARK-J, PLUTO and TASSO. Meanwhile the competing e^+e^- machine PEP at Stanford has also started to produce results and the combined data from about 7 different experiments has helped considerably to overcome the problem of the low cross sections. For the present, no data are available on polarisation (parity violating) effects, nor has anyone yet observed quark charge asymmetry effects analogous to the muon asymmetry.

6.1 Lepton Sector Coupling Constants v,a

In order to determine the lepton coupling constants, data from three processes have been analysed, ie.

$$\begin{aligned} e^+e^- &\to e^+e^- \\ &\to \mu^+\mu^- \\ &\to \tau^+\tau^- \quad . \end{aligned}$$

In all the analyses, the results have been presented as v^2 and a^2 although strictly, the measured quantities are v_e^2, $v_e v_\mu$, $v_e v_\tau$ etc. in the three processes. Assuming e-μ-τ universality allows data from the three processes to be combined in a single analysis. There is a justification for

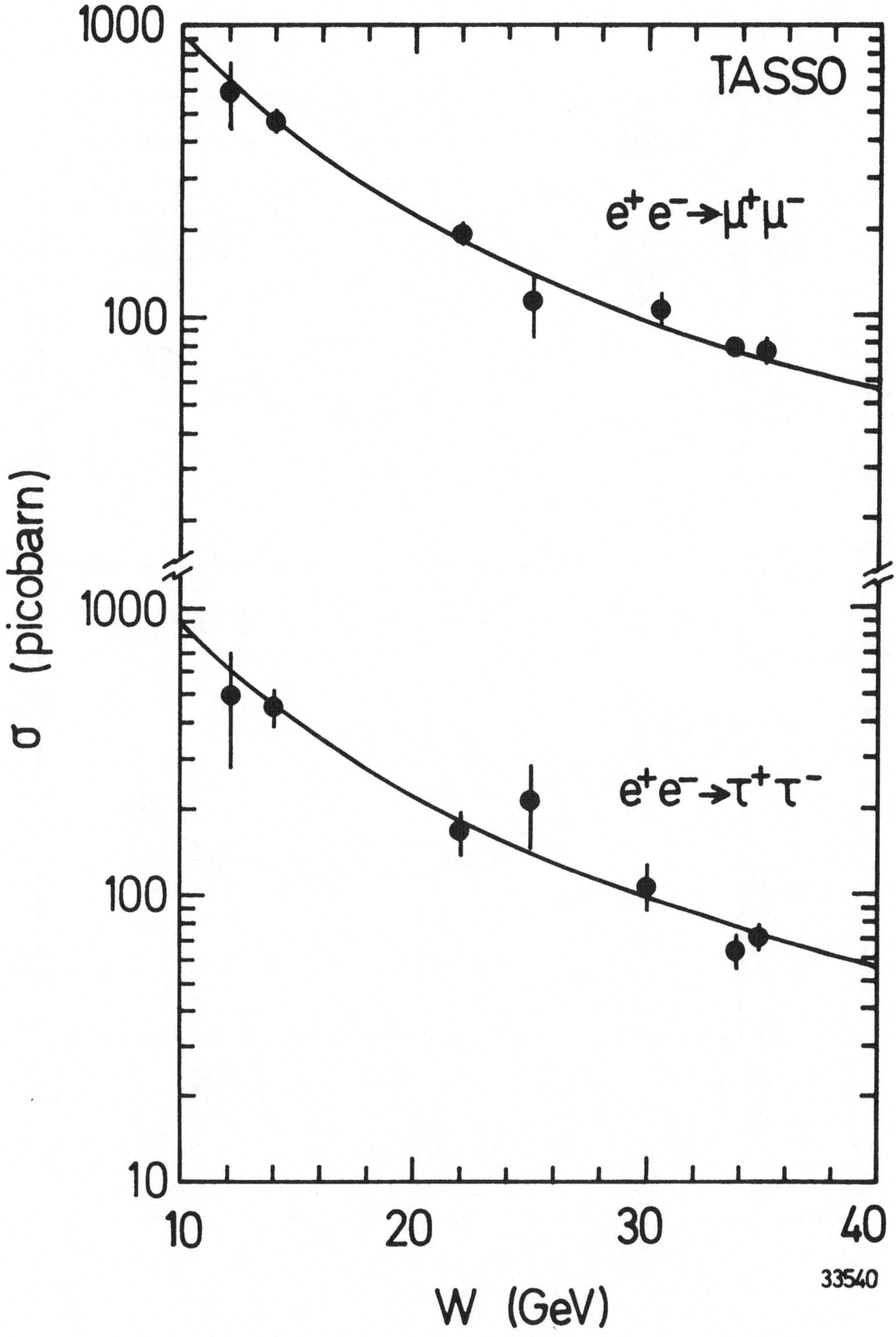

Fig. 11. The total cross sections for $\mu^+\mu^-$ and $\tau^+\tau^-$ production as a function of $W = \sqrt{s}$.

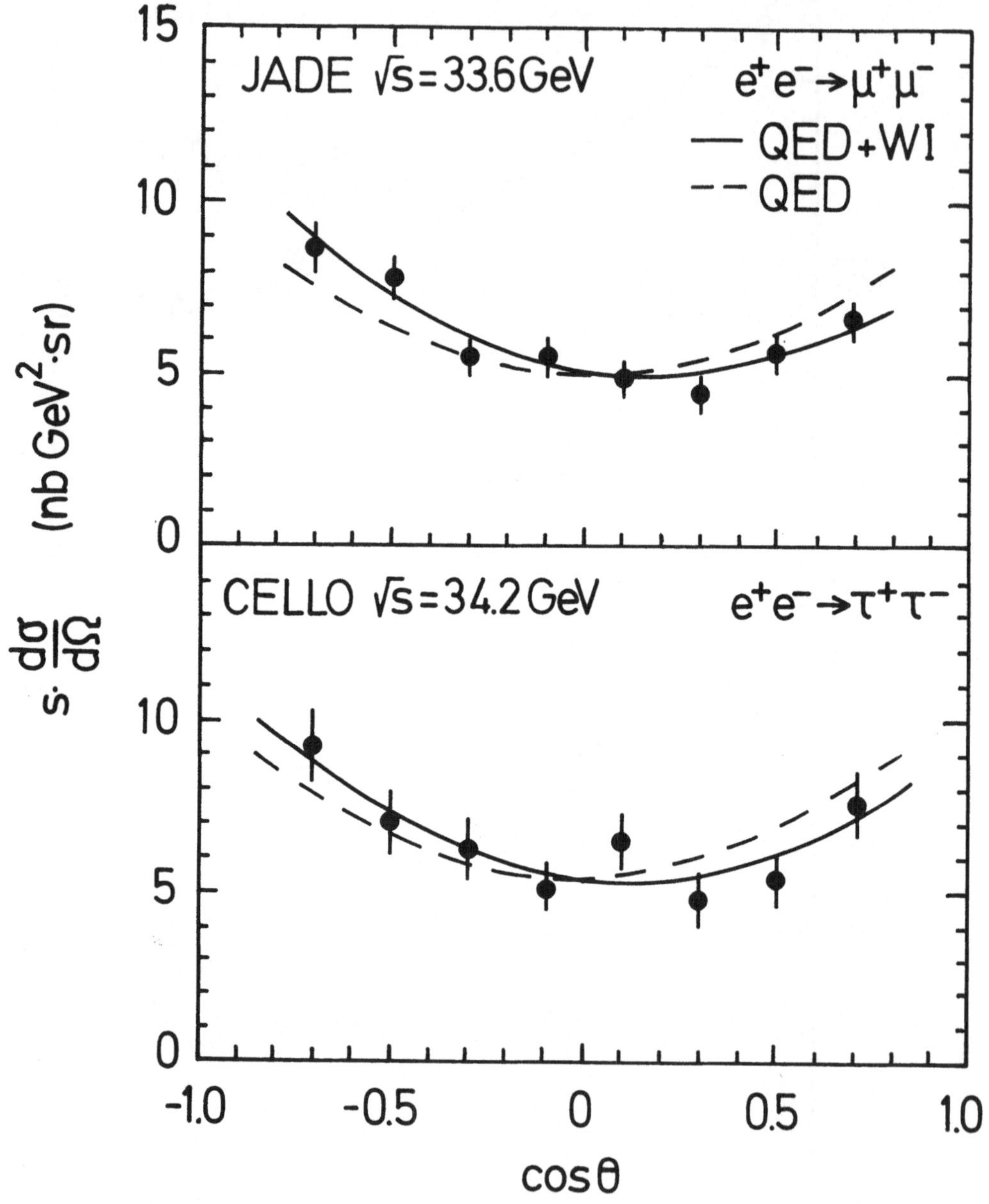

Fig. 12. The angular distributions for $\mu^+\mu^-$ and $\tau^+\tau^-$ production at $s \approx 1150\ \mathrm{GeV}^2$.

this assumption. Fig. 11 shows the measured total cross sections for $\mu^+\mu^-$ and $\tau^+\tau^-$ production which is sensitive to $v_e v_\mu$ and $v_e v_\tau$. The data support μ-τ universalityin that both $v_e v_\mu$ and $v_e v_\tau$ appear to be close to zero. However, the possibility: $v_e \approx 0$ and v_μ and v_τ non-zero <u>and</u> not equal to each other is not excluded. To test this latter possibility, one should measure the τ polarisation (via the $\mu\nu_\mu\nu_\tau$ decay mode). From equation 11, this measures $a_e v_\tau$ and the zeroing effect of v_e can be avoided.

Further evidence for μ-τ universality comes from the shape of the differential cross section - which is sensitive to $a_e a_{\mu,\tau}$. The measured $\frac{d\sigma}{d\Omega}$ are shown in fig. 12; they do not contradict each other. Most experiments quote a result for the forward-backward asymmetry since apparatus biases largely cancel out. The various experiments tabulate as follows:

	A_μ (measured)	A_μ (expected)	Average	Expected
CELLO	$-6.4 \pm 6.4\%$	-9.1%		
JADE	$-12.7 \pm 2.7 \pm 1\%$	-9.1%		
MARK-J	$-8.4 \pm 2.9 \pm 1.1\%$	-9.5%		
PLUTO	$+7 \pm 10\%$	-9.8%	$-9.4 \pm 1.5\%$	-8.1%
TASSO	$-16.1 \pm 3.2\%$	-9.2%		
MAC	$-0.9 \pm 5.7\%$	-5.0%		
MARK II	$-4.0 \pm 3.5\%$	-5.0%		

	A_τ (measured)	A_τ (expected)	Average	Expected
CELLO	$-10.3 \pm 5.2\%$	-9.2%		
JADE	$-4.7 \pm 4.2 \pm 1.2\%$	-7.5%		
MARK J	$-7.0 \pm 7.2 \pm 2.1\%$	-9.5%	$-5.8 \pm 2.8\%$	-8.5%
TASSO	$-0.4 \pm 6.6\%$	-9.1%		

(where two errors are quoted, the second is systematic).

The τ and μ results are in fairly good agreement and together provide a clear, unambiguous demonstration of the existence of neutral current effects in e^+e^- reactions. The reason why the expectation of the standard model is different for τ and μ is because the theory has been integrated over apparatus angular acceptance and this is slightly smaller in some experiments for τ than for μ.

There is always the possibility that the apparatus has introduced an asymmetry via instrumental biases. To investigate this, JADE tested for instrumental effects by measuring a large sample of cosmic muons and found A_{cosmic} = = -0.7 ± 1.1%, close to zero as expected. The $e^+e^- \rightarrow \mu^+\mu^-$ asymmetry has also been measured in the same experiments at low s where weak effects are expected to be negligible. No evidence for an experimental bias can be found.

Combining data for the various final states, a fit can be made to the measured angular distributions to find v^2 and a^2. Since data is continually being collected, these values are progressively being improved; the status quoted here corresponds to the spring 1982 conferences [19]. The results for the various experiments can be listed as follows:

	v^2	a^2
CELLO	-0.16 ± 0.32	1.12 ± 0.48
JADE	0.40 ± 0.28	1.24 ± 0.32
MARK-J	0.12 ± 0.24	0.88 ± 0.32
TASSO	-0.44 ± 0.52	2.12 ± 0.40
MAC		0.16 ± 0.88
MARK II	0.20 ± 0.40	0.96 ± 0.64.

They can be combined to obtain a result for the lepton couplings from all e^+e^- experiments:

$$v^2 = 0.10 \pm 0.14 \quad \text{and} \quad a^2 = 1.22 \pm 0.17$$

or

$$g_v^2 = 0.026 \pm 0.035 \quad \text{and} \quad g_a^2 = 0.305 \pm 0.043 ,$$

remembering they have been derived assuming e-μ-τ universality and factorisation. A further assumption in the v^2, a^2 analyses was to absorb propagator effects into the coupling constants, ie. the results are actually for

$$\frac{v^2 m_Z^2}{m_Z^2-s} \quad \text{and} \quad \frac{a^2 m_Z^2}{m_Z^2-s} \quad .$$

If $m_Z \approx 90$ GeV then the above numbers would have to be divided by a factor 1.16.

A comparison can now be made with the couplings determined in νe elastic scattering, also a purely leptonic process with no hadronic structure involved. Elastic νe scattering measures the product of the neutrino and electronic couplings (see Sakurai tetragon). It is customary however, to assume V-A coupling for the neutrinos and to use the data to determine g_a and g_v for the electron - see fig. 13. The cross sections depend on terms like $(g_a+g_v)^2$ and $(g_a-g_v)^2$; hence the ellipses from the measurement of a single number σ. All three elliptical regions coincide only in the small regions $(g_v, g_a) \approx (0, -\frac{1}{2})$ or $\approx (-\frac{1}{2}, 0)$ leaving a dominant axial or dominant vector ambiguity from νe reactions alone. A comparison with the results from e^+e^- can be made in the v^2-a^2 plane (remember $v = 2g_v$ and $a = 2g_a$) as shown in fig. 14. This is a comparison between all νe and all e^+e^- data under the assumptions:

1) ν couplings are assumed to be V-A
2) factorisation
3) e-μ-τ universality.

The comparison removes the νe ambiguity, the dominant axial solution alone survives and furthermore, both sets of purely leptonic processes are in good agreement with each other. It is worth noting that e^+e^- measurements are already competitive in determining the axial coupling but lag behind a little on the determination of the vector coupling.

The results for v^2 and a^2 overlap the line corresponding to the standard model ($a^2 = 1$, $v^2 = (-1+4\sin^2\theta_W)^2$) and it is

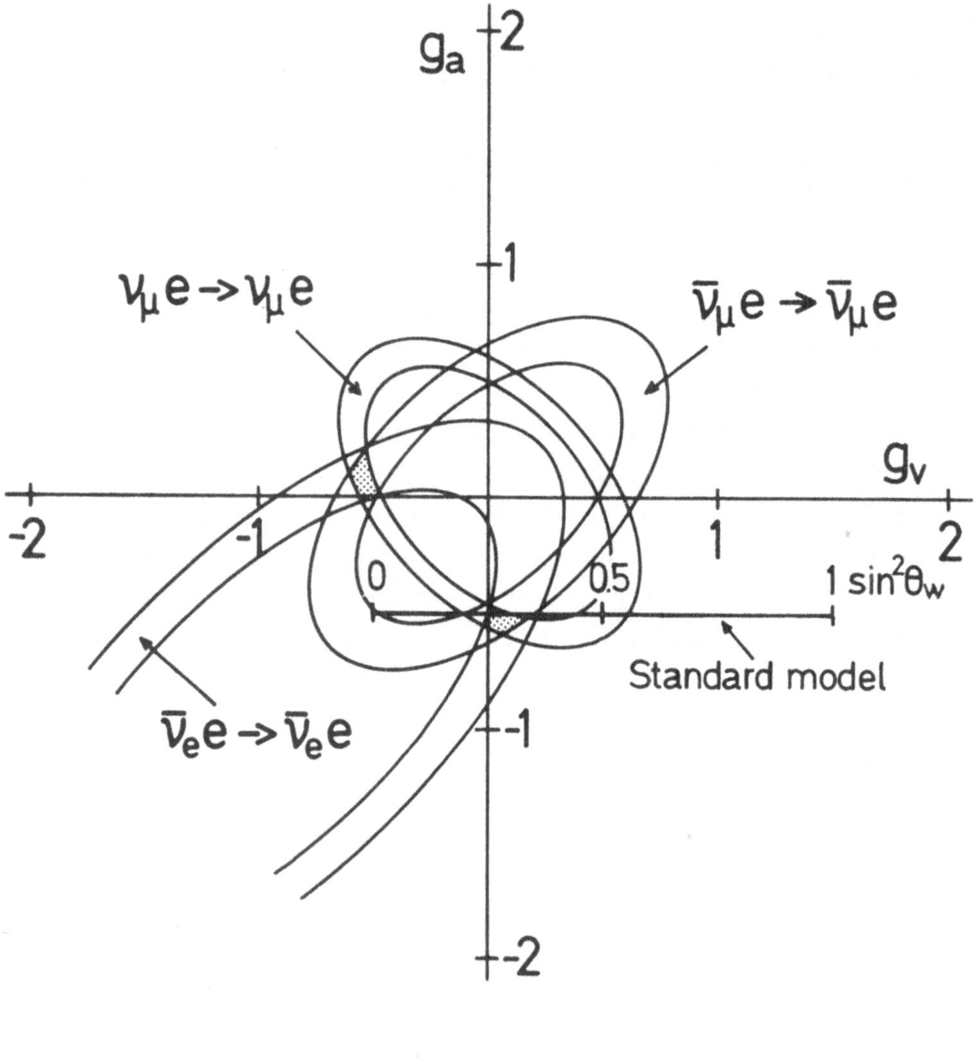

Fig. 13. The measurement of g_a and g_v in νe elastic scattering experiments.

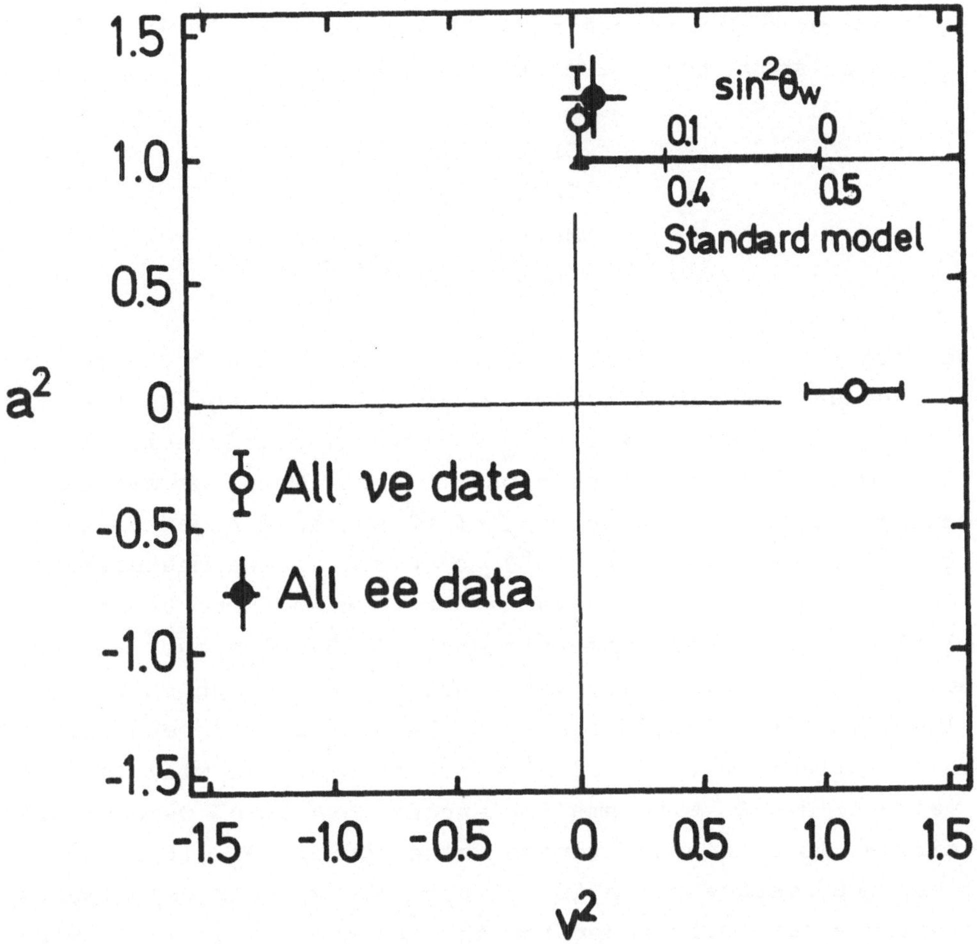

Fig. 14. A comparison of lepton weak couplings determined in e^+e^- and νe experiments.

thus reasonable to proceed to the next stage which is to determine $\sin^2\theta_W$. This is done in the fitting procedure by fixing $a^2 = 1$, substituting $v^2 = (-1+4\sin^2\theta_W)^2$ and including the propagator depencence $m_Z^2 = 37.3^2/(\sin^2\theta_W \cos^2\theta_W)$. Typical results as shown below are obtained:

CELLO	0.25 ± 0.12
JADE	0.25 ± 0.12
MARK J	0.25 ± 0.11
TASSO	$0.29 {+0.09 \atop -0.11}$.

The fact that the various results are all close to 0.25, even though the quoted errors would allow a large scatter is due to the fact that $\sin^2\theta_W$ is essentially determined via v^2 which happens to be close to zero. Then, as can be seen in fig. 15, a measurement of v^2 which is negative will yield a single value for $\sin^2\theta_W$ of 0.25. If the measurement gives a slightly positive value for v^2 then there are two solutions for $\sin^2\theta_W$ symmetric about 0.25 which overlap and merge to give again 0.25. (Propagator effects can shift the value slightly off 0.25.) Due to the quadratic dependence of the χ^2 surface, a value of v^2 which is negative will lead to a value of $\sin^2\theta_W$ with smaller errors than if v^2 were slightly positive. To obtain an average value for $\sin^2\theta_W$, it is safer to use the average value for v^2 from the various experiments. Assuming a parabolic error for the combined value and fixing $a^2 = 1$, then from all experiments on e^+e^-

$$\sin^2\theta_W = 0.25 \pm 0.06 \ .$$

6.2 Quark Sector Coupling Constants v_f, a_f

In this case, we have only one process

$$e^+e^- \to q\bar{q} \to \text{hadrons}$$

because a summation is carried out over all the quarks.

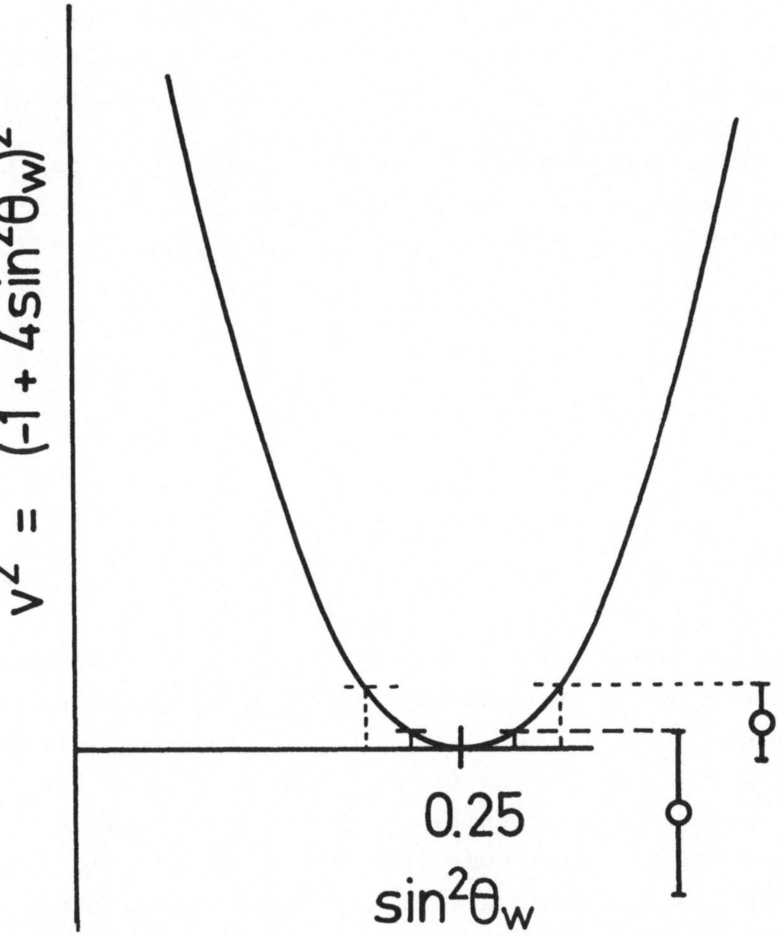

Fig. 15. The dependence of v^2 on $\sin^2\theta_W$ showing how $\sin^2\theta_W =$ $= 0.25$ arises, and errors which depend on v^2 as well as the error on v^2.

The cross section values used in the analysis by JADE cover the range from s = 144 GeV^2 to s = 1350 GeV^2, a variation of almost a factor 10 in s. The s range is just as important as the accuracy of the measurement of R since different values of $\sin^2\theta_W$ correspond to different s dependences.

A model independent analysis can be carried out as follows: From equation (8), summing over d, u, s, c and b quarks and multiplying by a factor 3 for colour,

$$R = 3\sum Q_f^2 + \frac{3}{2} gV + \frac{3}{16} g^2 W$$

where

$$V = v\sum Q_f v_f$$

$$W = (v^2 + a^2)\sum (v_f^2 + a_f^2)$$

and now

$$g \approx -\sqrt{2}G_F s/4\pi\alpha \ .$$

Neglecting propagator effects, the above equation can be fitted to the data to determine V and W, and the measurement then corresponds to a determination of these particular combinations of couplings. Including QCD effects following the prescription of Jersak et al.[11], the result for V and W shown in fig. 16 is obtained. Fig. 16 also shows the prediction for the V-W locus from the standard model. The l.s.d. ellipse overlaps part of the standard model and hence the measured V and W agree with the calculated value for certain values of $\sin^2\theta_W$. The next stage is to determine the value of $\sin^2\theta_W$. This is done by fitting the full expression for R, substituting the expressions for v_f and a_f from table 2 and including propagator effects as before. The effect of the overall normalisation uncertainty in the R values (assumed to be independent of s) is taken into account by including the normalisation as a fit parameter with measured value 1.00 ± 0.07. The result can be expressed as an l.s.d. contour in the $\sin^2\theta_W$ - normalisation plane - see fig. 17. The fitted normalisation of 0.986 is well within the 7% uncertainty.

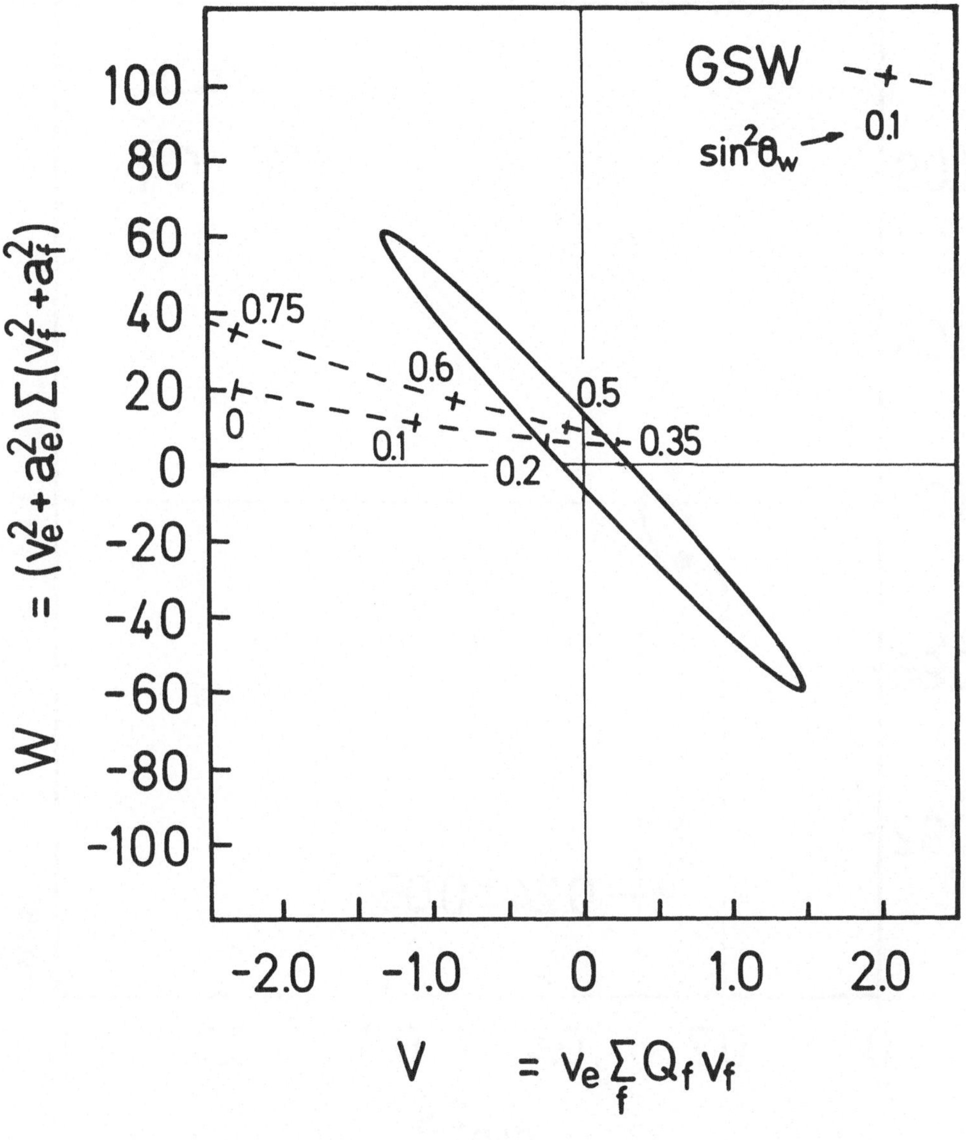

Fig. 16. A model independent fit to determine the V and W combinations of the quark weak couplings.

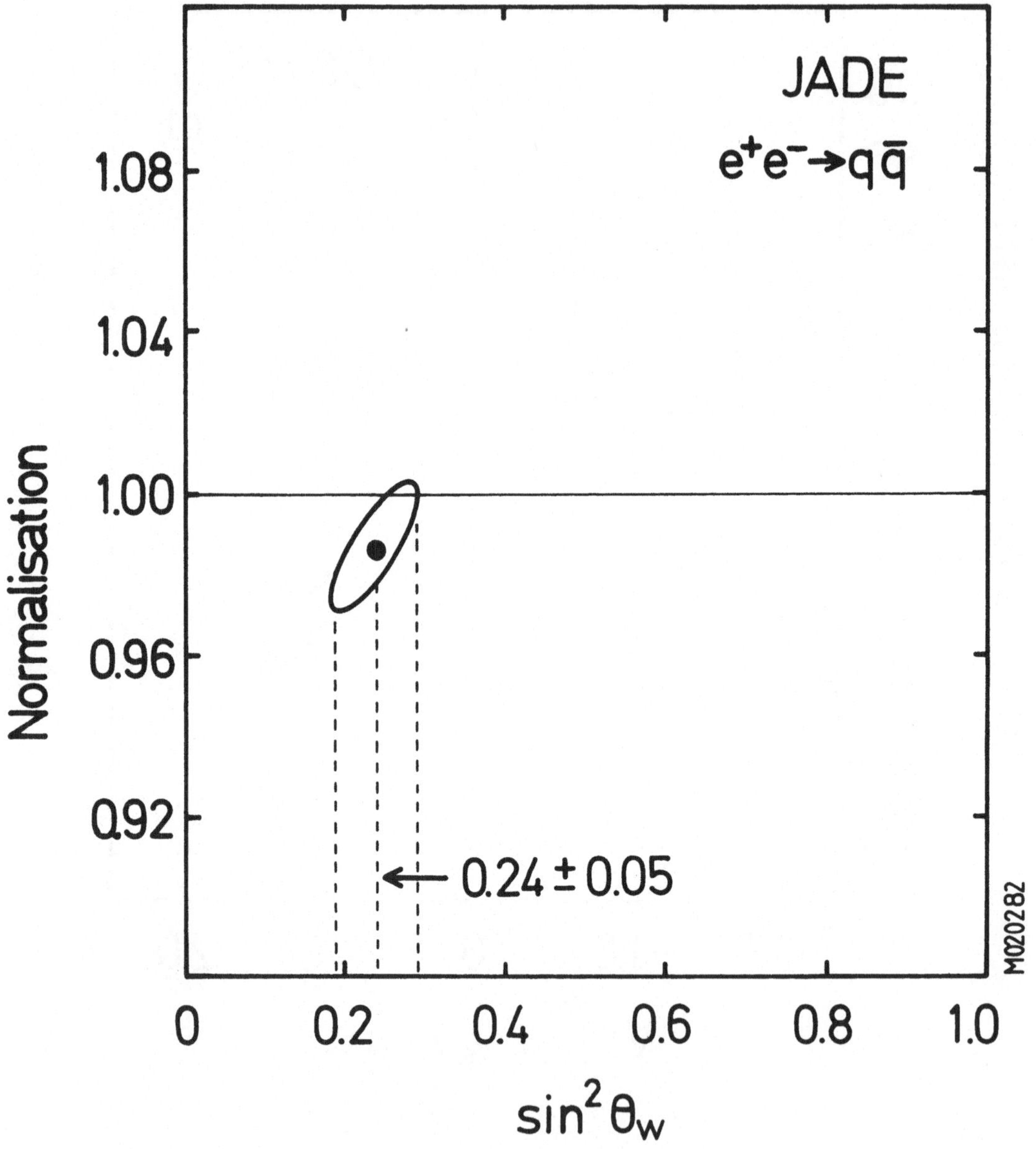

Fig. 17. A two parameter fit to determine the overall normalisation of the JADE data and a value of $\sin^2\theta_W$.

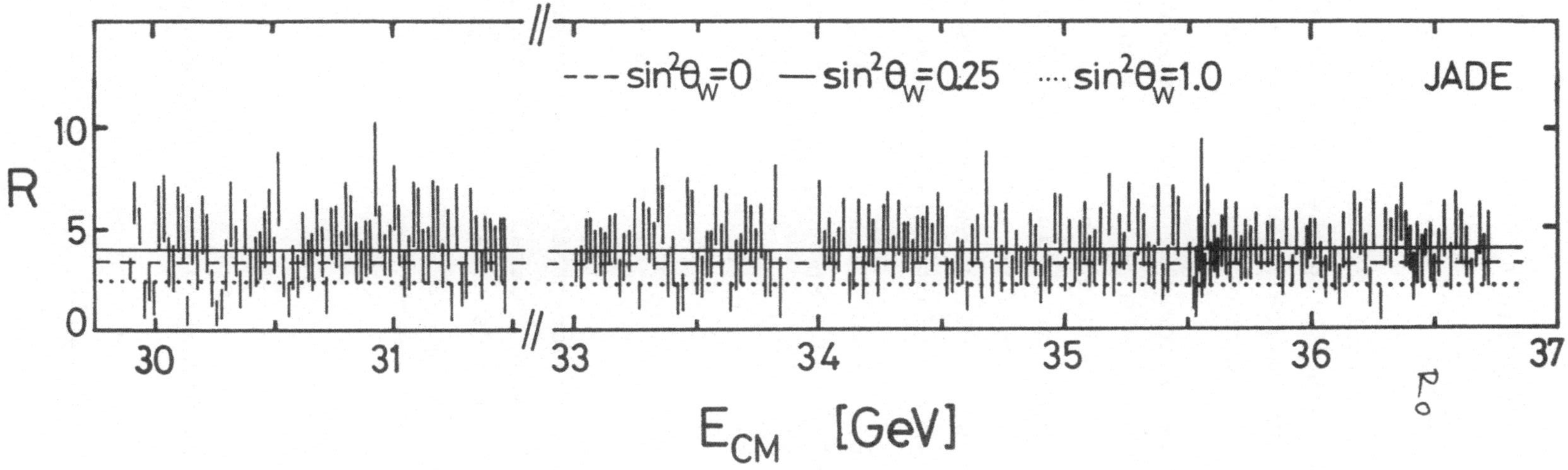

Fig. 18. A measurement of R in fine steps of energy. No evidence is seen for a peak whereas a Z^0 with conventional couplings would give an R value of several thousand.

The fitted value for $\sin^2\theta_W$ is

$$\sin^2\theta_W = 0.24 \pm 0.05 \text{ (d, u, s, c and b quarks)}$$

where the error contains both statistical and systematic effects. It is worth emphasising the fact that e^+e^- processes produce all quarks democratically and provide a unique opportunity to measure the couplings of s, c and b quarks.

A further source of error is in the uncertainty in α_s. Varying α_s by ±30% (ie. $\alpha_s = 0.18 \pm 0.05$) produced a change of ±0.03 in $\sin^2\theta_W$. An important conclusion is that $\sin^2\theta_W$ has the same value for a democratic mixture of all known quarks as it does for leptons (νe and e^+e^- results) or for light quarks (νq and eq results). Moreover, the e^+e^- measurementshave been made at much larger values of Q^2 than previously.

In the analysis, one can "freeze" out the d and u quarks by fixing $\sin^2\theta_W$ for d and u quarks at the world average of 0.23 and determine the mixing angle for s, c and b quarks alone. The result in this case is

$$\sin^2\theta_W = 0.25 \begin{smallmatrix}+0.12\\-0.06\end{smallmatrix} \text{ (s,c and b quarks).}$$

It seems that $\sin^2\theta_W$ has just that value which makes the lepton coupling to the weak neutral current almost pure axial. This value is close to the "special" value of $\frac{1}{4}$ displayed in table 3. Whether this is an approximate coincidence or is significant, remains to be seen.

6.3 What Is the Mass of the Z^0?

There are several ways to draw conclusions about the Z^0 mass from e^+e^- data. These can be listed as follows:

1. Search for peak in energy scan data.
2. Comparison of data with multi Z (low Z mass models).

3. Deviation from linear s (propagator) dependence (Sakurai model).
4. Have faith in the standard model, then $m_Z^2 = \frac{37.3^2}{\sin^2\theta_W(1-\sin^2\theta_W)}$.

6.3.1 Energy scan

Much of the data at PETRA has been taken in small steps of beam energy, scanning for narrow peaks. The results of this scan above 30 GeV are shown in fig. 18 where no significant peaks are observed. Since a Z^o with conventional couplings would produce an R value of several thousand, this can essentially be ruled out in the energy range covered.

6.3.2 Multi gauge boson models

The measured cross sections for lepton and hadron final states can be compared with the predictions of the extended gauge models discussed above. In the case of the two boson models of Barger et al. and de Groot et al. which each predict two bosons with masses m_1 and m_2, the data is compared with theory for all possible m_1 and m_2 value. Regions of m_1, m_2 space can be excluded if the fitted χ^2 is unacceptable (eg. $\chi^2 > \chi^2_{min} + 6$) leaving a restricted region of allowed m_1 and m_2 values. Fig. 19 shows the result for these two models. Disregarding the small pathological regions under the toes of the curves (where the bosons barely couple) one can say

$$\left.\begin{array}{ll} m_Z > 50 \text{ GeV} & \text{Barger et al.} \\ m_Z > 80 \text{ GeV} & \text{de Groot et al.} \end{array}\right\} \text{95\% CL limits.}$$

In terms of the C parameter, this corresponds to $C \lesssim 0.03$, 95% CL referring to equation 12, the coefficient of the extra piece of j^2_{em} is slightly smaller than the coefficient of j^2_{em} in the standard model ($\sin^4\theta_W$). Alternatively, this

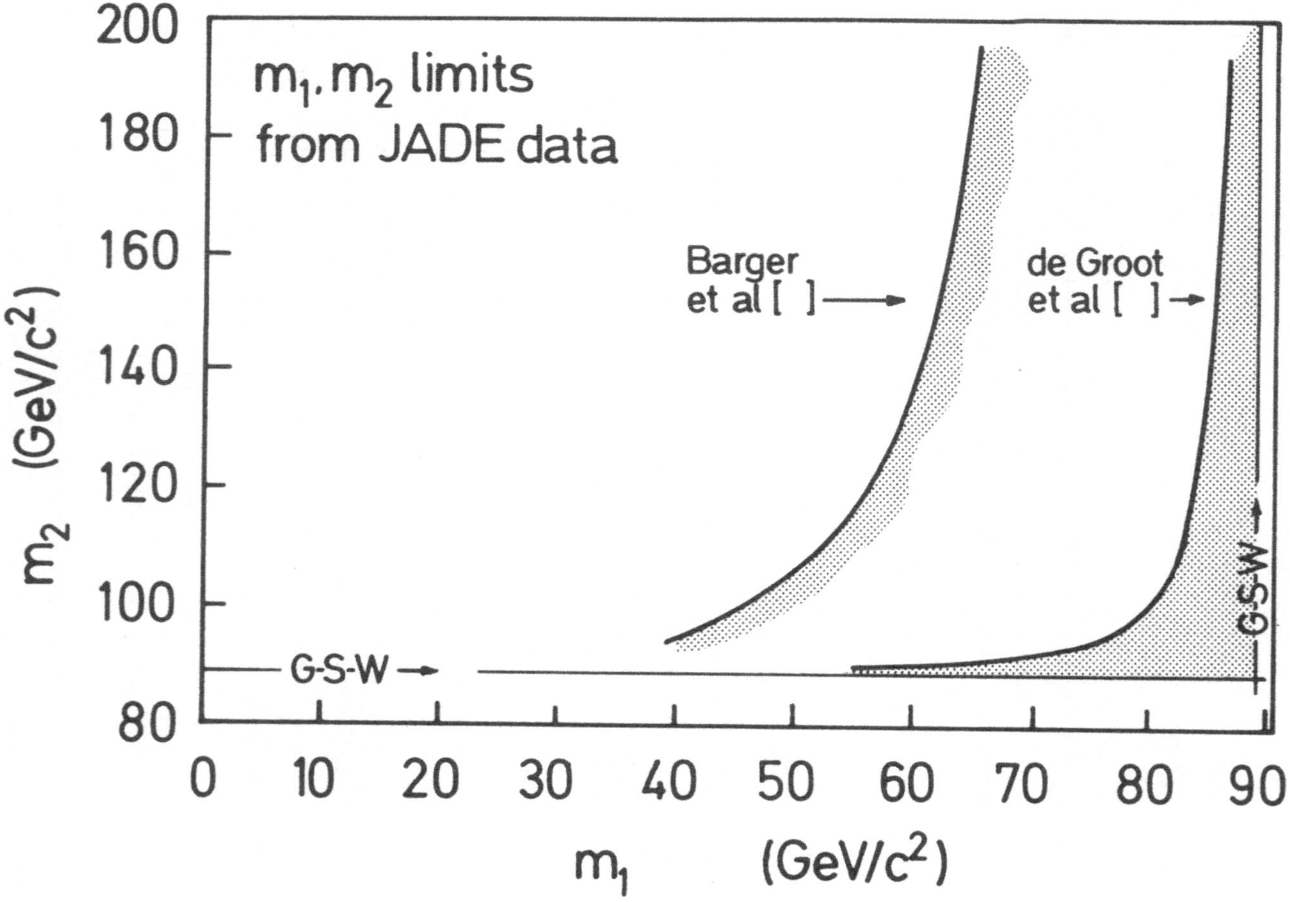

Fig. 19. The regions of m_1 and m_2 space excluded by e^+e^- data. The shaded areas correspond to values of m_1 and m_2 still allowed by the data.

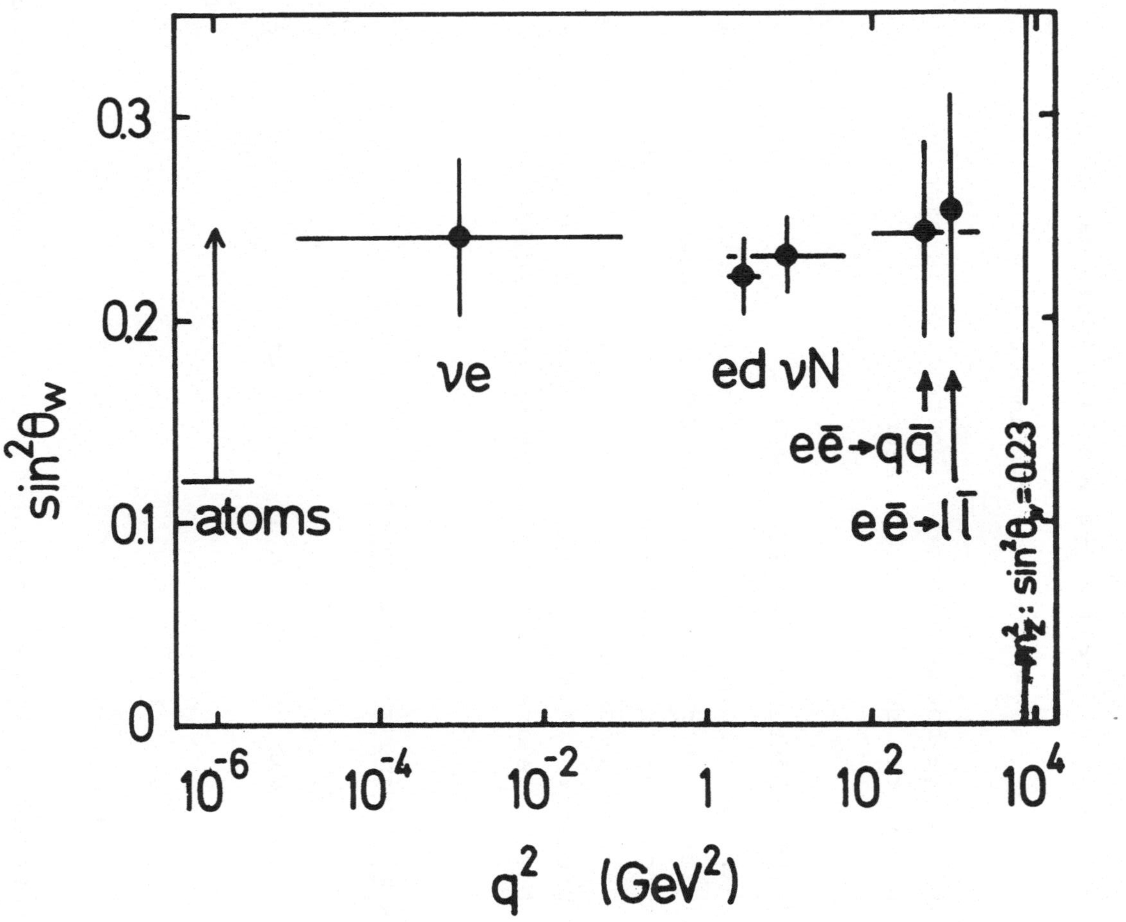

Fig. 20. The Q^2 dependence of $\sin^2\theta_w$ determined in various processes.

can be expressed as the sum rule

$$\frac{\int\frac{ds}{s}\sigma(e^+e^-\to\gamma,Z_1\ldots Z_N\to all)-\int\frac{ds}{s}\sigma(e^+e^-\to\gamma,Z(standard)\to all)}{\int\frac{ds}{s}\sigma(e^+e^-\to\gamma,Z(standard)\to all}<0.48$$

ie. multigauge bosons can still contribute as much to this sum rule as the standard model alone.

6.3.3 Comparison with Sakurai model

If the complete set of JADE data on lepton and hadron final states is compared with the Sakurai model where the v and a couplings are all fixed at the low Q^2 values but the m_Z is a free parameter, then $m_Z > 54$ GeV with 95% confidence. This procedure is an extended version of the method whereby the mass is deduced via propagator effects in the asymmetry A_μ.

6.3.4 m_Z in the standard model

The values of $\sin^2\theta_W$ determined from $e^+e^- \to$ lepton pair and hadrons can be inserted in the relation $m_Z^2 = \frac{37.3^2}{\sin^2\theta_W\cos^2\theta_W}$, then:

$$e^+e^- \to l^+l^- \quad 77 < m_Z < 116 \text{ GeV} \quad 95\% \text{ CL}$$

$$e^+e^- \to q\bar{q} \quad 78 < m_Z < 112 \text{ GeV} \quad 95\% \text{ CL}.$$

But note that this result depends heavily on the mass formula of the standard model (ie. Higgs structure) and is the most model dependent result. The results of these considerations on m_Z can be summarised as follows:

Method		Limit on m_Z	
Direct scan		> 36.8 GeV	
Extended models			
$SU(2)\times U(1)\times SU(2)'$		> 50 GeV	
$SU(2)\times U(1)\times \tilde{U}(1)$		> 80 GeV	
Sakurai model		> 54 GeV	
Standard model	$116 > m_Z$	> 77 GeV	all lepton data
	$112 > m_Z$	> 78 GeV	Jade hadron data

CONCLUSIONS

The data from experiments at PETRA and PEP now demonstrate the existence of neutral current effects in e^+e^- processes for the first time. Traditionally, this would have been termed a violation of QED in an electromagnetic process. Information is also available on the neutral weak couplings of leptons and quarks which complements the results from neutrino and electron scattering. In the lepton sector the information derived from νe and e^+e^- processes are of comparable quality.

The important conclusion from e^+e^- data is that the range of validity of the standard model has been extended to include μ and τ leptons and the second and third generation of quarks. Furthermore, the Q^2 values at which measurements have been taken are about an order of magnitude greater than in other processes. The situation is summarised in fig. 20 (borrowed from Perkins [20] with the e^+e^- data updated) which shows the Q^2 dependence of $\sin^2\theta_W$.

Although the data are consistent with the standard model, one should not lose sight of the fact that they are also consistent with the old fashioned four fermion point coupling with axial and vector couplings which happen to agree with the values predicted by the standard model. No propagator effects, which would indicate a finite range of the weak force, have yet been detected. But if the standard model is really correct, then there is a Z^o with a mass of about 90 GeV. Therefore one would expect that the $\bar{p}p$ collider should see it soon, or that the experiments at PETRA and PEP should see propagator effects in the $\mu^+\mu^-$ asymmetry which would be a factor 1.25 larger at the current highest energies than for point coupling. If the mass of the Z^o turns out to be very large then the $\mu^+\mu^-$ asymmetry will continue to rise linearly through the PETRA/PEP energy range and will thereafter rise less rapidly than linearly, eventually falling to zero due to the weak squared term, ie.

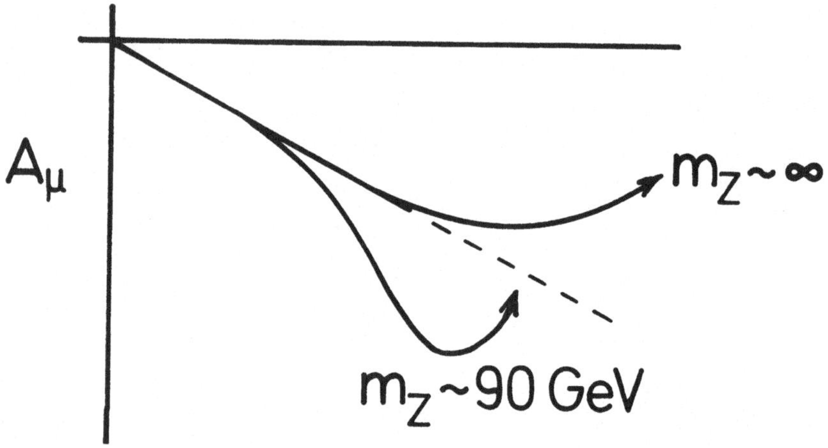

If the Z^o is discovered then attention can be turned to study the detailed properties of the Z^o (and W via $e^+e^- \rightarrow$ $\rightarrow W^+W^-$), the Higgs or techniparticle structure, a program which will probably be carried out at the SLC and LEP in a few years time.

In the meantime however, measurements of polarisation effects and quark charge asymmetries are likely to provide new and interesting information.

ACKNOWLEDGEMENTS

I am most grateful to Prof. Dr. H. Mitter for the invitation to the Schladming school, to work on and present these lectures in such pleasant surroundings. I also recognise the support of my colleagues in the JADE collaboration who expertly collect, analyse and discuss the experimental data. In particular, I thank Drs. D.P. Barber, J.B. Dainton, M.C. Goddard and B. Naroska for many useful discussions.

REFERENCES

1. R. Marshall, RL-80-029 and Proc. XV. Rencontre de Moriond, Les Arcs, France, 9-21 March 1980.
2. N. Cabibbo and R. Gatto, Phys. Rev. 124 (1961) 1577.
3. J. Ellis and M.K. Gaillard, Physics with very high energy e^+e^- colliding beams, CERN 76-18 (November 1976).
4. F.A. Behrends and R. Kleiss, Nucl. Phys. B177 (1981) 237.
5. J.J. Sakurai, UCLA/77/TEP/15 and Neutrino '77, Int. Conf. on Neutrino Physics and Neutrino Astrophysics, Elbrus, USSR, June 1977.
6. R. Budny, Phys. Lett. 55B (1975) 227.
7. L.M. Sehgal, PITHA 80/5. Int. School of Elementary Particles Physics, Kupari, Dubrovnik, Yugoslavia, 16-30 September 1979.
8. R.P. Feynman, Proc. Neutrino '72, Vol.II, Balatonfüred, June 1972.
   G. Farrar and J.L. Rosner, Phys. Rev. D7 (1973) 2747.
   R. Cahn and E. Colglazier, Phys. Rev. D9 (1974) 2658.
   S.J. Brodsky and N. Weiss, Phys. Rev. D16 (1977) 2325.
   C.J. Maxwell and M.J. Teper, Zeit. f. Phys. C7 (1981) 295.
9. G.F. Pearce, private communication.
10. A. Bartl et al., Zeit. f. Phys. C6 (1980) 335.
    J. Nieves, Phys. Rev. D20 (1979) 2775.
    J. Ranft and G. Ranft, Zeit. f. Phys. C12 (1982) 253.
11. J. Jersak et al., Phys. Lett. 98B (1981) 363.

12. V. Barger et al., Phys. Lett. D22 (1980) 727.
13. V. Elias et al., Phys. Lett. 73B (1978) 451.
14. E.H. de Groot et al., Phys. Lett. 90B (1980) 427 and 95B (1980) 128.
15. E.H. de Groot, G.J. Gounaris and D. Schildknecht, Zeit. f. Phys. C5 (1980) 127, E.H. de Groot and D. Schildknecht, Zeit. f. Phys. C10 (1981) 139 and BI-TP 80/32 December 1980.
16. M. Kuroda and D.Schildknecht, BI-TP 81/22 December 1981.
17. G.J. Gounaris and D. Schildknecht, Zeit. f. Phys. C12 (1982) 57.
18. P.Q. Hung and J.J. Sakurai, Nucl. Phys. B143 (1978) 81.
19. P. Steffen, XVII. Rencontre de Moriond. Les Arcs, France, March 1982. A. Wagner, Frühjahrstagung der Deutschen Phys. Ges. Karlsruhe, March 1982.
20. D.H. Perkins, Oxford Univ. preprint 81/037. Royal Society Meeting on Gauge Theories, London, April 1981.

Acta Physica Austriaca, Suppl. XXIV, 125–155 (1982)

THE CERN PROTON-ANTIPROTON COLLIDER+

by

M. SPIRO

Dep. de Phys. Part. Elémentaires

Gif-sur-Yvette

Since summer 1981 the CERN proton-antiproton collider produces collisons at energies at least 10 times higher than the highest energies ever achieved with accelerators. This gives a good chance to discover the weak bosons (W and Z^o boson), which mediate the weak forces, in the next coming years much before the new accelerators designed to study these bosons are built.

Table 1 and table 2 emphasize the main features in the comparison between the $p\bar{p}$ collider and existing machines, and between the $p\bar{p}$ collider and future planned machines.

I. HISTORY OF THE PROJECT

The origins of the project

The first CERN plans orchestrated particularly by Carlo Rubbia emerged in 1976 following crucial advances in accelerator physics. Indeed though antiprotons have the same properties as

+Lecture given at the XXI. Internationale Universitätswochen für Kernphysik, Schladming, Austria, February 25-March 6, 1982.

protons except for the sign of their electric charge and could be in principle accelerated and stored in the same magnet ring as protons, it was not possible until recently to produce well defined and concentrated antiproton beams of sufficient intensity and density to provide enough collisions in a reasonable amount of time for useful physics to be done.

This situation has changed with the invention of beam cooling. Beam cooling refers to a process which concentrates all the particle energies around this value which is cooled down.

Antiprotons can be produced in the collisions of protons with a metal target at relatively low rates (10^{-6}). But the antiprotons emerge with a wide range of momenta unsuited for a beam transfer line, an accelerator, or a storage ring which is designed to handle a well defined momentum.

A cooling technique was proposed in 1966 by Gersh Budker at Novosibirsk in USSR. It involves the use of electron beams travelling along with the antiproton beam at the same velocity. The electron beam is much more intense and well defined than the antiproton one; the antiprotons transfer their "heat" to the electron beam and the antiproton momenta are concentrated around the desired value. The scheme has successfully been tested in 1975 at Novosibirsk and subsequently also been demonstrated at CERN and Fermilab.

However, at CERN, an alternative idea called "stochastic cooling" was invented by Simon van der Meer in 1968. The distribution of particles across a beam is observed at pick up stations in the ring, and from this information corrections to the beam are calculated,Fig. 1. As the beam passes further around the ring, the same section of the beam is subjected to an electric field derived from the calculated corrections. Over millions of repetitions the beam is progressively cooled. The technique was first successfully demonstrated at the ISR in 1975. This led to the construction of a small storage ring aptly known as ICE (Initial Cooling Experiment) specificatly to study stochastic and electron cooling. The results from ICE, particularly on stochastic cooling, were so spectacular that in 1978 CERN was able to give the green light to the antiproton project. Table 3.

The role of the antiproton accumulator

The heart of the antiproton project is the Antiproton Accumulator (AA) where the stochastic beam cooling technique is applied to build up antiproton beams some 10000 times more intense than has ever been achieved before.

Protons are first accelerated in the Proton Synchrotron. Fig.2. When the protons reach 26 GeV they are ejected from the PS and strike a target in front of the AA ring. From the spray of secondary particles, a focussing system selects antiprotons of energy around 3.5 GeV for injection into the AA. This energy gives the maximum yield of antiprotons - for each pulse of 10^{13} protons on target, some 2×10^{7} antiprotons are injected. (In other words, for every million protons hitting the target only two antiprotons are collected!) This operation is repeated every 2.4 s.

Eventually the AA is required to provide beams containing 6×10^{11} particles, and about 30 000 PS pulses (representing about a full day's operation of the machine) will be needed to supply all these antiprotons.

The AA's vacuum chamber is unusually large (70 cm wide) to give space for all the necessary manoeuvres and is held at high vacuum (10^{-10} torr) to minimize loss of antiprotons due to collisions with residual gas molecules.

The first pulse,Fig. 3, is injected and bent by "kicker" magnets so that the antiprotons orbit on the outside of the vacuum chamber. During injection, this region is shielded from the rest of the chamber by mechanically operated shutter. This protects the antiprotons subsequently stored in the main body of the chamber from the magnetic fields of these kickers, and makes it possible to cool the low density injected beam without being swamped by the much stronger signals from the high density stack. The first injected pulse is monitored at pick-up stations, and other kicker magnets act upon it to cool the antiprotons.

In two seconds the pulse is precooled so that the momentum spread of the particles is reduced by a factor of ten. The shutter is then lowered and the precooled antiprotons move into the stack position in the main body of the chamber. The shutter rizes again and the second pulse is injected to receive the same treatment.

While this sequence of injection, precooling and transfer to the stack proceeds, cooling is applied to the stack. After fifty pulses are stacked, some two minutes after injection, about 10^9 antiprotons would be in the stack, being progressively cooled.

After about an hour and 1500 pulses, when some 3×10^{10} antiprotons would be orbiting in the stack, a core "tuned" by the cooling system forms near the inside of the vacuum chamber. Forty hours and 60 000 injected pulses later, some 10^{12} antiprotons would be orbiting in the stack, with the majority (6×10^{11}) of them concentrated in the core, thanks to the cooling system. It is this core which will provide the intense antiproton beam for colliding beam physics.

A residue of some 4×10^{11} antiprotons would remain in the AA stack to start the next core. Injection of antiproton pulses would continue so that 24 hours later another core of 6×10^{11} cooled antiprotons would be ready for ejection. Table 4.

Antiprotons in the CERN machines

All three of CERN's large machines are involved in the exploitation of these intense antiproton beams. The pulse from the AA is first sent to the Proton Synchrotron. Because of space limitations in the AA, the injection and ejection systems are located in the same section of the ring. The antiproton beam therefore has to be turned around in a loop of magnets before it can be injected into the PS, where it circulates in the opposite direction to the protons. The antiprotons are then accelerated in six bunches from 3.5 to 26 GeV, when they can be sent either to the SPS or the ISR.

At the SPS, the antiprotons and protons circulate in the same ring, but in opposite directions. Ultimately, the protons would be injected in one pulse from the PS and made to circulate in six bunches evenly spaced around the ring, with 10^{11} protons in each bunch. The antiprotons would be injected in two pulses from the PS to build up an intensity of 10^{11} antiprotons in each of six bunches around the ring. The two sets of bunches could then be accelerated simultaneously to 270 GeV. Two beam collision regions are being equipped with large and sophisticated particle detectors to observe the very high energy interactions.

At the ISR, the antiprotons can orbit in one ring at 26 GeV while the PS sends 26 GeV protons to orbit the other ring in the opposite direction. The energy of each beam can be increased up to 31 GeV by using the ISR's own acceleration system. Ultimately, five injections of antiprotons, each containing 6×10^{11} antiprotons, are envisaged to achieve a luminosity of 10^{30}. The ISR detectors have already taken their first proton-antiproton data.

At the PS, the 3,5 GeV antiprotons from the AA can be used for a very different range of experiments. It is planned to decelerate the antiprotons to 0.3 GeV, and eject them into a small storage ring (about 20 m across), called LEAR (Low Energy Antiproton Ring). In this case, the aim is not initially to have proton-antiproton collisions (although this may follow later), but to build up low energy anti-proton beams of a quality and intensity never before achieved. To fill LEAR, the intricate beam gymnastics necessary for higher energy antiprotons are no longer required. The antiprotons can be drawn from the AA in more modest pulses of over 10^9 particles. The LEAR antiproton energy will cover the range 0.1-2 GeV, and stochastic beam cooling systems will help preserve the beam quality.

II. PRESENT STATUS AND NEAR FUTURE OF THE MACHINE

The main parameter which characterizes the performances of the machine is the luminosity. The luminosity is defined by

$$L_{cm^{-2}sec^{-1}} = \frac{N(\text{number of events per second})}{\sigma(\text{total cross section in } cm^2)} .$$

L can in turn be calculated from beam parameters:

$$L \sim \frac{N_p N_{\bar{p}}}{N_B \sqrt{\varepsilon_x \beta_x^* \varepsilon_y \beta_y^*}}$$

where

$N_p (N_{\bar{p}})$	total number of (anti-)protons
N_B	number of bunches
$\varepsilon_{x,y}$	horizontal (vertical) emittance
$\beta^*_{x,y}$	horizontal (vertical) beta value in the crossing region.

N_p/N_B or $N_{\bar{p}}/N_B$ (whichever is bigger) determines the beam-beam tune shift. See Appendix 1 for β and ε definition. Table 5 gives the performance of the collider in november and december 1981 compared with the designed values. The expected gains for 1982 are also shown in this table.

One has to compare these figures with the following predictions:

$$m_W = 78 \text{ GeV} \qquad m_{Z^o} = 89 \text{ GeV}$$

$$\sigma(p\bar{p} \rightarrow W + X) \sim 2 \times 10^{-33} \text{ cm}^2$$

and

$$\sigma(p\bar{p} \rightarrow Z^o + X) \sim 10^{-33} \text{ cm}^2 \quad .$$

However $W \rightarrow e\nu$ or $\mu\nu$ and $Z^o \rightarrow e^+e^-$ or $\mu^+\mu^-$ are the only modes with clear signature, and $B(W \rightarrow e\nu) = 8\%$, $B(Z^o \rightarrow e^+e^-) = 3\%$; so for $L = 10^{29}$ $cm^{-2}sec^{-1}$ we get 10 $W \rightarrow e\nu$ or $\mu\nu$ per day, 1 $Z^o \rightarrow e^+e^-$ or $\mu^+\mu^-$.

III. EXPERIMENTS AT THE COLLIDER

To study the collisions detectors are installed at two of the long straight sections (LSS4 and LSS5) on the SPS ring. At both locations, extensive excavation was necessary to make room for the big experimental set ups and to allow them to be moved in and out of the ring. Table 6.

At LSS5 is the detector of the UA1 experiment involving about a hundred physicists from Aachen/Annecy/Birmingham/CERN/Queen Mary College London/College de France/Riverside/Rome/Saclay/Vienna. The apparatus weights well over 2000 tons and is housed in an experimental area which was excavated from the surface, and consists of two cylindrical pits, each 20 m in diameter, roofed over and joined by a connecting chumber.

The detector can be rolled out of the SPS ring on rails into the garage pit for modification, or while the SPS is running for fixed target physics. The large central magnet weights over 800 tons and provides a field of 0.7 T in a volume of 85 m^3.

Inside this magnet and surrounding the beam intersection is the central detector constructed by CERN consisting of six shells of drift chambers with image read out. To minimize shower production in this region of the detector a light honey comb supporting structure is used.

Surrounding the central detector are the photon and electron detectors consisting of 48 "gondolas-like" lead scintillator sandwich stacks. These calorimeters were constructed at Saclay while Vienna took the responsibility for the electronics. This part of the detector should sign the electrons coming from W or Z decays. The "bouchons" electromagnetic calorimeters have been built at Annecy. The British contingent of the UA1 collaboration had a large investment in the hadron calorimeter which is installed in the return yoke of the magnet. End caps close the effective volume. Sophisticated electronic trigger logic for UA1 has been developed at Rutherford.

Muons traversing all this apparatus are picked up by outer layers of drift tubes constructed by Aachen. A lot of hadron energy is released in narrow forward cones around the beam pipe. This is covered by both College de France and INFN Rome calorimeters and drift chumbers. Even further out some 23 m on either side of the main detector are UA1 detectors supplied by CERN + Riverside to monitor beam luminosity. UA1 is then a 4π general purpose detector which can measure the momentum of almost all charged particles and is also optimized to identify the electrons, muons and neutrinos (the last ones through transverse energy imbalance) coming from the weak boson decays.

Incorporated in UA1 is the Annecy/CERN UA3 experiment using a modest detector to search for magnetic monopols. For the other experimental area at LSS4, a large underground "cathedral" was excavated. It houses the UA2 experiment (Bern/CERN/Copenhagen/Orsay/Pavia/Saclay).

The detector of UA2 (Fig. 4) is largely dedicated to the observation of the leptonic (electron) decay modes of the weak vector bosons. The design provides for large solid angle electromagnetic and hadronic calorimetry with a high degree of segmentation. However there is no magnetic field in the central region unlike UA1. Neutrinos can be detected through missing transverse energy. In addition, a small azimuthal wedge in the central region is instrumented as a magnetic spectrometer to cover additional aspects of pp collisions.

The central part of the UA2 detector is also to be used in the UA4 experiment (Amsterdam/CERN/Genova/Naples/Pisa).

UA5 contains two long (7.5 m) streamer chambers in which the tracks of emerging charged particles can be photographed (Fig. 5). The detector has already recorded proton-antiproton collisions in the ISR and now at the collider.

Recently approved is the UA6 experiment, to use an internal hydrogen jet target to study inclusive electromagnetic final states and lambda production at 22.5 GeV center of mass energy by a CERN/Lausanne/Michigan/Rockefeller collaboration.

IV. FIRST RESULTS

(See Physics Letters 107B, Dec. 1981)

These are results from UA1 and UA5 collaborations on data collected in November. The main features of these first data are:

1) Average charged multiplicity increases from ISR to the collider following a log s extrapolation (Fig. 6).
2) Rapidity plateau has risen and extended in rapidity range (Fig. 7). However it does not extend as much as it could in width. This together with colarimeter measurements suggest that $\langle p_T \rangle \sim 500$ MeV. At ISR $\langle p_T \rangle = 300$ MeV. More precise using the central detector of UA1 and the magnetic field on december data are coming very soon.

It has to be pointed out that both results 1) and 2) agree with cosmic rays experiments.

3) Using the variable $Z = n/\langle n \rangle$ where n is the observed multiplicity and $\langle n \rangle$ the average multiplicity, the shape of the distribution does not vary from PS energies to ISR energies and then to collider energies (i.e. over two orders of magnitude in energies), Fig. 8. This is the so-called KNO scaling. This means also that we get exponential tails for high multiplicity unlike Poisson distributions (Fig. 9). Large multiplicity are not rare events. Jets may not be obvious at all in hadron collisions.
4) Correlations between p_T and multiplicity as suggested by comsic rays experiments are not seen (Fig. 10).
5) Centauro type of events as observed in some cosmic ray experiments, with a very copious production of charged hadrons (~ 100) and no or almost no gammas from π^0 production and decays do not show up. They might simply be explained by the very large fluctuations observed in multiplicity distributions.

V. HOW W'S AND Z^0 SHOULD SHOW UP IN UA1 and UA2

The signature of the W's and Z^0 is most likely to be through their leptonic decay modes:

UA1	$W^{\pm} \to e^{\pm}\nu$ $\mu^{\pm}\nu$	UA2	$W^{\pm} \to e^{\pm}\nu$
UA1	$Z^{o} \to e^{+}e^{-}$ $\mu^{+}\mu^{-}$	UA2	$Z^{o} \to e^{+}e^{-}$

Signature for W

1) A well identified direct lepton with $p_T > 20$ GeV.
2) Clear imbalanced p_T transverse momentum, corresponding to a neutrino which escapes.
3) Statistically a jacobian peak for the p_T distribution of the lepton near $m_W/2$.
4) Statistically a characteristic asymmetry in the lepton angular distribution coming from the weak decay of W.

Signature for Z^o

It is much easier but requires a very high luminosity. A peak at the Z^o mass should appear in the invariant mass distribution of the lepton pairs.

CONCLUSION

If everything goes right 1982 might hopefully be the W year.

APPENDIX 1

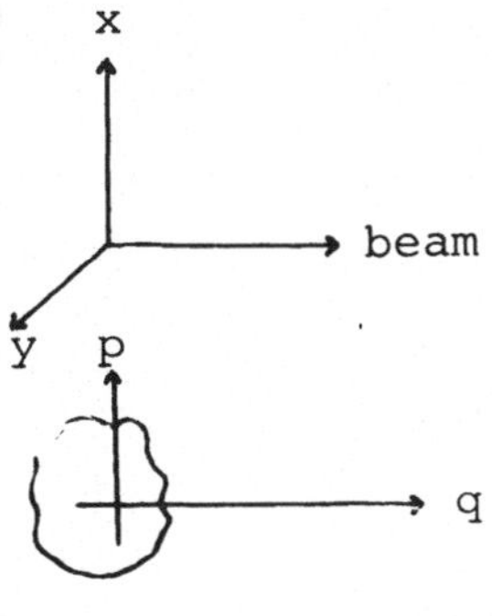

β and ε in accelerators theory.
We define the two variables p and q as: $q = x$; $p = \frac{m\dot{x}}{\sqrt{1-v^2/c^2}} = mc\gamma\beta_x$,
where $\beta_x = \frac{\dot{x}}{c}$ and

$\gamma = [1 - (\dot{x}^2 + \dot{y}^2 + \dot{z}^2)]^{-1/2}$.

A group of particles, the beam, can be thought of as an area or closed curve which includes the p and q coordinates of every particle. The shape and position of this area changes as the motion proceeds and each particle follows its own trajectory. But the area $\int pdq$ is conserved. This is called Liouville's theorem.

Now, in accelerator theory we are interested in:

$$\text{divergence} = x' = \frac{dx}{ds} \quad \text{or} \quad y' = \frac{dy}{ds}, \text{ and}$$

$$\text{displacement} = x \ .$$

Liouville's theorem becomes: $\int pdq = mc\int \gamma\beta_x dx = mc\beta\gamma\int x'dx = \text{const.}$

$$\cdot\int x'dx = \varepsilon \ .$$

As acceleration proceeds, ε decreases like $\frac{1}{\beta\gamma}$, e.g. inversely proportional to the energy of the proton above a few GeV.

In addition the x,x' plot looks in most of the places like an ellipse. We define the β value at this place (do not confuse with v/c) to be proportional to the square of the maximum displacement in x.

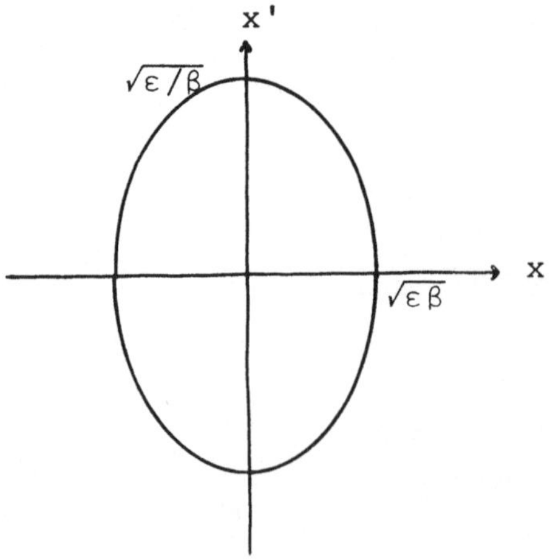

MACHINE TYPE	PLACE	COLLISIONS	$E_{C.M.}$	ADVANTAGES	DISADVANTAGES
PROTON SYNCHROTRON	SPS CERN FNAL (Chicago) 1975	400 GeV protons on fixed targets	30 GeV max	Luminosity ($10^{37}cm^{-2}/s$) Secondary beams	low energy C.M. proton is made of 3 quarks
PROTON PROTON COLLISION RINGS	ISR(CERN) 1971	31 GeV p 31 GeV p	62 GeV	Luminosity 10^{32}, high energy	need 2 rings, only pp, proton is made of 3 quarks
e^+e^- COLLISION RING	PETRA (Hamburg) 1978 PEP (Stanford)	20 GeV e^- 20 GeV e^+	40 GeV	Energy "fully used", one ring only	low cross sect. Bremsstrahlung
$p\bar{p}$ COLLIDER	CERN (Geneva) 1981	270 GeV p 270 GeV $\bar{p}$	540 GeV	huge energy, one ring only	low luminosity, proton and $\bar{p}$ are made of quarks

Table 1. Comparison between varicus existing machines and the CERN collider. (Note that luminosity ($cm^{-2}sec^{-1}$) = number of event per sec./total cross section (cm^2).)

MACHINE TYPE	PLACE	COLLISIONS	$E_{C.M.}$	COMMENTS
PROTON SYNC.	DOUBLER (FNAL) 1983	1000 GeV p on fixed targ.	45 GeV	no weak boson produced
$p\bar{p}$ COLLIDER	FNAL 1985	1000 GeV p 1000 GeV $\bar{p}$	2000 GeV	design: a few thousand W,Z per day hidden into 5×10^{10} int./day
pp COLLISION RINGS	ISABELLE (Brookhaven) 1986	400 GeV p 400 GeV p	800 GeV	a few W and Z per sec hidden in 5×10^{7} int./day
$^{+}\,^{-}$ COLLISION RINGS	LEP (CERN) 1989	50 GeV e^+ 50 GeV e^-	100 GeV	a few Z^o per sec., no background
$p\bar{p}$ COLLIDER	CERN existing	270 GeV p 270 GeV $\bar{p}$	540 GeV	a few hundred W,Z/ day hidden into 5×10^{9} int./day

Table 2. Comparison between the CERN Collider and future machines.

Idea	1976	Search for the IVB using existing accelerators (C.Rubbia)
Ice Approval	1978	Construction of AA transfer lines to CPS, SPS, Experiment UA1
	End 1978	Transfer line CPS-ISR Experiments UA2-UA5
	5/1980	Lear
Operation	14/7/1980	$\bar{p}$ in AA
	3/4/1981	Collisions in ISR
	7/8/1981	Collisions in SPS (also observable by UA1)
	1983	Lear

Table 3. The CERN $p\bar{p}$ enterprise.

Parameters	
Central Momentum	3.5 GeV/c
Circumference	151.08 m at injection (CPS/4) 155.84 m at stack centre
Working Point	$Q_x = 2.284 \quad Q_y = 2.276$ at stack centre
Total $\Delta p/p$	$\pm 30 \times 10^{-3}$
Vacuum Pressure	10^{-10} torr
Target	Tungsten Rod 110 mm long, 3 mm Ø, followed by magnetic horn
Protons/Pulse on target	1×10^{13} every 2.4 sec
Antiprotons/Pulse	2.5×10^{7} in AA acceptance
Acceptance (Injection)	100 hor, 100 vert mm rad. at $\Delta p/p = 7.5 \times 10^{-3}$
(Transfer to CPS)	7.6 hor, 4.5 vert. mm rad. at $p/p = 1.1 \times 10^{-3}$
First fill (1×10^{12} $\bar{p}$)	50 000 pulses, 33 hrs.
Refill (6×10^{11} $\bar{p}$)	30 000 pulses, 20 hrs.

Table 4. The proton-antiproton accumulator (R. Billinge, S. v.d. Meer).

Performance of SPS $\bar{p}p$

	Design	Nov.	Dec.
Number of p/bunch stored	10^{11}	7×10^{10}	7×10^{10}
Number of $\bar{p}$/bunch stored	5×10^{10}	2×10^{9}	5×10^{9}
Proton bunches	6	2	2
Antiproton bunches	2×6	1	1
Tune shift/ crossing	.003	.003	.003
Lifetime of p's	>24 hrs	∿150 hrs	∿150 hrs
Lifetime of $\bar{p}$'s	>24 hrs	*≤9 hrs	∿20 hrs
Low β	1.0 m	∿60 m	V 3.5 m H 7.0 m
Luminosity	10^{30}	2×10^{25}	$\sim10^{27}$

*RF for $\bar{p}$'s not yet tuned

Expected gains:

Low β	× 2	V = 0.75 m H = 1.5 m
m^o = of $\bar{p}$'s	× 10	
bunches	× 3	
emittances	× 2	

→ $\sim10^{29}$ cm^{-2} sec^{-1}

Table 5

Approved Experiments

UA1	4π Detector (Rubbia) Aachen, Annecy, Birmingham, CERN, CdF, QMC, Riverside, Rome, Rutherford Saclay, Vienna
UA2	Large detector Bern, CERN, Copenhagen, Orsay, Pavia, Saclay
UA3	Monopole search Annecy, CERN
UA4	Elastic and total cross section (Matthiae) Amsterdam, CERN, Genova, Napoli, Pisa
UA5	Streamer Chamber (Rushbrooke) Bonn, Brussels, Cambridge, Stockholm

Table 6

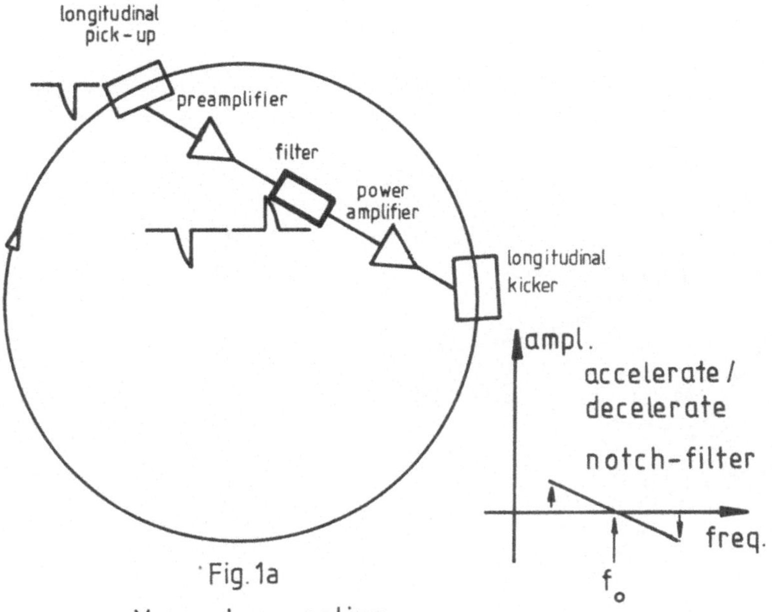

Fig. 1a
Momentum cooling

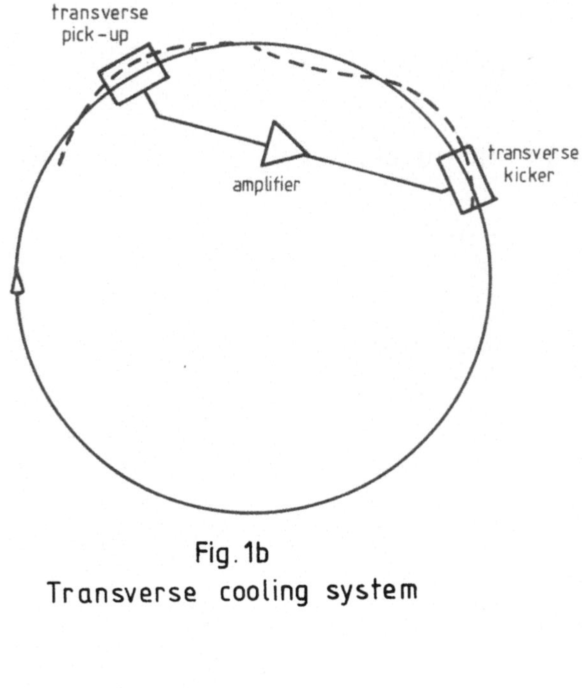

Fig. 1b
Transverse cooling system

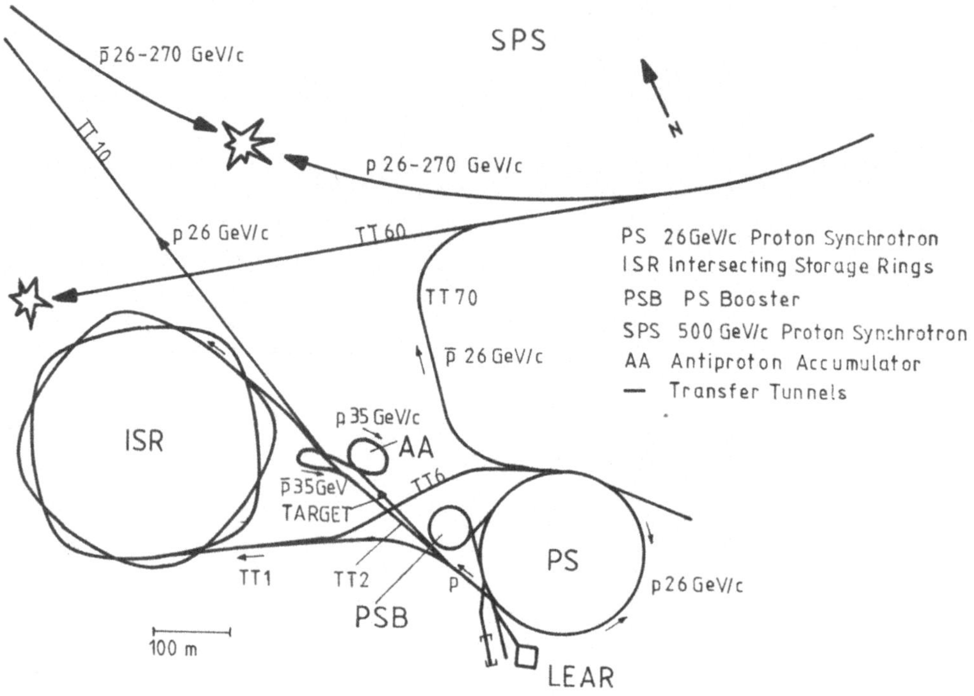
SPS
p̄ 26-270 GeV/c
p 26-270 GeV/c
N
TT 10
p 26 GeV/c
TT 60
TT 70
p̄ 26 GeV/c
ISR
p 35 GeV/c
AA
p̄ 35 GeV
TARGET
TT6
TT1
TT2
PSB
PS
P
p 26 GeV/c
LEAR
100 m
PS 26GeV/c Proton Synchrotron
ISR Intersecting Storage Rings
PSB PS Booster
SPS 500 GeV/c Proton Synchrotron
AA Antiproton Accumulator
— Transfer Tunnels

Fig. 2

Over-all site layout

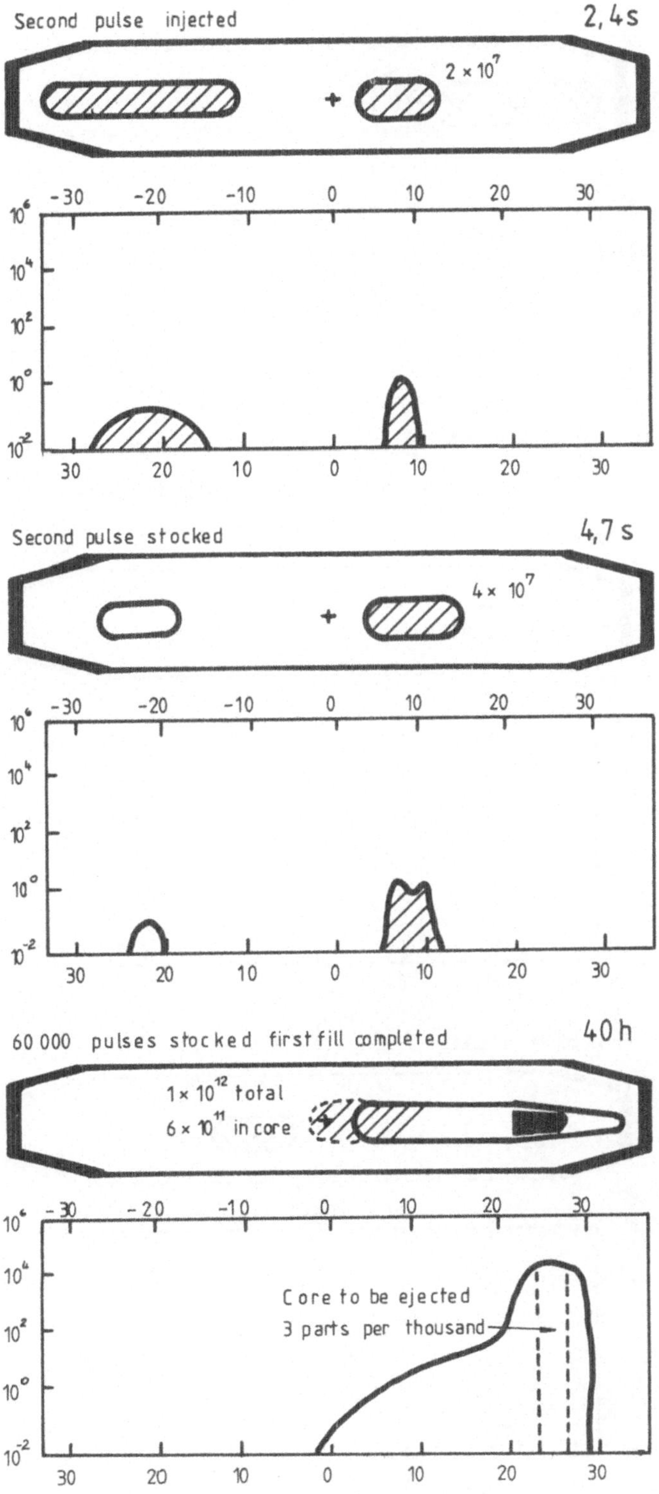

Second pulse injected
2,4s
2 × 10^7
-30 -20 -10 0 10 20 30
10^6
10^4
10^2
10^0
10^-2
30 20 10 0 10 20 30
Second pulse stocked
4,7 s
4 × 10^7
60 000 pulses stocked first fill completed
40 h
1 × 10^12 total
6 × 10^11 in core
Core to be ejected
3 parts per thousand

Fig.3a

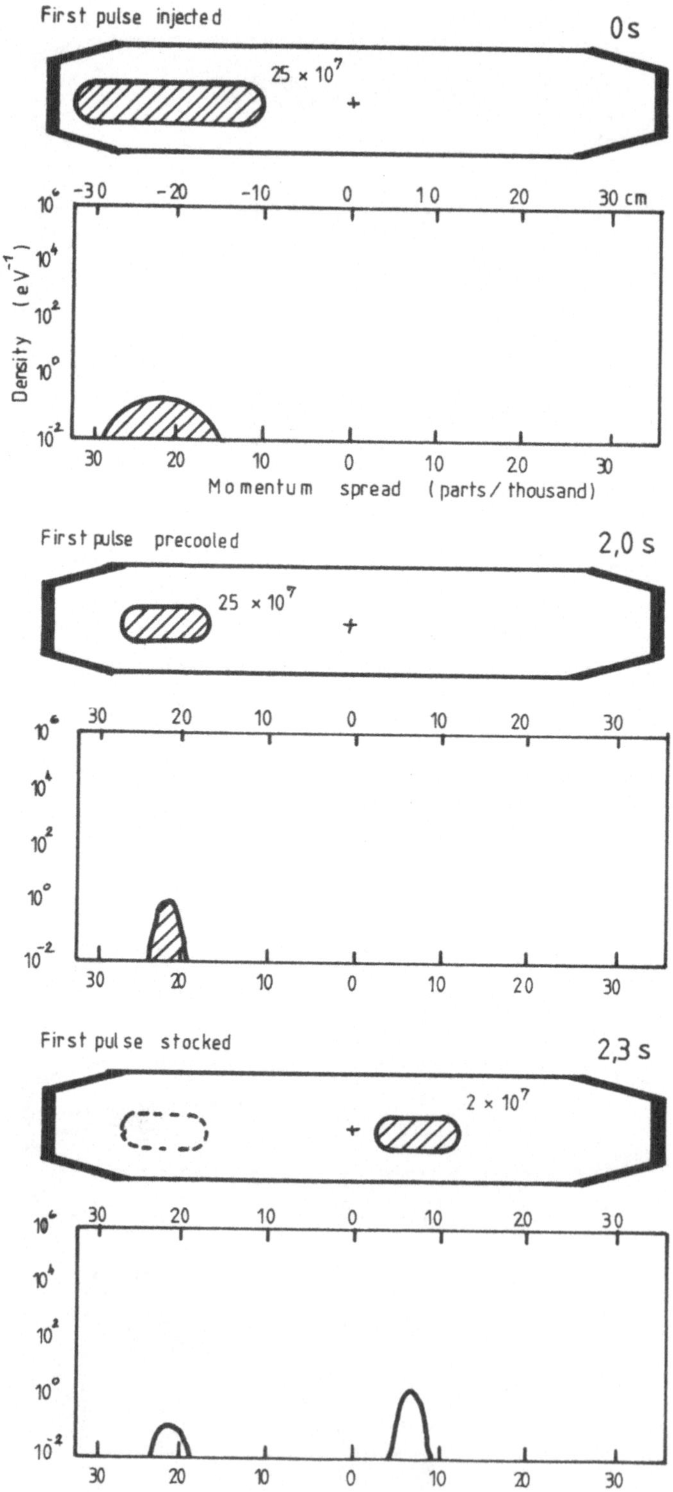
First pulse injected
0s
25 × 10^7
-30
-20
-10
0
10
20
30 cm
Density (eV^-1)
Momentum spread (parts / thousand)
First pulse precooled
2,0 s
25 × 10^7
First pulse stocked
2,3 s
2 × 10^7

Fig. 3b

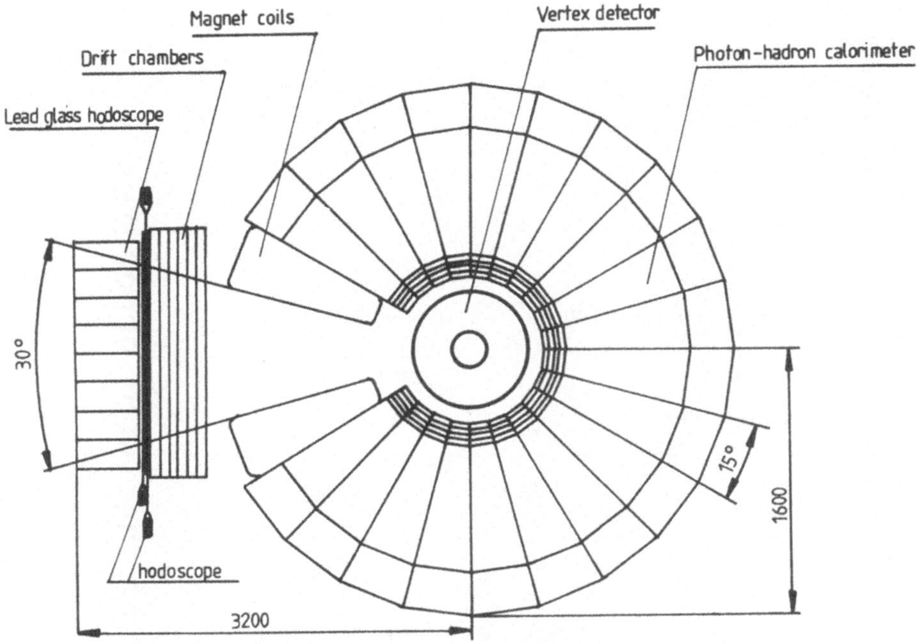

Fig.4: Cross-section normal to the beam of the UA2 detector in the "wedge" configuration

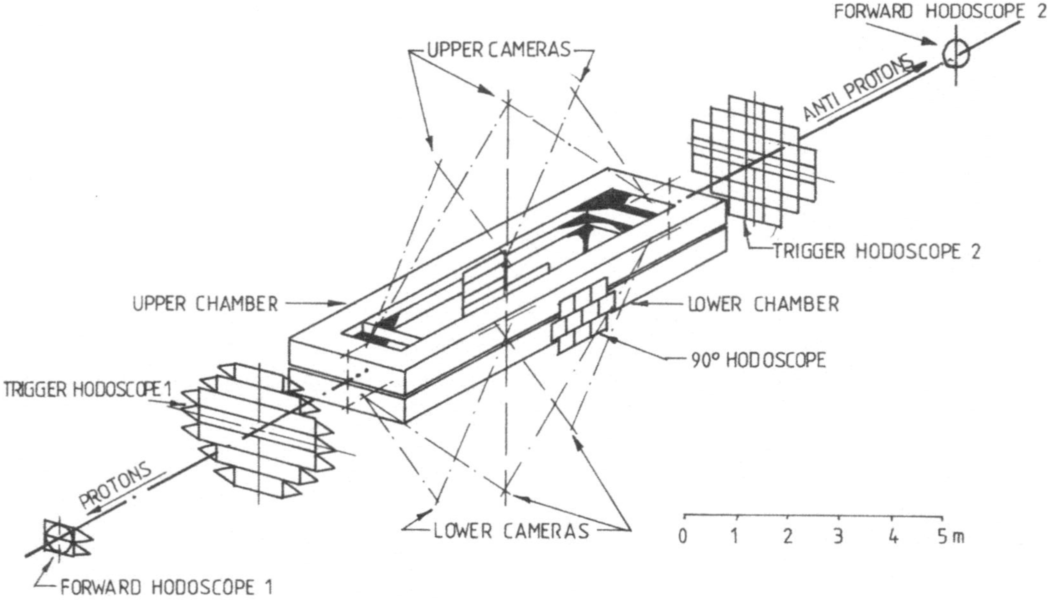

Fig.5: Schematic layout of the UA5 apparatus

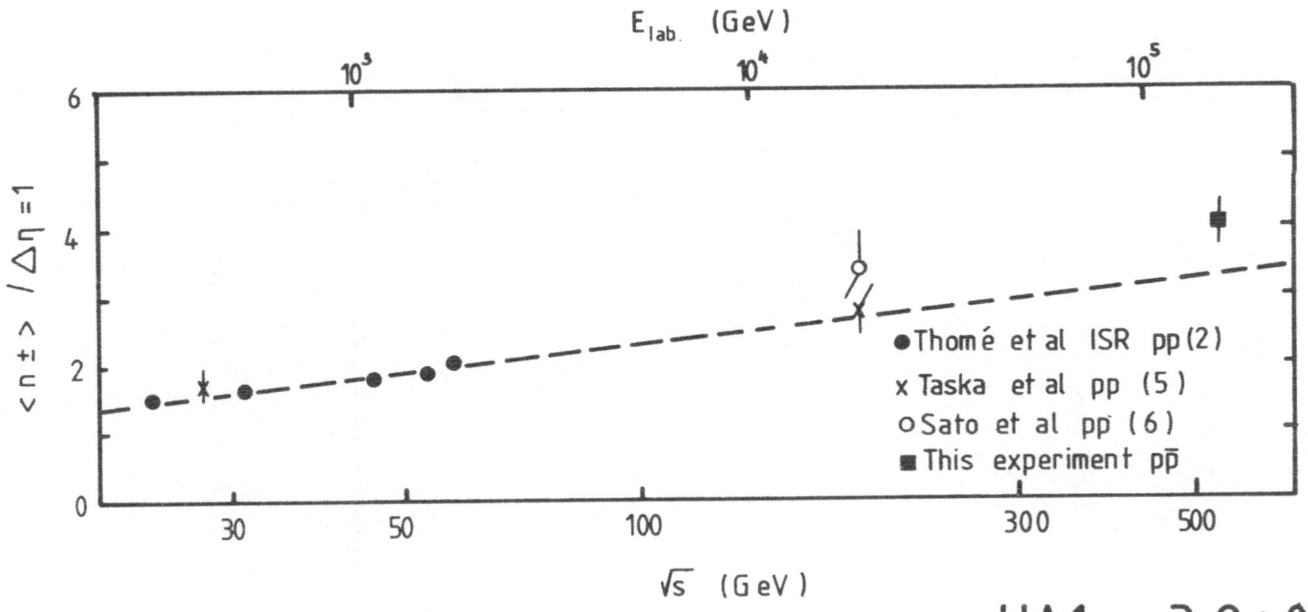

UA1 3,9±0,3

Fig. 6a

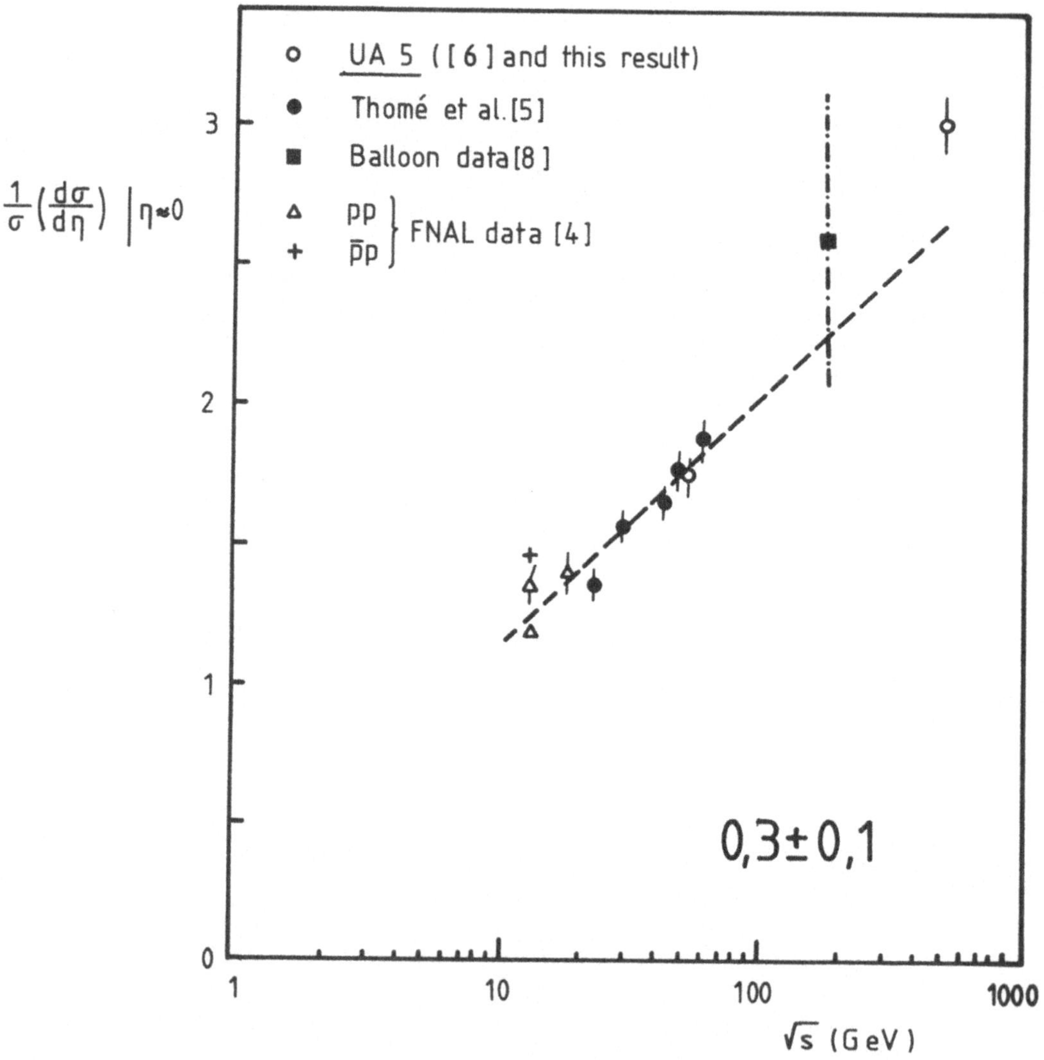
UA 5 ([6] and this result)
Thomé et al.[5]
Balloon data[8]
pp
p̄p
FNAL data [4]
$\frac{1}{\sigma}\left(\frac{d\sigma}{d\eta}\right)\Big|_{\eta\approx 0}$
0,3±0,1
√s (GeV)
0
1
2
3
1
10
100
1000

Fig. 6b

UA 5

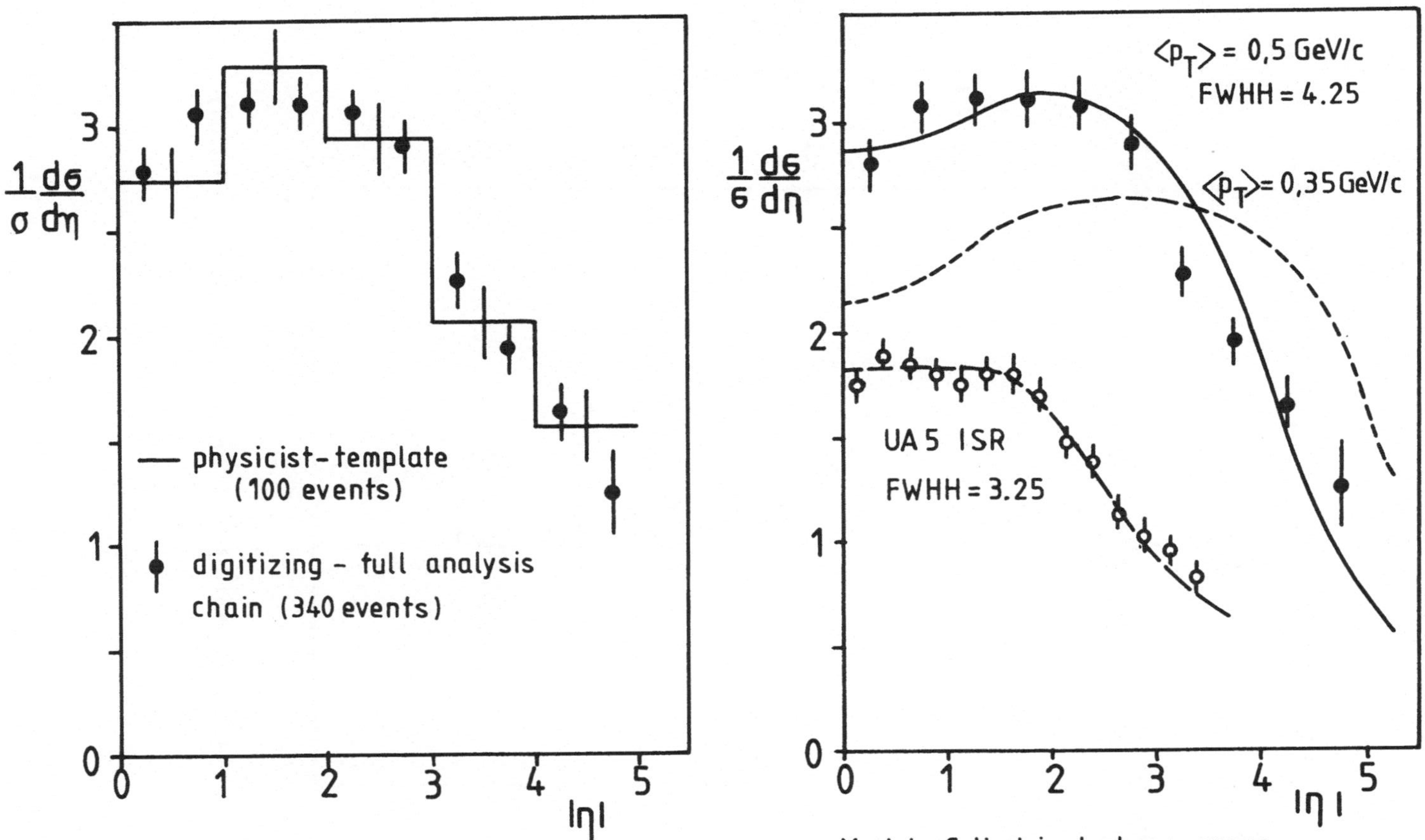

Fig. 7a

Model: Cylindrical phase space;
limiting p_T leading particle effect

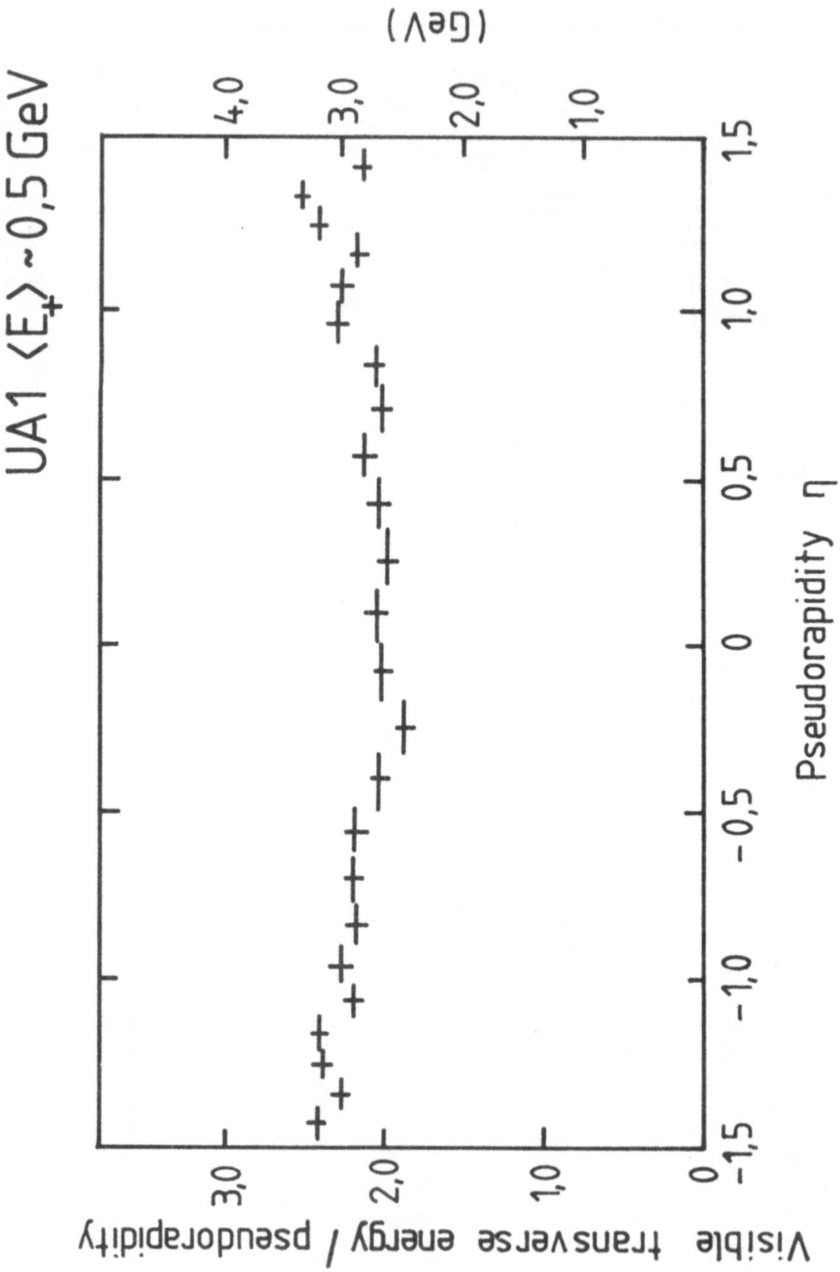
UA1 ⟨E_T⟩ ~ 0,5 GeV
Visible transverse energy / pseudorapidity
(GeV)
0
1,0
2,0
3,0
4,0
−1,5
−1,0
−0,5
0
0,5
1,0
1,5
Pseudorapidity η

Fig. 7b

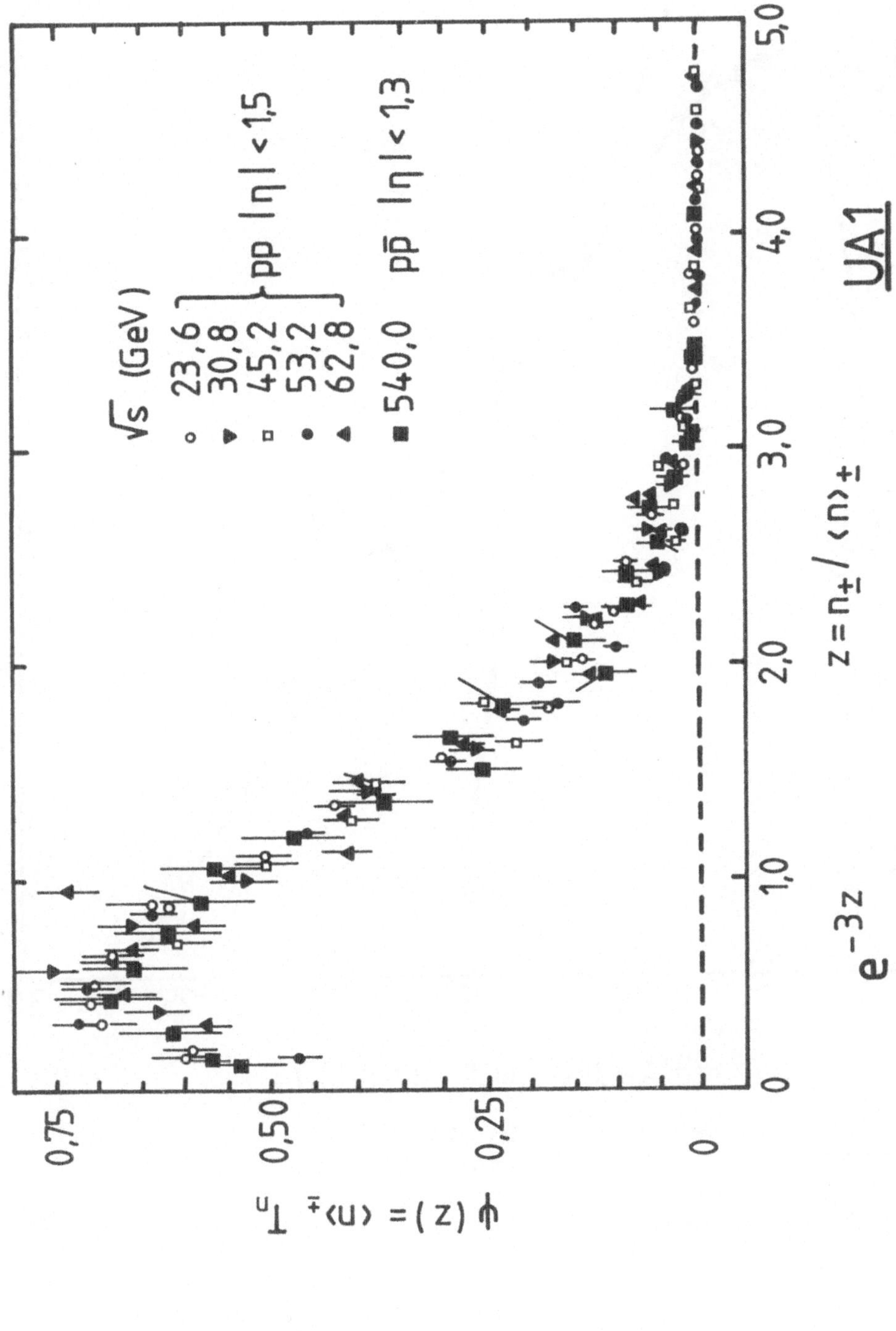

√s (GeV)
23,6
30,8
45,2
53,2
62,8
pp |η| < 1,5
540,0 pp̄ |η| < 1,3
UA1
e^-3z
z = n± / ⟨n⟩±
ψ (z) = ⟨n⟩± T_n
0
0,25
0,50
0,75
1,0
2,0
3,0
4,0
5,0

Fig. 8

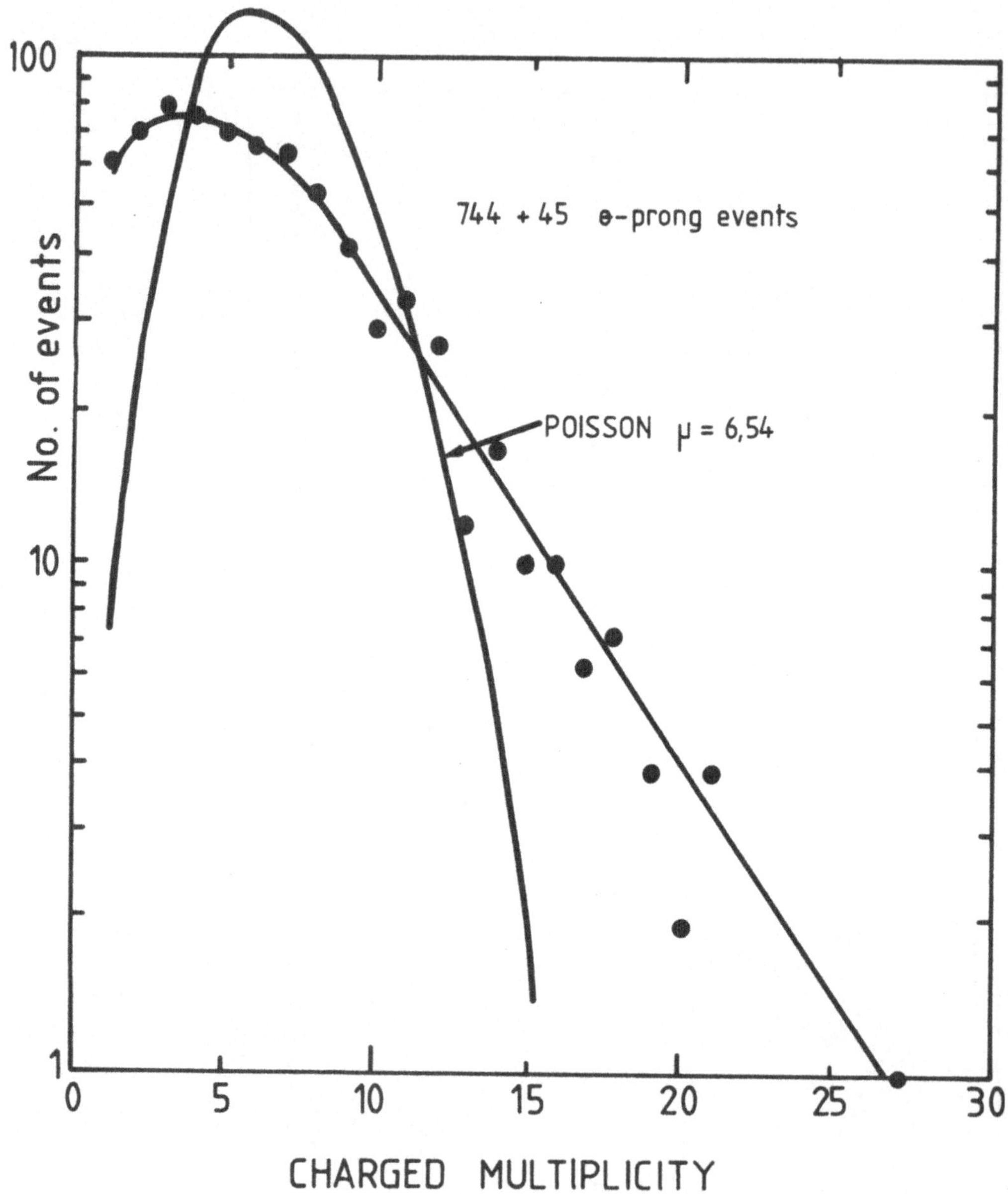
100
10
1
No. of events
744 + 45 θ-prong events
POISSON μ = 6,54
0
5
10
15
20
25
30
CHARGED MULTIPLICITY

Fig. 9

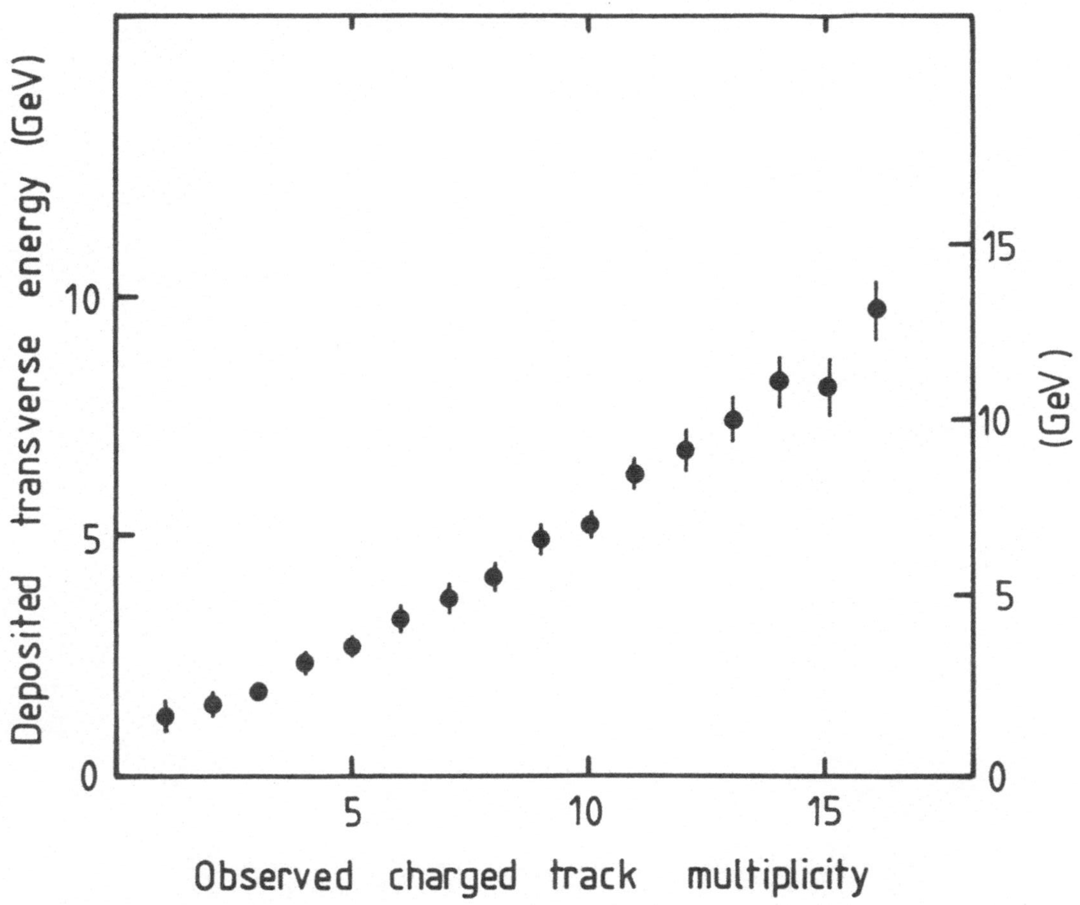
Deposited transverse energy (GeV)
(GeV)
0
5
10
15
Observed charged track multiplicity

Fig. 10

Acta Physica Austriaca, Suppl. XXIV, 157–201 (1982)

WEAK INTERACTIONS OF QUARKS+

by

F. WAGNER
Max-Planck-Institut für Physik und Astrophysik
Werner Heisenberg-Institut für Physik
München, Fed. Rep. of Germany

INTRODUCTION

Our present believe is that interactions between the fundamental constituents of matter are described by renormalizable gauge theories. In the present lecture the question will be discussed to what extent weak interactions of quarks can be described by the so-called standard model. The theoretical background has been covered by the preceding lecture of Professor Ecker. Only for leptons the low energy extrapolation of the standard model can be directly confronted to the data (Part 2). Since free quarks cannot be observed, one has to make additional assumptions about their behaviour inside a hadron. Various possibilities are discussed in Part 1. These assumptions can be tested together with the standard model for quarks in semileptonic processes if one assumes the model works for the leptonic part of the interactions (Part 3). For decays involving

+Lectures presented at the XXI. Internationale Universitätswochen für Kernphysik, Schladming, Austria, Feb. 25-March 6, 1982.

only quarks (nonleptonic processes) we have much less predictive power as far as the weak interactions are concerned since strong final state interactions play an important role. These problems are discussed in the last Part (4).

1. BASIC ASSUMPTIONS

1.1 Effective Hamiltonian

As you have learned in the lecture from Professor Ecker one interpretation of electroweak interactions is in terms of a SU(2) × U(1) gauge theory (Weinberg-Salam [1]) or the so-called Standard Model (SM). At present energies we can neither test the symmetric version for momentum transfer $Q^2 \gg m_W^2$ nor the effective field theory after symmetry breaking valid for $Q^2 \simeq m_W^2$. Only the low energy extrapolation for $Q^2 \ll m_W^2$ is accessible for present-day experiments. All alternatives to the SM are constructed such that they lead to the same low energy limit [2]. So we restrict ourselves to the following effective interaction of the current-current form

$$H_I = -\frac{e^2}{Q^2} J_\mu^{el} J^{el\mu} + \frac{G}{\sqrt{2}} [J^{+\mu} J_\mu^- + 2\rho J_\mu^N J^{N\mu}] . \qquad (1.1)$$

The first term is the usual electromagnetic interaction, the second describes the V-A charged currents of weak interaction known since the sixties, and the last gives the neutral current whose discovery in 1972 lead to a new picture of weak interaction. All what is left from the original SU(2) × U(1) model is the universal coupling G between all currents. The strength ρ of the neutral current depends on the way SU(2) × U(1) is broken. For a Higgs doublet we have

$$\rho = \frac{m_W^2}{m_Z^2 \cos^2\theta_W} = 1 \quad .$$

Since the empirical value $\rho = 0.985 \pm 0.026$ [3] is compatible with 1, we will assume $\rho = 1$ in the following. Left-handed fermions form $(q_1, q_2)_L$. Massive right-handed fermions are put in isospin singlets. Each doublet i contributes to the charged current a term

$$J_\mu^+ = \sum_{ij} \bar{q}_1^{(i)} \gamma_\mu (1-\gamma_5) q_2^{(j)} C_{ij} \qquad (1.2)$$

where C describes possible mixing between different families. Empirically for leptons we have C = 1 and for quarks we have to insert the Cabibbo rotation respectively the KM matrix [4]. More about these mixings you will hear in Professor Sakurai's lecture. Whereas the charged current (1.2) contains only left-handed fermions, the neutral current will involve both right- and left-handed fermions:

$$J_\mu^N = \sum_{ik} \bar{q}_k^{(i)} \gamma_\mu (g_V - g_A \gamma_5) q_k^{(i)} \quad . \qquad (1.3)$$

In the SM the couplings can be expressed by the isospin I_3 and charge Q of a fermion by

$$g_V = I_3 - 2Q \sin^2\theta_W \qquad (1.4)$$

$$g_A = I_3 \quad . \qquad (1.5)$$

The experimental value of $\sin\theta_W$ is close to $\frac{1}{2}$, which has the consequence that charged leptons have no vector coupling to J^N. Therefore some typical weak effects like R-L asymmetries will be absent in e^-e^+ interactions. As a typical prediction of the SM we get for quarks $g_V \neq 0$ due to their non-integer charge. Very often we will denote H_I from Equ. (1.1) by a diagram H_I =  to emphasize the connection between spin and flavor.

Our H_I can be confronted directly with the data for leptons. Since free quarks have not been seen yet, we have to make additional assumptions about their behaviour inside hadrons. A successfull comparison of weak properties of quarks with data will be interpreted by the true believer of the Quark Model (QM) as success of SM, whereas the heretic may say one has found just the right assumptions. You will see in Sections 3 and 4 that life for the heretic has become increasingly harder in the last years.

1.2 Quark Model Assumptions

The notion "QM for hadrons" is rather a compendium of rules found mostly by trial and error in 20 years of experience in hadronic reactions mixed with elements of a field theory (QCD) and results of a naive nonrelativistic potential calculation. Therefore in the absence of a real theory any comparison of weak properties with experiment needs assumptions. Obviously life will be easier if one of the currents in (1.1) contains only leptons:

1.2.1 Semileptonic Processes

Here we need to compute the current between two hadronic states A and B. One always assumes additivity

$$\langle A|J|B\rangle = \sum_{ij} \langle A|i\rangle \langle q_i|J|q_j\rangle \langle j|B\rangle \qquad (1.6)$$

that the current is a sum of quark contributions multiplied with transition amplitudes whose squares

$$p_i(A) = |\langle A|i\rangle|^2 \qquad (1.7)$$

give the probability to find quark i in A. Equ. (1.6) itself is not very useful since the p_i^A are, in general, unknown. Often the state B will be any number of hadrons (inclusive state). In analogy to statistical mechanics interference terms in a many-body state can be neglected and the sum over the states contained in B leads to

$$\sum_X |\langle i|X\rangle|^2 = 1 \quad . \qquad (1.8)$$

Dropping the interference terms and inserting Eq. (1.8) into Equ. (1.6) we arrive at the assumption of "incoherent additivity" (or I),

$$I: \quad \sum_X |\langle A|J|X\rangle|^2 = \sum_{ij} p_i(a) |\langle q_i|J|q_j\rangle|^2 \quad . \qquad (1.9)$$

If states A and B contain only one hadron we can restrict ourselves to diagonal terms and take the probabilities from the simplest flavor spin wave functions modified by form factors at higher Q^2; this we call coherent additivity (shortly C),

$$C: \quad \langle A|J|A\rangle = \sum_i p_i(A) \, \langle q_i|J|q_i\rangle \quad . \qquad (1.10)$$

Obviously at $Q^2 = 0$ (1.10) is true for conserved quantities (vector-like charges or $\partial_\mu A^\mu$). In these cases (1.10) needs only an extrapolation to $Q^2 \neq 0$, whereas for nonconserved quantities (axial charges or moments of vector-charges) it is a rather drastic assumption. Also assumption I is equally drastic since dropping the interference terms means the constraints from conservation laws (energy momentum, angular momentum, isospin,...) cannot be taken into account. Note that both assumptions (1.9) and (1.10) involve the probabilities $p_i(A)$, but for completely different physical reasons. Especially one does not expect them equal in both cases.

1.2.2 Nonleptonic Decays

One main difference between the processes with only quarks and the semileptonic ones is the occurence of final state interactions. The flavor flow diagram  will be changed. For example, the charged current Hamiltonian $J_{s\to c} \cdot J_{u\to d}$ can be changed by gluon exchange into a form of neutral current $J_{c\to u} \cdot J_{s\to d}$ according to the following diagram:

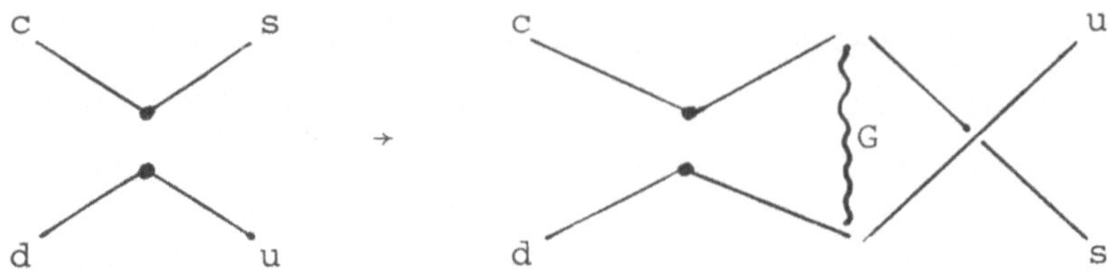

Therefore the effective Hamiltonian had to be replaced

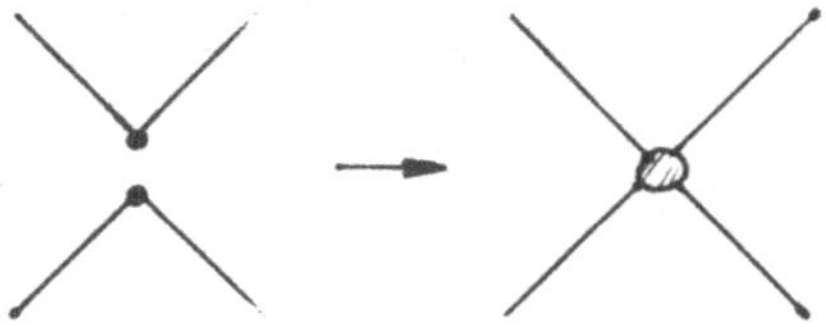

due to gluon exchange and/or final state interaction .

A second complication arises, namely that two competing processes are possible. If we consider, for example, the decay of a meson $Q\bar{q}$ this can proceed by analogy to the semi-leptonic case by the decay diagram:

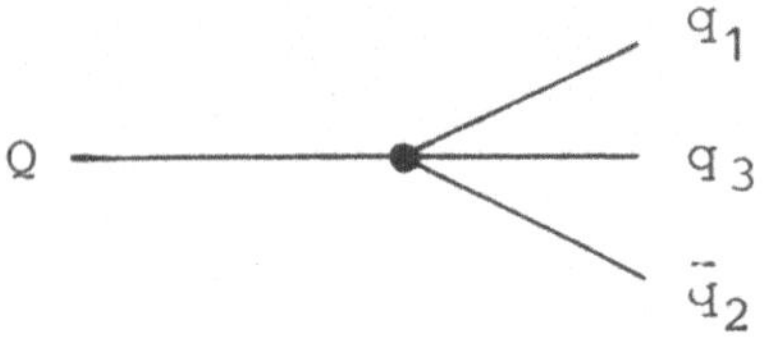

(1.11)

D:

or the W exchange (sometimes also called annihilation)

W:

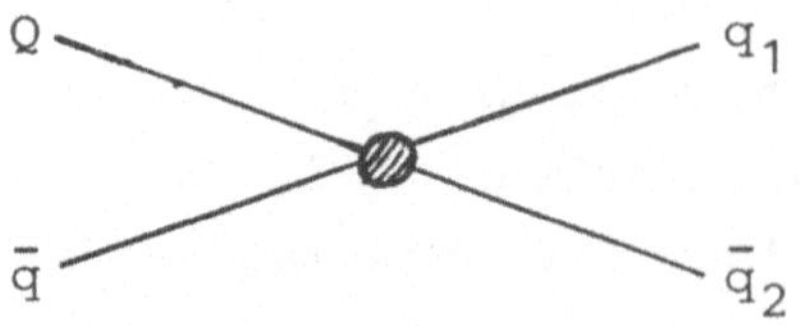

(1.12)

Later we will refer to (1.11) or (1.12) as assumption D,

respectively W. For the hadronization of the final quarks both can be combined with the two additivity assumptions I or C.

1.2.3 Examples

It might be useful to show the additivity assumptions in some examples. For application of (1.10) we need the probabilities resulting for the proton with spin up (↑) in an S-wave three quark state. They are given by

$$p_i(p\uparrow) = \frac{5}{3}\, u\uparrow,\ \frac{2}{3}d\downarrow,\ \frac{1}{3}d\uparrow,\ \frac{1}{3}u\downarrow, \qquad (1.13)$$

which means $\frac{5}{3}$ for a u-quark spin up, $\frac{2}{3}$ for a d-quark spin down, a.s.o.. By interchanging quarks we get the probability for $\Sigma^{\pm}$, Ξ^{o}_{-}, n. It is better to remember the numbers in (1.13) than the proof we will give in the Appendix.

As a first application we apply (1.10) to the magnetic moment of the proton

$$\mu_p = \sum_i p_i(p\uparrow)\ \langle q_i|\mu|q_i\rangle\ . \qquad (1.14)$$

The quark moments are in terms of the quark magneton $\mu_o = \frac{e}{2m_q}$ given by

$$\langle q_i|\mu|q_i\rangle = \mu_o \cdot Q_i \cdot (2S_Z)_i \qquad (1.15)$$

where Q_i and $(S_Z)_i$ denote charge and spin of quark i. Inserting the numbers (1.13) and Equ. (1.14) into (1.15) we obtain

$$\mu_p = \mu_o\ (\frac{5}{3}\,\frac{2}{3} + \frac{2}{3}\,\frac{1}{3} - \frac{1}{3}\,\frac{2}{3} - \frac{1}{3}\,\frac{1}{3}) - \mu_o\ .$$

The neutron magnetic moment is obtained by interchanging u ↔ d in Equ. (1.13) which leads to $\mu_n = -\frac{2}{3}\mu_o$. Therefore we have proven the famous SU(6) relation [5]

$$\mu_p = -\frac{3}{2}\mu_n \ .$$

Next we apply assumption I from Equ.(1.9) to the ratio D of deep inelastic e^-n to e^-p scattering

$$e^- \, {}^{p}_{n} \rightarrow e^- X \ . \qquad (1.16)$$

In this case the probabilities are called structure function which depend on the function x of the p(n) momentum carried by the quark. With $P_i(p) = q_i^p(x)$ we obtain from (1.19) for the ratio between p and n cross sections

$$D = \frac{\sum q^n(x) Q_q^2}{\sum q^p(x) Q_q^2} \ . \qquad (1.17)$$

If the ratio of probabilities finding u or d in the proton is $(u/d)_p = 2$ independent of x (this implies $u/d = \frac{1}{2}$ for n) we find for D

$$D = \frac{2\,\frac{1}{9} + \frac{2}{9}}{2\,\frac{4}{9} + \frac{1}{9}} = 0.67 \ . \qquad (1.18)$$

If we count only quarks with the same spin as the proton (which is motivated for $x \sim 1$ from QCD [6]) we have $(u/d)_p =$ 5 and $D = 0.43$. Finally, we can use the proton as a flavor tester for u-quarks near $x = 1$ (likewise the n counts only d-quarks), as it is assumed in the so-called Feynman-Field [7] parametrization. This gives for D

$$D = \frac{\frac{1}{9}}{\frac{4}{9}} = 0.25 \ . \qquad (1.19)$$

The data [8] shown in Fig. 1 indicate that near $x = 1$, $d^p = 0$ may be a useful assumption.

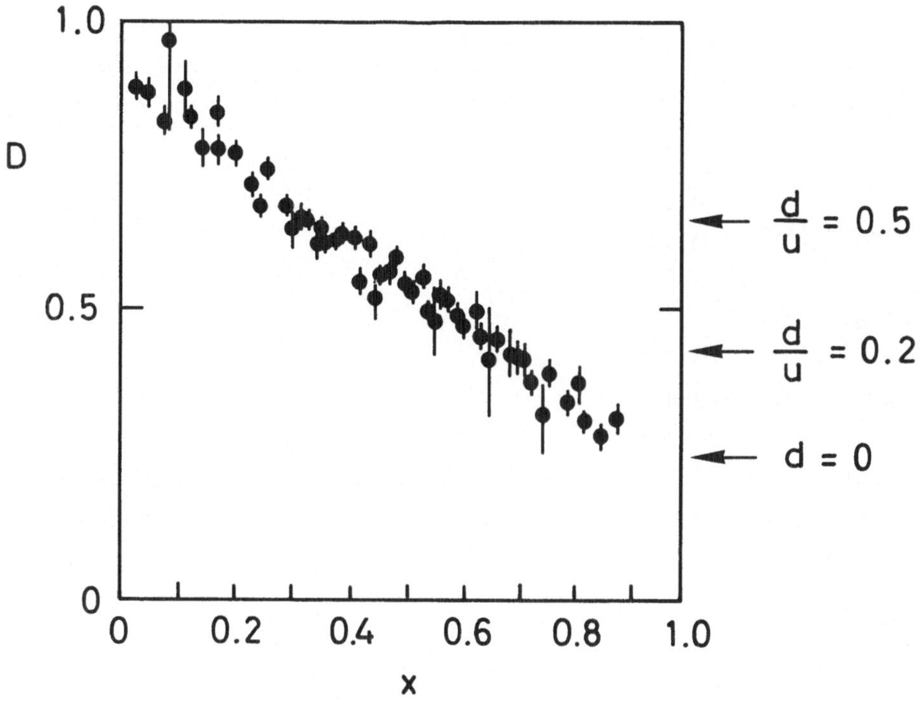

Fig. 1

2. STANDARD MODEL FOR LEPTONS

The strategy for testing the weak properties of quarks will be that one assumes SM works for the leptons and we can test the quarks with semileptonic processes together with the quark model assumption. Therefore it will be useful to look how the SM works for leptons.

2.1 Charged Currents

The oldest indication for V-A comes from the μ-decay

$$\mu^- \to e^- \bar{\nu}_e \, \nu_\mu \, . \qquad (2.1)$$

The electron spectrum for unpolarized μ is described by the so-called Michel parameter ρ. Historically it tooke a very long time before experiments reached the value 0.75

given by V or A (see Fig. 2). Even with polarized μ the present limits [9] on other interaction than V-A (or $g_A = g_V$ according to Equ. (1.2)) are very unsatisfactory for one of the basic facts of weak interactions:

$$g_A/g_V = 0.86 \begin{matrix} +0.33 \\ -0.11 \end{matrix}$$

$$|g_{S,P}| < 0.33|g_V|$$

$$|g_T| < 0.28|g_V| \quad . \tag{2.2}$$

Recently another test became possible [10]. For V or A interactions the linear polarization of μ^+ produced in the reaction

$$\bar{\nu}\ Fe \rightarrow \mu^+\ X \tag{2.3}$$

is the same as for the right-handed $\bar{\nu}$, whereas for S,P and T interaction the opposite is true (note, γ_5 anti-commutes with V,A but commutes with S,P and T). From the measured value of polarization of μ^+ the authors conclude

$$\sigma_{S,P} < 0.15\ \sigma_{ALL} \quad . \tag{2.4}$$

The upper limit (2.4) is not better than (2.2) but allows a test with $Q^2 \sim 100\ GeV^2$ much larger than $Q^2 \lesssim m_\mu^2$ in the decay (2.1).

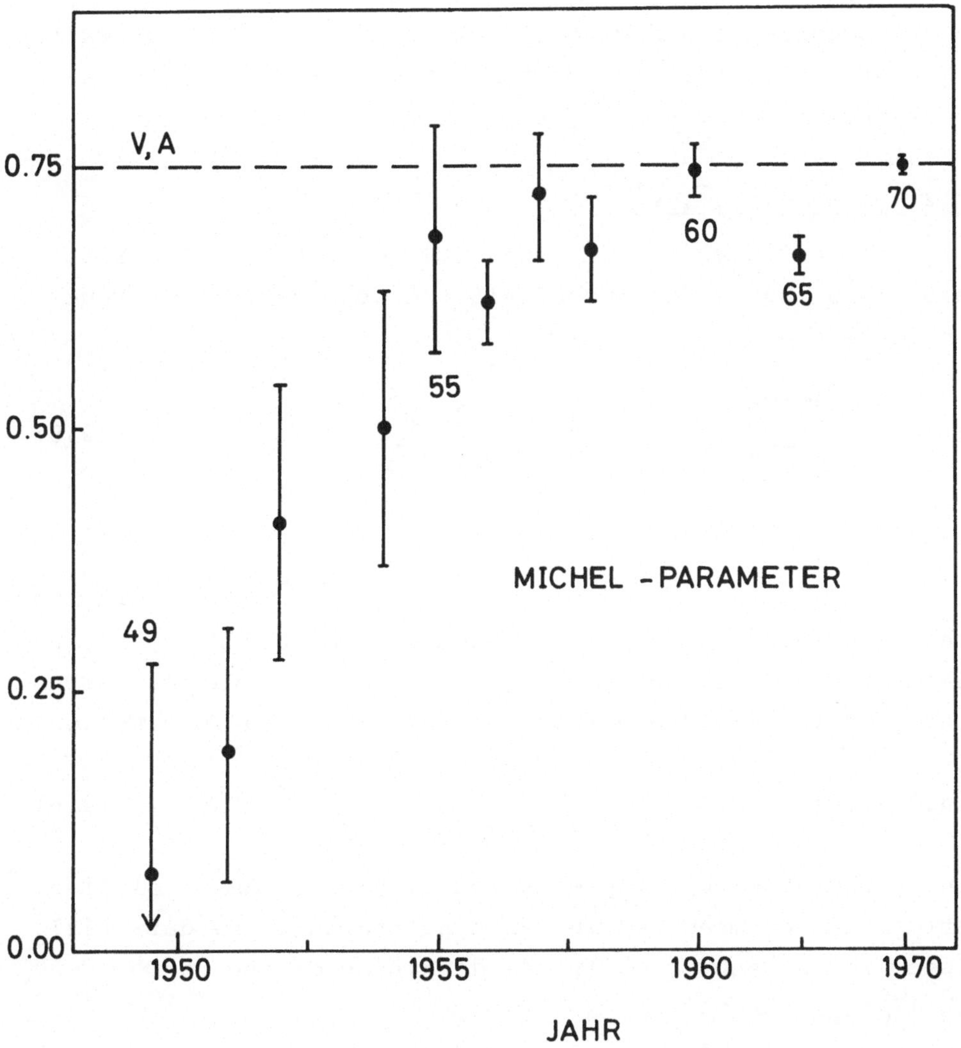
V, A
0.75
0.50
0.25
0.00
49
55
60
65
70
MICHEL - PARAMETER
1950
1955
1960
1970
JAHR

Fig. 2

2.2 Neutral Lepton Current

Occurrence and size of the neutral current interaction is a typical prediction of the Hamilton (1.1). This can be tested in $e^+e^- \to \mu^+\mu^-$, e^+e^- or $\nu e \to \nu e$ scattering.

2.2.1 $e^+e^- \to \mu^+\mu^-$, e^+e^-

In addition to the usual QED contribution the reaction $e^+e^- \to \mu^+\mu^-$ can also proceed via the weak current-current

H = 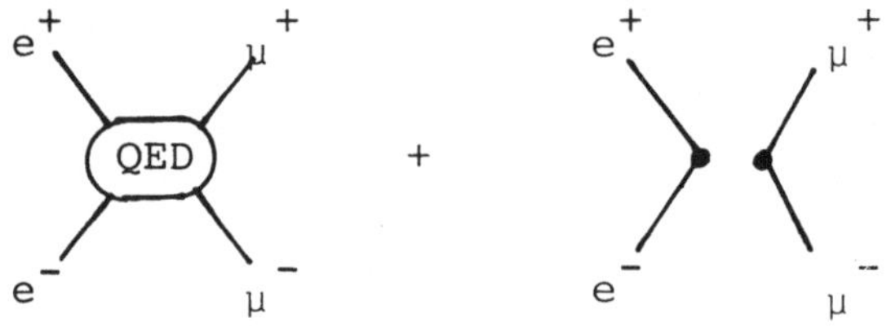 (2.5)

The main effect is the interference term leading to a forward-backward asymmetry $\sim -\frac{sG}{\pi\alpha} \cdot g_A^2 \cos\theta$. In various PETRA experiments [11] the effect has been measured leading to

$$|g_A| = 0.50 \pm 0.09 \quad , \quad g_V \sim 0 \quad , \tag{2.6}$$

in agreement with the prediction (1.5) from SM. A similar effect can be measured in Babha scattering. The data [12] (see Fig. 3) show clearly the deviation of the ratio $\frac{d\sigma}{d\Omega}/(\frac{d\sigma}{d\Omega})_{QED}$ from 1.

The other possible reaction consists in:

2.2.2 νe Scattering

The differential cross section for the reaction $\nu_\mu e^- \to \nu_\mu e^-$ can be written

$$\frac{d\sigma}{dy} = \frac{G^2 . s}{4\pi} \cdot f(y) \tag{2.7}$$

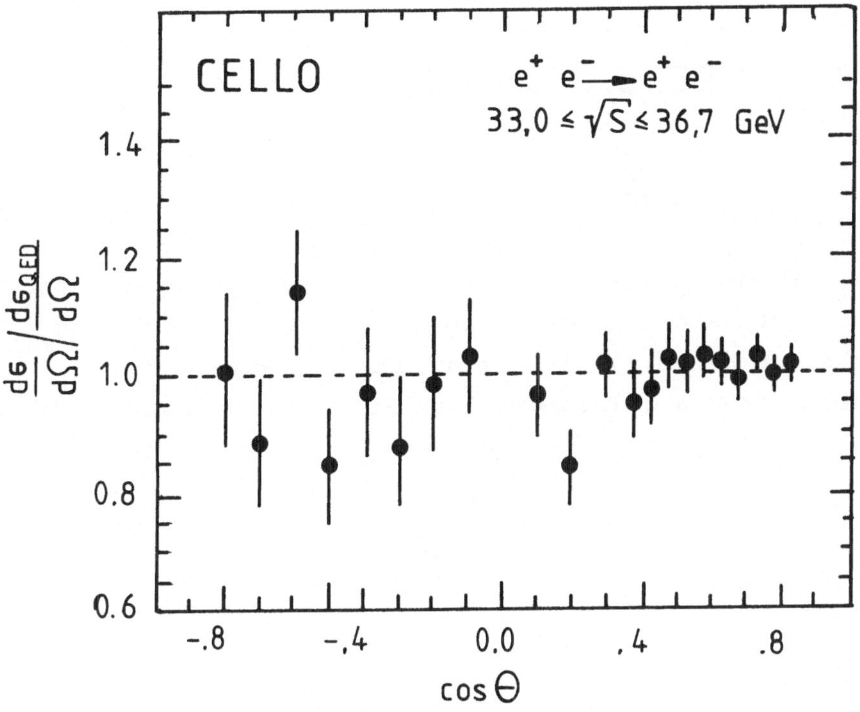

Fig. 3

where we use $y = \frac{1-\cos\theta}{2}$ instead of the center of mass angle θ to make the connection to deep inelastic scattering on hadrons more clearly which we will discuss in Chapter 3. The factor G^2s must be there for dimensional reasons if we neglect the electron mass (s denotes the total invariant mass squared). The factor f depends only on y and is quadratic in the neutral coupling constants g_A and g_V. Since the ν is purely left-handed, the scattering of ν_L on e_L is proportional to $(g_A + g_V)^2$ and on e_R to $(g_A - g_V)^2$.

$$\nu_L\, e_L \;:\; f = (g_A + g_V)^2 \tag{2.8}$$

$$\nu_L\, e_R \;:\; f = (g_A - g_V)^2\,(1-y)^2 \quad . \tag{2.9}$$

The different y behaviour in (2.8) and (2.9) we can understand in the following way:

$\nu_L \; e_L$:

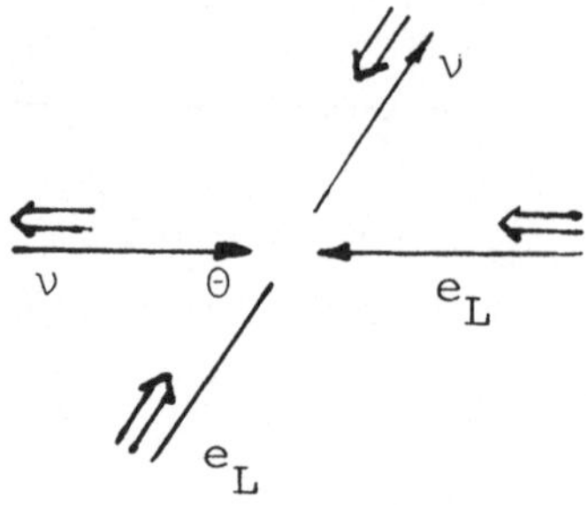

For $\nu_L e_L$ the helicities of the incoming leptons and of the outgoing leptons are always opposite. For any scattering angle the net helicity flip is always 0, and we get no suppression from angular momentum. For $\nu_L e_R$, however,

$\nu_L \; e_R$:

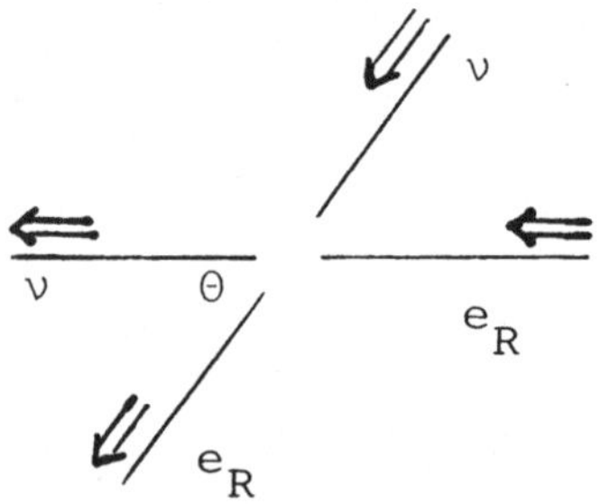

the helicity of the incoming and outgoing particles is 1. Only at $\theta = 0$ the net helicity flip is zero, therefore at $\theta = 0$ or $y = 0$ there is no suppression, whereas at $\theta = \pi$ there are two units of helicity net flip implying a double zero at $y = 1$, which explains the y dependence in Equ.(2.9). Inserting (2.8) and (2.9) into (2.7) we get

$$\frac{d\sigma}{dy} \left(\nu_\mu e \rightarrow \nu_\mu e \right) = \frac{G^2 s}{4\pi} \left[(g_A + g_V)^2 + (1-y)^2 \, (g_A - g_V)^2 \right]. \quad (2.10)$$

Only the proof of the innocent-looking $\frac{1}{4\pi}$ requires a long calculation of traces and two-body phase space. If we integrate (2.10) over y, the $(1-y)^2$ leads to a factor 1/3 and we get in units of $\frac{G^2 s}{12\pi}$ the cross section listed in the first line of Table 1 :

Table 1

	$\sigma \cdot \frac{12\pi}{G^2 s}$	$g_A = \frac{1}{2}$, $g_V = 0$
$\nu_\mu e \to \nu_\mu e$	$3(g_A + g_V)^2 + (g_A - g_V)^2$	1
$\bar{\nu}_\mu e \to \bar{\nu}_\mu e$	$3(g_A - g_V)^2 + (g_A + g_V)^2$	1
$\bar{\nu}_e e \to \bar{\nu}_e e$	$3(g_A - g_V)^2 + (2 + g_A + g_V)^2$	7

To obtain $\bar{\nu}_\mu e$ from (2.10) we apply CP to the ν side of the reaction. Under CP ν changes into $\bar{\nu}$. Since the vector current is even and the axial current is odd under CP, we get the second line in Table 1. For $\bar{\nu}_e e$ we have to add the contribution from the charged current with $g_A = g_V = 1$. The measured cross sections can be represented by ellipses in the (g_A, g_V) plane. As we see from Fig. 4, all three intersect in two points. The solution $g_A = 0$ is excluded from e^+e^- data. The second solution falls into the prediction of the SM model (the line $g_A = \frac{1}{2}$) leading to the following value of $\sin^2\theta_W$ [13]:

$$\sin^2\theta_W = 0.24 \pm 0.04 \; ; \qquad (2.11)$$

so we conclude that the SM works for leptons, and we can test weak properties of the quarks as well as the assumptions in semileptonic reactions.

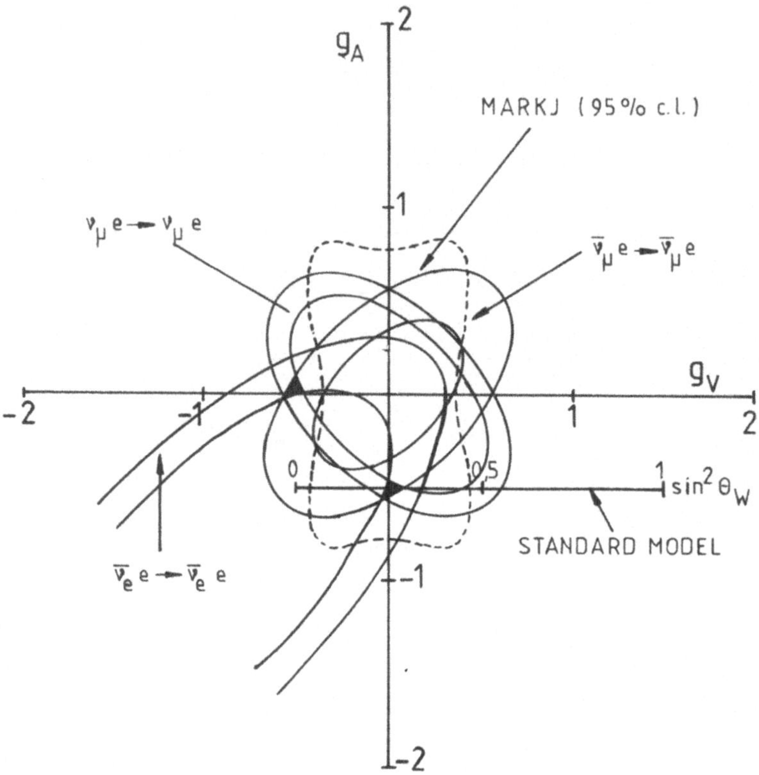

Fig. 4

Finally we would like to mention that $\sin\theta_W = \frac{1}{2}$ corresponds to ideal mixing in SU(2) which is broken by a high mass vector boson into SU(2) × U(1). If one adds [14] to the right-handed doublet $e^+ \bar{\nu}$ the charged conjugate of a left-handed lepton (μ, τ or a mixture) (normally the SU(2) singlet), one has a right-handed triplet (essentially the old Mahmoud-Konopinski scheme [15]). The charge is given by the Gell-Mann relation

$$Q = \frac{1}{2}(\lambda_3 + \sqrt{3}\ \lambda_8) \ . \qquad (2.12)$$

Comparing this with the ansatz of the SM [16],

$$Q \sim \sin\theta_W I_3 + \cos\theta_W \ Y \ . \tag{2.13}$$

We read off that $tg\theta_W = \frac{1}{\sqrt{3}}$ or $\sin\theta_W = \frac{1}{2}$. Unfortunately quarks require an extra hypercharge, due to their non-integer charge. Therefore the scheme is not possible in a gauge theory with universal coupling. In composite models universality is not mandatory but in the way these models are constructed [17] a weak SU(3) cannot be accomodated.

3. SEMILEPTONIC PROCESSES

In these processes one always assumes additivity. Only one quark is active, the others are simply spectators.

3.1 β-Decays

A particularly simple decay is the β-decay $n \to p e \bar{\nu}_e$. Even if $g_A = g_V = 1$ on the quark level we do not expect $G_A = G_V$, since right-handed protons contain also left-handed quarks according to Equ.(1.13). Applying coherent additivity to the β-decay we get (calling spin up ↑ left)

$$G_{A,V} = \cos\theta_c [\langle p\uparrow | j_+ | n\uparrow\rangle \mp \langle p\downarrow | j_+ | n\downarrow\rangle] \tag{3.1}$$

where j_+ denotes the left-handed (↑) quark current. First we use isospin rotation to express $\langle p|j|n\rangle$ as a diagonal element:

$$\langle p|j_+|n\rangle = 2 \cdot \langle p|j_3|p\rangle \ . \tag{3.2}$$

The factor 2 is the product from a Gordon coefficient $\sqrt{2}$ and a $\sqrt{2}$ from our non-conventional normalization of the charged current in (1.2). $2j_3$ is nothing else but the difference between u- und d-quarks. The second term in (3.1) we can rotate in spin space to express in terms of a right-handed quark current. So we get

$$G_{A,V} = \cos\theta_c[\langle p|(n_{u\uparrow} - n_{d\uparrow}) \pm (n_{u\downarrow} - n_{d\downarrow})|p\rangle] \,. \tag{3.3}$$

Inserting the probabilities (1.13) for the proton we find

$$G_V = \cos\theta_c \left(\frac{5}{3} - \frac{1}{3} + \frac{1}{3} - \frac{2}{3}\right) = \cos\theta_c \tag{3.4a}$$

$$G_A = \cos\theta_c \left(\frac{5}{3} - \frac{1}{3} - \frac{1}{3} + \frac{2}{3}\right) = \frac{5}{3}\cos\theta_c \,. \tag{3.4b}$$

The predicted value of $\frac{G_A}{G_V} = \frac{5}{3}$ is remarkably close to the measured ratio of 1.254 ± 0.007 [18]. There exists a large literature [19], which additional effects can bring down $\frac{5}{3}$ to the measured value. If we now consider the strangeness changing decay $\Sigma^- \to n e^- \bar{\nu}_e$, it involves the matrix element

$$\langle\Sigma^-|j^+_{su}|n\rangle = 2 \cdot \langle\Sigma^-|(j_3)_{su}|\Sigma^-\rangle = \langle\Sigma^-|n_s|\Sigma^-\rangle \,. \tag{3.5}$$

j^+_{su} can be rotated into $2(j_3)_{su}$ by an s-u isospin rotation (sometimes called V-spin) which counts the number of s-quarks, since there are no u-quarks in Σ^-. Therefore we get

$$G_V(\Sigma^- \to n) = \sin\theta_c \langle n_{s\uparrow} + n_{s\downarrow}\rangle$$

$$G_A(\Sigma^- \to n) = \sin\theta_c \langle n_{s\uparrow} - n_{s\downarrow}\rangle \,. \tag{3.6}$$

s plays the same role for Σ^- as d does for the p. Therefore we take just the d-quark probabilities in (1.13) leading to

$$G_V = \sin\theta_c \left(\frac{2}{3} + \frac{1}{3}\right) = \sin\theta_c \tag{3.7}$$

$$G_A = \sin\theta_c \left(\frac{1}{3} - \frac{2}{3}\right) = -\frac{1}{3}\sin\theta_c \,. \tag{3.8}$$

The (G_A/G_V) -ratio of $-\frac{1}{3}$ agrees marginally with the measured number ±0.39 ± 0.07 [18]. We should point out that an extraordinary thing happens in the β-decays. Instead of the quark model we can use flavor SU(3), assuming the currents transform as octets. Any matrix element depends on two couplings (F and D). Therefore the three best-measured decays $n \to pe\nu$, $\Sigma^- \to ne^-\nu$ and $\Lambda \to pe\nu$ must satisfy a sum rule which can be written as

$$3 \left(\frac{G_A}{G_V}\right)_{\Lambda\to p} - \left(\frac{G_A}{G_V}\right)_{\Sigma^-\to n} = 2 \cdot \left(\frac{G_A}{G_V}\right)_{n\to p} . \qquad (3.9)$$

Empirically [18]we have for both sides

$$2.30 \pm 0.17 = 2.51 \pm 0.02 .$$

Whereas the quark model predictions(3.3) and (3.4) work only qualitatively, (3.9) is in good agreement with the data. Alternatively one can fit all decays within flavor SU(3), using different θ_c in axial and vector part. Again both agree [18] within a few per cent ,

$$(\sin\theta_c)_A = (\sin\theta_c)_V \cdot (1.01 \pm 0.03) , \qquad (3.10)$$

Cabibbo universality we can test for vector currents since there assumption C involves only an extrapolation in Q^2 which is small. Before doing this, one has to apply radiative correction. For our effective H_I(1.1) they are infinite. You have to use the broken effective field theory [20,21]. This is the only glimpse we have in the moment to the field theoretical basis of H_I. Taking G from μ-decay by universality we find

$$\frac{G \cdot G_V}{G} = \cos\theta_c , \quad \theta_c = 13.1 \pm 0.5^\circ . \qquad (3.11)$$

Both from $\Sigma^-$and K_{e2} decay we get [22]

$$(G_V)_\Sigma = (G_V)_{K_{e2}} = \sin\theta_c , \quad \theta_c = 12.7 \pm 0.5^\circ . \qquad (3.12)$$

The agreement of θ_c in both cases tests coherent additivity together with V-A doublet structure of quarks.

3.2 Semileptonic Decays of D- or B-Mesons

Additivity implies for semileptonic decays of a heavy quark Q(c,b,t...) that they should proceed according to the diagram

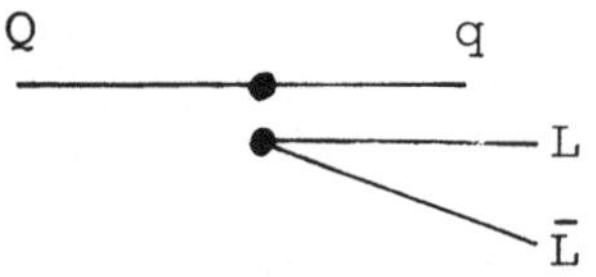

(3.13)

which gives the rate

$$\Gamma_{SL}(Q) = \Gamma_o \cdot f^2$$

$$\Gamma_o = \frac{G^2}{192\pi^3} m_Q^5$$

where f^2 depends on the mixing angles involved. There are many predictions from isospin flavor SU(3) [23] but they cannot be tested at the moment since only branching ratios are measured and the life time is not known accurately enough.

3.3 Deep Inelastic Lepton Quark Scattering

Most of our information about weak properties comes from e,ν scattering on p or nuclei N. Incoherent additivity says the cross section for $\nu p \to \nu x$ is related to $\sigma(\nu q \to \nu q')$ by

$$\sigma(\nu p \to \nu x) = \sum_q \sigma(\nu q \to \nu q')\, p_q(p) \ . \tag{3.14}$$

We assume that Q^2, m_x^2 (mass of the produced hadronic system) are much larger than 1 GeV^2. As in the $\nu e \to \nu e$ case we use as one variable y (which is the energy loss of the ν in the laboratory frame). As second variable it turns out that the inelasticity

$$x = \frac{Q^2}{Q^2+m_x^2} \tag{3.15}$$

is a very convenient choice. One simply takes the νe formulae (2.10), multiplies it with x from phase space and the probability a(x) for finding a quark at inelasticity x, and sums over all quarks,

$$\frac{d\sigma}{dxdy} = \frac{G^2 s}{4\pi} \cdot x\,[(g_A + g_V)^2 + (g_A - g_V)^2\,(1-y)^2]q(x)\,. \tag{3.16}$$

In general q can also depend on Q^2, however not for structureless quarks or partons. QCD radiative corrections give rise to a Q^2 dependence. Since this is mainly a strong interaction problem we do not consider this in the present lecture. Also, present-day statistics are not sufficient to test in detail the x,y and Q^2 dependence at the same time. Now we compare Equ.(3.16) with the general formula obtained by current-current interactions with ν or $\bar{\nu}$,

$$\frac{d\sigma}{dxdy}^{\nu,\bar{\nu}} = \frac{G^2 s}{2\pi}\,[(1-y)F_2 + y^2\,x\,F_1 \pm x(y-\frac{y^2}{2})F_3]\,. \tag{3.17}$$

The $F_i(x,Q^2)$ are called the structure functions. The $\bar{\nu}\bar{q}\to\bar{\nu}\bar{q}$ cross section is the same as (3.16) by CPT. By CP we obtain, as in the νe case, the $\bar{\nu}q\to\bar{\nu}q$ by reverting the sign of g_A. If we compare (3.17) with (3.16) for q or $\bar{q}$ we have three functions F_i for two quark probabilities q(x) and $\bar{q}(x)$. Therefore there must be one relation between the F_i, the so-called Callan-Gross relation [24]

$$2x\,F_1 = F_2\,, \tag{3.18}$$

and $F_{2,3}$ can be expressed in terms of $q,\bar{q}(x)$ by

$$F_2 = \sum_q (g_{A,q}^2 + g_{V,q}^2)\,(q(x) + \bar{q}(x)) \tag{3.19}$$

$$F_3 = \sum_q 2\,g_{V,q}\,g_{A,q}(q(x) - \bar{q}(x))\,. \tag{3.20}$$

That F_i are independent of Q^2 (scaling) is due to the input of structureless quarks. The nice feature of our formulae is the fact that they hold also for charged current $\nu p \to \bar{\mu} X$ and electromagnetic scattering $\mu p \to \mu X$, if we make the following specializations:

Neutral currents

$\nu p \to \nu X$ $$g_{A,q} = I_3 \qquad (3.21)$$

$\nu N \to \nu X$ $$g_{V,q} = I_3 - 2Q_q^2 \sin^2\theta_W \qquad (3.22)$$

Charged currents (pure left-handed)

$\nu p \to \mu^- X$

$$g_A = g_V = 1$$

$\nu N \to \mu^- X$ (3.23)

Electric currents

$$g_A = 0$$

$\mu p \to \mu X$

$$g_V = Q_q \qquad (3.24)$$

In the last case we also have to replace in (3.16) and (3.17) the factor

$$\frac{G^2}{2\pi} \to \sigma_{coul} = \frac{4\pi\alpha^2}{(Q^2)^2} \quad . \qquad (3.25)$$

3.3.1 Callan-Gross Relation

Relation (3.18) essentially tests the spin $\frac{1}{2}$ of the quarks. From Fig. 5 the ratio $\frac{2xF_1}{F_2}$ is plotted as function of x in various Q^2 intervals for $e^- p \to e^- X$ [25]. It is compatible with 1, which tells us that within the assumption I the partons are spin $\frac{1}{2}$ particles. Note that the magnetic term F_1 would zero for spin 0 partons.

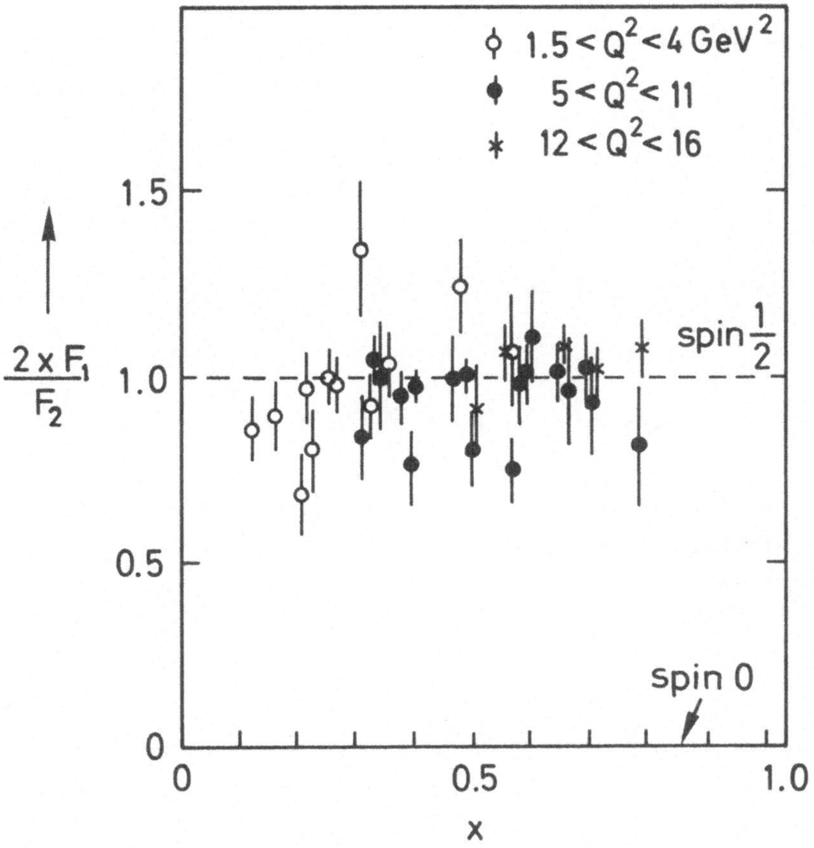

Fig. 5

3.3.2 Ratio of Charged Currents and Electromagnetic Current

If we consider $\nu N \rightarrow \mu X$ and $e^- N \rightarrow eX$, the coupling constants are given by Equ.(3.23) and (3.24). Since in both cases the same probabilities occur, there must be a relation between the structure function in both cases. The ν sees only d-quarks or $\bar{u}$-quarks since only the reactions

$$\nu d \rightarrow \mu^- u \qquad \text{or } \nu\bar{u} \rightarrow \mu^- \bar{d} \tag{3.26}$$

are possible. One uses a target with equal number of protons and neutrons. Therefore the protons give to the sum (3.19) a contribution $d + \bar{u}$. Since the probability of finding a d-quark in the neutron is equal the probability of finding a u-quark in the proton, we get from the neutrons a term $u + \bar{d}$. Adding both together in Equ.(3.19), we get with $g_A = g_V = 1$

$$F_2^{\nu N} = 2x(d + \bar{u} + u + \bar{d}) \ . \tag{3.27}$$

For e^--scattering we get in the same way with $g_A = 0$, $g_V = Q_q$,

$$F_2^{eN} = x[\tfrac{4}{9}(u + \bar{u} + d + \bar{d}) + \tfrac{1}{9}\ (u + \bar{u} + d + \bar{d})] \ . \tag{3.28}$$

Again the contribution with "wrong charge" ($\frac{4}{9}(d + \bar{d})$) describes the contribution of neutrons in N. Taking the ratio of Equ.(3.27) and (3.28) we get

$$F_2^{\nu N} = \frac{18}{5}\ F_2^{eN} \ . \tag{3.29}$$

The prediction (3.29) is shown in Fig. 6. The eN data [26] are parametrized by the curve, which is in very good agreement with ν data from [27]. Therefore the partons have the charges one expects from SM.

3.3.3 Absence of Right-Handed Quarks

The difference between ν and $\bar{\nu}$ allows an upper limit for right-handed coupled quarks. From Equ.(3.16) we get for $\nu N \to \mu^- X$

$$\sigma(\nu N) \sim q_L(x) + (1-y)^2\ q_R(x) \ . \tag{3.30}$$

Left-handed quarks (or right-handed $\bar{q}$) have $g_A = g_V = 1$, whereas right-handed quarks (or left-handed $\bar{q}$) have $g_A = -g_V = 1$, which gives the second term in (3.30). For

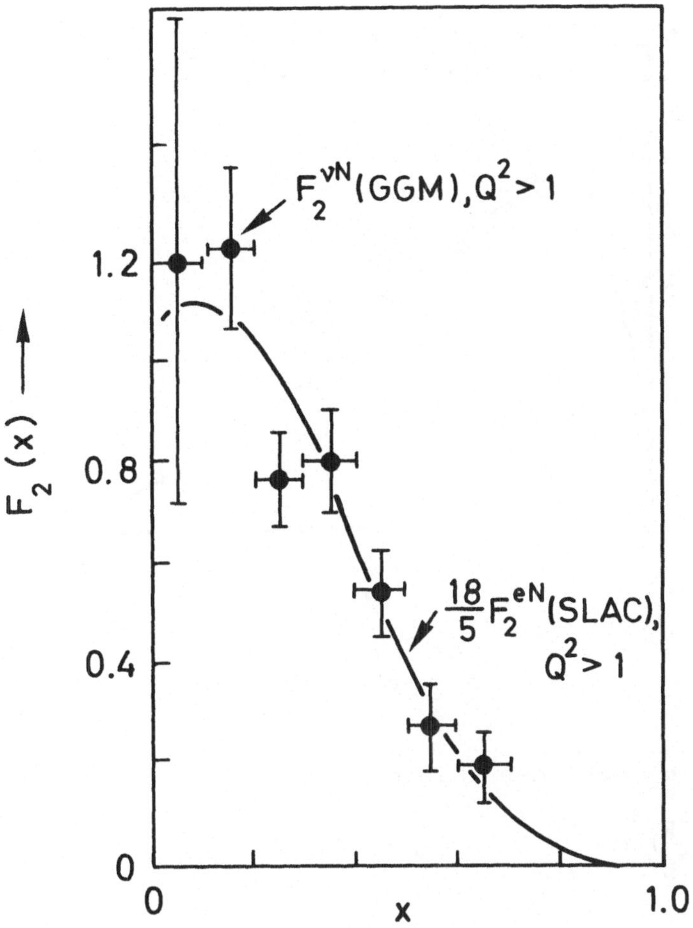

Fig. 6

$\bar{\nu}$ we interchange by CP R and L and obtain

$$\sigma(\bar{\nu}\ N) \sim q_R(x) + (1-y)^2\ q_L(x) \quad . \tag{3.31}$$

From the two measured cross sections q_R and q_L can be extracted [28], which are shown in Fig. 7.

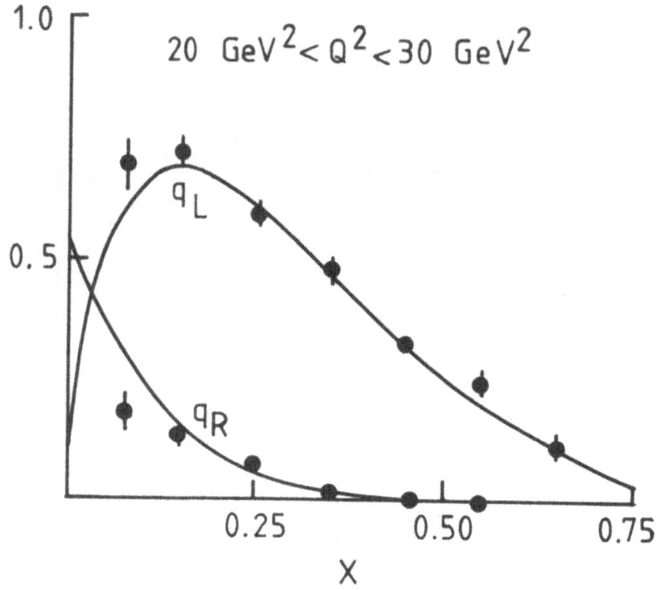

Fig. 7

At $x > 0.5$, q_R vanishes. Since

$$q_R = \bar{q} + \left(\frac{g_R}{g_L}\right)^2 q \tag{3.32}$$

where g_R denotes a possible right-handed coupling which is, of course, 0 in the SM. Since both terms in (3.32) are positive, one gets an upper limit for $\left(\frac{g_R}{g_L}\right)^2 q$. Taking q from q_L we can determine an upper limit for $\left(\frac{g_R}{g_L}\right)^2$,

$$(g_R/g_L)^2 < 0.10\ . \tag{3.33}$$

We conclude that, assuming incoherent additivity, there are no right-handed currents for quarks with the same accuracy as for μ. At small x the sea $q\bar{q}$ pairs dominate. We can test those, if we use a D-meson as a flavor tester. One considers the di μ-events, which are assumed to come from $\nu N \to \mu^- D X$, $D \to \mu^+$. The D must be produced by the subreaction

$$\nu s \rightarrow \mu^- c \tag{3.34}$$

where s is a strange quark in the sea. From the y dependence one can decide whether the s-quark in N was right- or left-handed, leading to

$$\left(\frac{g_R}{g_L}\right)^2 \begin{cases} < 0.10 & \text{CDHS [29]} \\ = 0.15 \pm 0.10 & \text{CHARM [30]} \end{cases} .$$

Again, the experimental values are compatible with no right-handed coupling also for the sea quarks.

3.3.4 Neutral Current Couplings

A very peculiar feature of the SM are the neutral current couplings of quarks

$$g_{L,R} = \frac{1}{2}(g_A \pm g_V) = \begin{matrix} I_3 - Q \sin^2\theta_W \\ Q \sin^2\theta_W \end{matrix} \tag{3.35}$$

where, in contrast to charged leptons, g_V does not vanish for $\sin\theta_W = \frac{1}{2}$. The same analysis which we did in the case of charged currents has been performed also in the neutral case. For simplicity let us assume no antiquarks and only d,n-quarks in the target. The numbers quoted later on are corrected for their small amounts. If we take the probabilities from charged ν reactions, we have only four coupling constants g_{UL}^2, g^2_{UR}, g_{dL}^2 and g_{dR}^2 to determine. There are four possible experiments: $\sigma(\bar{\nu}N)$, $\sigma(\nu N)$, $\sigma(\bar{\nu}p)$ and $\sigma(\nu p)$. $\bar{\nu}$ and ν data distinguish between R and L, whereas the difference between I = 0 targets (N) and p-data are used to separate u and d. The signs can never be measured in incoherent processes. One has to use exclusive reactions, like $\nu p \rightarrow \nu p$ or $\nu p \rightarrow \nu\Delta^+$. Those processes need the current matrix element

$$<p|J_\mu^N|p> \tag{3.36}$$

which involve linearly the coupling constants, if one uses coherent additivity (1.10). The Q^2 dependence is taken from the analogous reactions $e^-p \to e^-p$ or $\nu n \to \mu^-p$. For the normalization at $Q^2 = 0$ one uses, as in the β-decay, the proton probabilities (1.13). Since we want to distinguish only signs, this should be a reasonable assumption. The results [31] of such an analysis, together with the prediction of the SM with $\sin^2\theta_W = 0.23$, are listed in Table 2 ,

Table 2

	g_{UL}	g_{UR}	g_{dL}	g_{dR}
EXP.	0.351±0.037	-0.179±0.032	-0.415±0.055	-0.010±0.046
Theory				
$\sin^2\theta_W$=0.23	0.345	-0.155	-0.423	0.077

which shows that quarks have the right properties in the SM, provided the additivity assumptions are true.

Another way to measure the Weinberg angle comes from the ratio between charged and neutral currents

$$\frac{[\sigma(\nu N) - \sigma(\bar{\nu}N)]_{NC}}{[\sigma(\nu N) - \sigma(\bar{\nu}N)]_{CC}} = \sum_{u,d} g_{A,q}\, g_{V,q} = \frac{1}{2} - \sin^2\theta_W \; . \tag{3.37}$$

(3.37) can be obtained by inserting (3.21) to(3.23) into (3.20) and (3.20) into (3.17).

From the measured ratio one obtains [32] $\sin^2\theta_W = 0.19\pm$ ± 0.04.

3.3.5 Right-Left Asymmetry in $e^-d \to e^-X$

Up to now we used only ν data to test SM for quarks. Any two-component mass-less ν imposes automatically a V-A structure on the theory. This is different in scattering of linear polarized e on deuterium

$$e^-_{R,L}\, d \to e^- X \,. \tag{3.38}$$

One observes the left-right asymmetry

$$A = \frac{\sigma_R - \sigma_L}{\sigma_R + \sigma_L} \,. \tag{3.39}$$

To compute (3.39) we assume incoherent additivity. For simplicity we take the deuterium as an equal mixture of u- and d-quarks. The asymmetry arises from interference of the electromagnetic interaction with $A_N(e)V^N(q)$ where A(V) denote axial (vector) currents. The analogous term with $V^N(e)A(q)$ can be neglected since $g^e_V \sim 0$. Both terms are proportional to V(q) which is the same for u and d. Therefore the quark vector current factors out in H_I,

$$H_I = \frac{e^2}{Q^2} \sum_q V_\mu(q)\,[Q_q V^\mu(e) + \frac{Q^2}{2}\gamma \cdot g^V_q\, A^\mu(e)]\,, \tag{3.40}$$

where $\gamma = \frac{G^2}{\pi\alpha\sqrt{2}} \sim 2.10^{-4}\ \mathrm{GeV}^{-2}$. The factorization implies that A is independent of x or y. The numerator is the interference term and the dominator essentially the square of the first term.

$$A = -\gamma Q^2 \frac{\sum Q_q g^V_q}{\sum Q^2_q} = -\gamma Q^2 (\frac{9}{10} - 2\sin^2\theta_W) \,. \tag{3.41}$$

From the measured [33] asymmetry we get $\sin^2\theta_W = 0.224 \pm 0.020$ in agreement with previous determinations which are listed in Table 3.

Table 3

Process	$\sin^2\theta_W$
A in $e\uparrow d$	0.224 ± 0.020
NC	0.23
NC/CC	0.19 ± 0.04
νe	0.24 ± 0.04

At least one can say, with incoherent additivity for quarks, SM is compatible with the data.

4. NONLEPTONIC DECAYS

Nonleptonic weak processes can be only observed in decays forbidden by strong interaction. Historically the first attempts to understand strange particle decays did not use quarks. PCAC and current algebra, together with some assumption about factorization of H_I, give many restrictions on the decay amplitudes [34]. The main successes are ratios (for example the $\Delta I = \frac{1}{2}$ rule) between rates, but the absolute value of $\Gamma(K_s \to 2\pi)$ fails by two orders of magnitude. In the beginning the quark model suffered from the same desease, but over the last years it has been realized that strong final state interaction can qualitatively explain the rates.

First we have to decide whether in a given decay of a heavy quark Q, quark-decay or W-exchange dominates.

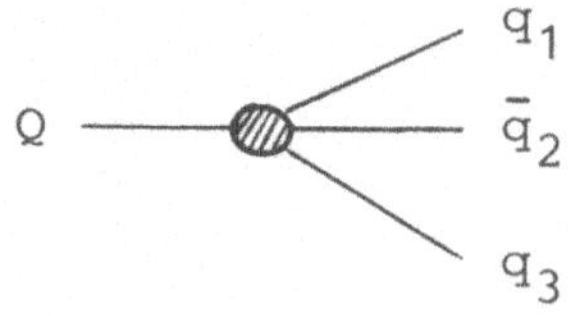

(4.1)D

or

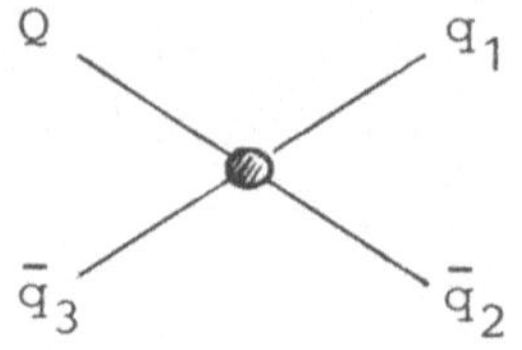

(4.2)W

For a rough estimate we assume incoherent additivity for hadronization of the quarks. Diagram D involves a three-body and diagram W a two-body final state. From phase space only we get for the ratio

$$\frac{\Gamma_D}{\Gamma_W} = \frac{N_c}{2} \left(\frac{Rm_Q}{4\pi}\right)^2 \tag{4.3}$$

where N_C is the number of colors and R a radius describing the ratio of the two-body and three-body amplitude. Using $R = 6\ \text{GeV}^{-1}$ we get the following numbers for Γ_D/Γ_W a function of m_Q:

m_Q	m_B	m_D	m_K	
Γ_D/Γ_W	3.2	0.4	small	(4.4)

From this very rough estimate we expect that heavy quarks (b,t...) mainly decay via diagram (4.1) whereas light quarks prefer (4.2).

4.1 The Effective Hamiltonian

The first problem we encounter is the estimate of strong interaction. As an example we use a vector gluon potential between the quarks, $q_1q_2q_3q_4$ written in momentum space [35]:

$$H_s = \sum_a \lambda^a_{13}\, \lambda^a_{24}\, \gamma^\mu_{13}\, (\gamma_\mu)_{24}\, v(q^2)\ . \tag{4.5}$$

For elementary gluon exchange we have $v = 4\pi\alpha_s/q^2$. Now we rewrite our weak Hamiltonian of charged current in the following way,

$$H_I = \frac{G}{2\sqrt{2}} \{(f_+ + f_-)(\bar{u}\ d)(\bar{s}\ c) + (f_+ - f_-)(\bar{u}\ c)(\bar{s}\ d)\}\ . \tag{4.6}$$

In Equ.(4.6) we took as example the reaction cd → us, other

decays can be obtained by replacing the appropriate quark flavors. The bracket $(\bar{u}d)$ is an abbreviation for the V-A interaction $(\bar{q}q) = \bar{q}\gamma_\mu(1-\gamma_5)q$. If there is no strong interaction, $f_+ = f_- = 1$, then the second term (a flavor-changing neutral current interaction) is absent. The important point is that the quarks in (4.6) should obey the Pauli principle, which means H_I must be antisymmetric under the exchange of $d \leftrightarrow c$. In spin space it is already antisymmetric, since it changes sign under a Fierz transformation $s \leftrightarrow d$. In flavor space the f_+ term (f_- term) is symmetric (antisymmetric) by construction. Therefore the Pauli principle requires that f_+ involves the color indices in a symmetric, and f_- in an antisymmetric way. This symmetry is not yet put into (4.6) since $\bar{u}d$ contains a unit matrix in color space. To exhibit the color (anti-)symmetry explicitly we replace now

$$f_+ \, 1.1 \rightarrow f_+\left(\frac{2}{3}\, 1.1 + \frac{1}{4}\,\lambda^a \cdot \lambda^a\right) \tag{4.7}$$

$$f_- \, 1.1 \rightarrow f_-\left(\frac{1}{3}\, 1.1 - \frac{1}{4}\,\lambda^a \cdot \lambda^a\right) \; . \tag{4.8}$$

$\lambda^a \cdot \lambda^a$ in (4.7) means $\sum \bar{q}_1\lambda^a\gamma_\mu(1-\gamma_5)q_2 \cdot \bar{q}_3\lambda^a\gamma^\mu(1-\gamma_5)q_4$. We have made yet no real change, we only exhibited the Pauli principle explicitly. With (4.7) and (4.8) H_I becomes

$$H_I = \frac{G}{2\sqrt{2}}\{f_+[\tfrac{2}{3}(\bar{u}d\;\bar{s}c + \bar{u}c\;\bar{s}d) + \tfrac{1}{4}(\bar{u}\lambda^a d\;\bar{s}\lambda^a c + d\leftrightarrow c)]$$
$$+ f_-[\tfrac{1}{3}(\bar{u}d\;\bar{s}c - \bar{u}c\;\bar{s}d) - \tfrac{1}{4}(\bar{u}\lambda^a d\;\bar{s}\lambda^a c - d\leftrightarrow c)]\} \; . \tag{4.9}$$

If we now calculate the first order correction due to the potential (4.5) according to the diagram

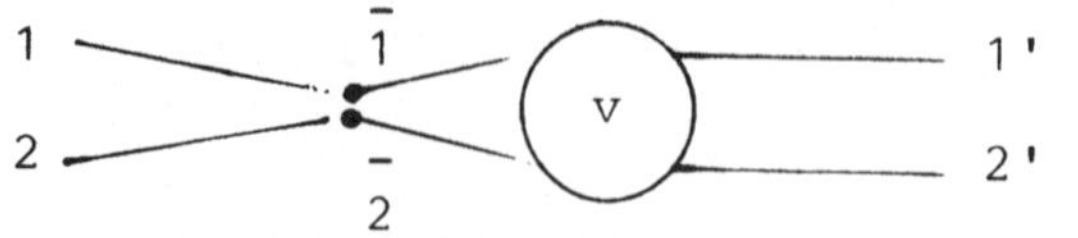

(4.10)

one finds that the combination of λ matrices occurring in (4.7) and (4.8) are eigenstates of λ^a_{11}, $\cdot\lambda^a_{22}$, with eigenvalues $\frac{4}{3}$ and $-\frac{8}{3}$ for the f_+ respectively f_- term. The $\gamma_\mu\gamma^\mu$ term in (4.5) does not change the V-A structure up to mass terms. Denoting the space integral in (4.10) by ρ we get the change

$$f_+ \rightarrow f_+(1 - \tfrac{4}{3}\rho)$$

$$f_- \rightarrow f_-(1 + \tfrac{8}{3}\rho) \quad . \qquad (4.11)$$

The signs in (4.11) must be such that it is repulsive or negative in the color symmetric part f_+ (by analogy to Coulomb's law) and attractive or positive for the color antisymmetric part f_-. It is well-known that the Born approximation works only for small Q^2. The average Q^2 in the loop integral (4.10) is of the order m_W^2 much larger than a typical hadronic energy. Therefore (4.11) can be considered only as the first term of an infinite series There are two ways to compute higher orders. First one assumes a geometrical series by interpreting (4.10) as the first term of ladder diagrams (similar to the K matrix expansion discussed in the lecture of Professor Dalitz).

$$f_+ = (1 + \tfrac{4}{3}\rho)^{-1} \ , \qquad f_- = (1 - \tfrac{8}{3}\rho)^{-1} \quad . \qquad (4.12)$$

The main point of this approximation is to show you there could be other ways to include strong than by summing up hard gluon contributions by renormalization group techniques [36]. For v we take one-gluon exchange. (4.11) is the first term of an exponential series, which means that the factor -2 in the first order contribution of v in Equ.(4.11) gets exponentiated, leading to the relation

$$f_+^2\, f_- = 1 \ . \qquad (4.13)$$

Solving the Callan-Symanzik equation [36] one finds

$$f_- = [1 + \gamma\, \alpha_s(Q^2) \log \frac{m_W^2}{Q^2}]^{1/\gamma} \qquad (4.14)$$

where $\gamma = \frac{1}{4}(11-\frac{2}{3}n_f)$ and $\alpha_s(Q^2)$ are the running coupling constants. Both (4.12) and (4.13), (4.14) have the property that by strong interaction $f_- > 1$ and $f_+ < 1$. This behaviour is not a particular property of the renormalization group equations but rather due to assumption that the strong interaction potential is proportional to $\lambda^a \cdot \lambda^a$. In the following Table 5 we compute the values of f for various Q^2 from (4.13), (4.14).

Table 5

	Q^2	f_+	f_-
	m_W^2	1	1
	m_b^2	0.80	1.56
	m_c^2	0.70	2.10
$\alpha_s \sim 1$,	m_K^2	~ 0.5	$\sim 4.$

Even $m_K^2/m_W^2 \sim 5.10^{-5}$ the changes are not very dramatic, which expresses again the fact that the log is essentially a constant.

4.2 Test of the Quark Decay Assumption (D)

With assumption D and the Hamiltonian (4.9) the amplitude for $Q \to q_1 q_2 \bar{q}_3$ can be computed analogous to the μ-decay. You can, of course, sit down and evaluate all the necessary traces over γ and λ matrices to find the result [37]:

$$\Gamma(Q \to q_1 q_2 \bar{q}_3) = \frac{G^2 m_Q^5}{122\pi^3}(f_-^2 + 2f_+^2) \quad . \qquad (4.15)$$

There is,however, a much simpler argument. The first factor in (4.13) is nothing else than the rate $\Gamma_o(\mu \to e\nu\bar{\nu})$ at $m^2 = m_Q^2$. (4.15) has to be quadratic in f_+ and f_-. Therefore we multiply Γ_o with

$$\Gamma(Q) = \Gamma_o(\alpha f_+^2 + \beta f_+ f_- + \gamma f_-^2) \quad . \tag{4.16}$$

For f_+ the quarks are in a symmetric color sixtet state, for f_- they are in an antisymmetric color triplet state. If we sum over all color indices those states cannot interfere, which means $\beta = 0$. From the multiplicity the weights of f_+ and f_- must appear in the ratio 6:3 or $\alpha = 2\gamma$. Putting $f_+ = f_- = 1$ we must recover the usual color factor 3 which proves Equ.(4.15). From the total rate (4.15) and incoherent additivity for the semileptonic rate we get the semileptonic branching ratio:

$$B(Q \to eX) = (N_L + 2f_+^2 + f_-^2)^{-1} \tag{4.17}$$

where N_L is the phase space corrected number of active leptons. Inserting the numbers f from Table 5 we get:

Table 6

Meson	B_{TH}	$B_{EXP.}$
B	15%	14 ± 5% for average over B^-, B^o [38]
D	13%	$B(D^o) \sim 4\%$, $B(D^+) \sim 20\%$ [39]
K	6%	$B(K^o) \sim 0.1\%$, $B(K^+) \sim 28\%$ [40]

From this table we see that quark decay may work for B but is absolutely disastrous for D(not to speak of K). These numbers fail partly because they had been predictions [41].

After the experiments one did neither throw away the QM nor the quark assumptions. The crucial assumption for this bad result was the assumption of incoherent quark decay where interference terms have been neglected. Their importance can be demonstrated at best in the K^+ decay. The dominant hadronic channel is $\pi^+\pi^o$. By angular moment conservation the $\pi\pi$'s are in an S-state. Mr. Bose tells that the isospin had to be even, which means I = 2 for a charged state. The f_- term in the decay $s \to ud\bar{u}$ is flavor antisymmetric in ud, and it carries therefore only $I = \frac{1}{2}$, which can never reach I = 2 with one spectator. In contrast to Equ. (4.17), f_- can enter only by the phase space-suppressed 3π decay mode. To cure the quark decay model Guberina et al. [42] assumed that f_+ and f_- are free parameters, since the shape of $v(q^2)$ is unknown. A decay proceeds in two steps: First, two hadronic color clusters $D \to AB$ according to (4.9) are formed, which subsequently decay incoherently. Since the first step is coherent, constraints from f.e. isospin can be imposed. As a result they get $f_+ = 0.80 << f_- = 6.6$ and the "post diction"

$$\Gamma_H(D^+) \cong \frac{1}{10} \Gamma_H(D^o) . \tag{4.18}$$

$f_+ << f_-$ also will lead to the $\Delta I = \frac{1}{2}$ rule for K. The very different hadronic width explains the different semileptonic branching ratio in Table 6 . A similar effect leads in the B-decay to the prediction [43]

$$\Gamma_H(B^o) \sim 1.3\ \Gamma_H(B^-) . \tag{4.19}$$

This should demonstrate how careful one should be with QM prediction in non-leptonic decays.

4.3 W-Exchange Model

Again we will assume the Hamiltonian (4.9). Due to

gluon corrections both K^+ and K^O can proceed via exchange. For K the $\Delta I = \frac{1}{2}$ rule will require $f_- \gg f_+$. For strange baryons f_+ cannot contribute since f_+ involves color symmetric quarks whereas the color baryon wave function is antisymmetric. Some people call this simple observation a theorem [44]. Dominance of $\Delta I = \frac{1}{2}$ is automatic. The $\Delta I = \frac{1}{2}$ rule has been generalized to four quarks in the D-decay called 20 plet dominance [45]. In W-exchange model D^+ cannot decay since $c\bar{u} \to s\bar{d}$ is possible but $\bar{d}c \not\to s\bar{u}$ by charge conservation. If W-exchange is dominant we expect the 20 plet rule

$$\Gamma(D^+) \ll \Gamma(D^O) \ . \tag{4.20}$$

Empirically $\Gamma(D^O) \sim 3\text{-}4\Gamma(D^+)$ [46]. Since the final state in $c\bar{u} \to s\bar{d}$ has $I = \frac{1}{2}$, we get

$$\frac{\Gamma(D^O \to K^O\pi^O)}{\Gamma(D^O \to K^-\pi^+)} = 0.5 \tag{4.21}$$

in agreement with the measured value of 0.73 ± 0.35 [47]. The same rule applies to ρ, which may be a case of future trouble [48].

The most serious objection against W-exchange comes from the so-called helicity suppression. If one looks at the decay rate of K in lepton neutrino

$$\Gamma(K^+ \to \mu\nu) = \frac{G^2}{4\pi} \frac{m_K^2 - m_\mu^2}{m_K} m_\mu^2 \cdot f_K^2 \ , \tag{4.22}$$

it is proportional to the lepton mass squared. Physically the leptons have to have the same helicity for a spin 0 state whereas V or A connects in the limit $m_L \to 0$ only leptons with opposite helicity. If we use the same argument for quarks the transition $c\bar{d} \to u\bar{s}$ should be proportional to the light quark masses, which are very small numbers

(∿5 MeV). To overcome this difficulty, several proposals have been made:

(i) Gluon Emission

Gluonic exchange does not change the V-A structure, but emission of real gluon does. It flips the spin of one quark according to the diagram (→ denoting the helicities):

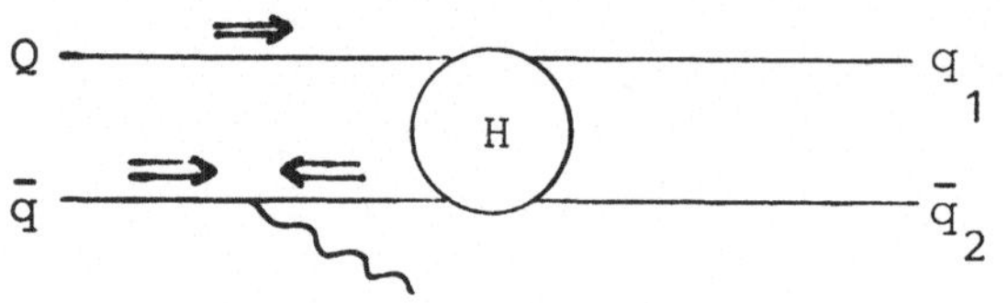

(4.23)

The original spin 0 color singlet $\bar{Q}q$ state is turned into a spin 1 color octet state which can decay into $q_1\bar{q}_2$ without helicity suppression. The rate can be estimated [48] by

$$\Gamma_W = \frac{G^2 m_D^5}{192\pi^3} \; r \; \frac{F_D^2}{(m_q)^2_{eff}} \; \alpha_s \left(\frac{f_+ + f_-}{2}\right)^2 \quad . \tag{4.24}$$

The first factor accounts for the phase space of the 3-particle final state (q_1. q_2, G), r corrects for finite masses, $\alpha_s \cdot F_D^2 (m_q)^{-2}_{eff}$ takes into account the magnetic emission of the gluon, which can be estimated analogously to $\pi \to \rho\gamma$. Note that $(m_q)_{eff}$ must be the large (300 MeV) constituent mass. $\frac{1}{2}(f_+ + f_-)^2$ replaces the factor $\frac{1}{3}(2f_+^2 + f_-^2)$ since f_+ and f_- will interfere after emission of a color octet. Putting in numbers [48] for D-decay, one gets the ratio

$$\Gamma_W / \Gamma_D \sim 1 - 2 \, ,$$

which is not enough to explain the difference between D^+ (which has only Γ_D) and D^0 life times.

(ii) Gluon as Part of the Wave Function

The preceding discussion is a first order QCD calculation which becomes highly questionable if gluonic effects are large ($f_- \gg 1$). Most likely the gluons become part of the wave function. In this case a D^o has besides the spin singlet ($\bar{u}c$) configuration also a ($\bar{u}c$) G component, where the quarks have to be in a color-octet spin-triplet state to make with the gluon a spin-0 color-singlet,

$$\psi_d = \psi_o(\bar{u}c) + \psi_1(\bar{u}\ c\ G)\ . \tag{4.25}$$

In the decay ψ_o can be neglected because of helicity suppression. Treating the gluon as a spectator the decay of the ψ_1 part of Equ.(4.25) is essentially the decay $(\bar{c}u)_{Triplet} \to \bar{s}d$ which we compute analogously to $\rho \to e^+e^-$,

$$\Gamma(\rho \to e^+e^-) = \frac{4\pi\alpha^2}{3} \frac{F_\rho^2}{m_\rho}\ . \tag{4.26}$$

For weak interactions we replace α by $\frac{Gm_\rho^2}{4\pi\sqrt{2}}$ and multiply with the factor $(\frac{1}{2}(f_+ + f_-))^2$ in order to get

$$\Gamma_W(D \to \text{HADRONS}) = \frac{G^2}{6\pi} m_D^3 F_D^2 \left(\frac{f_+ + f_-}{2}\right)^2\ . \tag{4.27}$$

F_D in Equ.(4.26) need not be the same as in Equ.(4.24). To estimate F_D we use the experimental observa that

$$\Gamma(V \to e^+e^-) = \langle Q_q\rangle^2 \Gamma_o \tag{4.28}$$

is true for any vector meson decay ($\rho, \omega, \phi, \psi, y$) with an universal Γ_o. Comparing (4.28) with (4.26) this can be only true if $F_V^2 = \frac{m_V}{m_\rho} F_\rho^2$ holds [48]. Assuming F_D to be the same as for vector meson of the same mass, Fritzsch et al. [49] get for the D-meson with a 25 % admixture of ψ_1 in the D-wave function

$$\Gamma_W \sim 3\Gamma_D \tag{4.29}$$

in reasonable agreement of my own interpretation of the

rather confused situation of the data [46] .

We have seen that gluonic effects may save the W-exchange picture from helicity suppression. On the other side one has opened a box of Pandora. If D contain already 25 % gluon mixture, light mesons and baryons will have even more. For mesons this is not of immediate danger since most of the predictions of the quark model depend only on the flavor content and not on the color content. However, for baryons the porbabilities for the proton from Equ.(1.9) crucially depend on the totally antisymmetric color wave function, which will change for quark color octet configurations. If one assumes that the only additional term for gluon admixture in the Hamiltonian H_o describing the spectrum is of the form [50]

$$H = H_o + \lambda . \bar{B}^c \psi^+ \bar{\sigma} \lambda^c \psi \tag{4.30}$$

where the color quark magnetic moment $\psi^+ \bar{\sigma}\lambda^c\psi$ interacts with a semiclassical color magnetic field, one can show that the ground state has the same flavor properties as the usual 3q ground state and that for p,Σ ,Λ the 3q G configuration actually dominates. Also the particular color magnetic interaction (4.30) solves the problem, why the so-called $\vec{L}.\vec{s}$ terms [51] are absent in the P-wave excited states.

We close this section by listing some more predictions of the W-exchange model. D^o and F^+ can decay via W-exchange, D^+ not. From the color properties of H_I,Equ.(4.9), we get the ratio [49] between D^o and F^+

$$\frac{\Gamma_H(F^+)}{\Gamma_H(D^o)} = \left(\frac{f_+-f_-}{f_++f_-}\right)^2 \sim \frac{1}{4} \quad ;$$

so we expect:

$$\tau(D^+) \quad > \quad \tau(F^+) \quad > \quad \tau(D^o)$$

$$\text{Exp.}[44][8\pm 2 \qquad 2^{+2}_{-1} \qquad 3\pm 2].10^{-13}\ \text{sec} \tag{4.31}$$

which needs better data to be checked. The Cabibbo suppressed D^+ decay can go via W-exchange ($c\bar{d} \to u\bar{d}$ is allowed). Therefore we expect

$$\Gamma_H(D^+ \to X(s=0)) > tg^2\theta_c\ \Gamma(D^+ \to X(s=-1))\ . \tag{4.32}$$

At the B-mass W-exchange is still important. Inserting the empirical rule $F_B^2 = \frac{m_B}{m_D} F_D^2$ into (4.27) we find for the relative importance of Γ_W and the quark decay rate Γ_D at the B-meson mass

$$\Gamma_W(D) : \Gamma_D(D) = \frac{m_B}{m_D} (\Gamma_W(B) : \Gamma_D(B))\ . \tag{4.33}$$

From $m_D/m_B \sim 0.4$ we see that $\Gamma_W(B) \sim \Gamma_D(B)$. A very decisive test, whether Γ_D or Γ_W dominates, comes from the F^+ decay. With W-exchange $F^+ = c\bar{s}$ changes into $u\bar{d}$, which will decay into π's, whereas quark decay leads to an $s\bar{s}\ u\bar{d}$ final quark state which should lead to a surplus of K and η.

From this whole discussion it should have become clear that in nonleptonic decays we mainly test our prejudices about the quark model and not the basic weak interaction. The main progress over the last years seems to be that at least qualitatively we understand the large difference in life times between D^o, K^o and D^+, K^+. By that, unfortunately, we ruined the $\Delta I = \frac{1}{2}$ rule as group theoretical property because it works in different cases for different reasons. On the other side one day we have to understand how hadrons are made out of quarks. For that, as repeatedly advertised by Lipkin , "Drink nonleptonic" may be a useful tool.

APPENDIX : SU(6) Quark Probabilities

The probabilities (1.13) can be derived from simple linear equations. From charge and isospin for the proton we must have

$$p_{u\uparrow} + p_{u\downarrow} = 2 \tag{A.1}$$

$$p_{d\uparrow} + p_{d\downarrow} = 1 \quad . \tag{A.2}$$

The third component of the total spin is $\frac{1}{2}$, leading to

$$p_{u\uparrow} + p_{d\uparrow} - p_{u\downarrow} - p_{d\downarrow} = 1 \quad . \tag{A.3}$$

A fourth equation comes from the total spin. Due to the symmetry of the wave function the two identical u-quarks are in a spin s = 1 state. For d↑ its third component s_3 has the value $s_3 = 0$, whereas d↓ requires $s_3 = 1$. In a total spin $I = \frac{1}{2}$ state the ratio of $p_{d\uparrow}/p_{d\downarrow}$ must be equal to the square of Clebsch-Gordon coefficients

$$\frac{p_{d\uparrow}}{p_{d\downarrow}} = \left| \frac{\langle \frac{1}{2}\ \frac{1}{2}\ 1 0 | \frac{1}{2}\ \frac{1}{2} \rangle}{\langle \frac{1}{2}\ -\frac{1}{2}\ 1 1 | \frac{1}{2}\ \frac{1}{2} \rangle} \right|^2 = \frac{1}{2} \quad . \tag{A.4}$$

(A.1), (A.2), (A.3) and (A.4) are four linear equations whose solution gives the probabilities Equ.(1.13). The corresponding probabilities for $^{\pm}$, Ξ^0 and N can be obtained by interchanging the quarks. $p_i(\ ^0)$ are given by the isospin sum rule

$$p_i(\ ^0) = \frac{1}{2}(p_i(\ ^+) + p_i(\ ^-)) \quad . \tag{A.5}$$

For the Λ we use the fact that the ud pair is in an I = O state, leading to

$$p_{s\uparrow} = 1, \qquad p_{s\downarrow} = 0, \qquad p_{u\uparrow} = p_{u\downarrow} = p_{d\uparrow} = p_{d\downarrow} = \frac{1}{2} \quad . \tag{A.6}$$

REFERENCES

1. S. Weinberg, Phys. Rev. Lett. 19, (1967) 1264 .
A. Salam, In "Elementary Particle Theory", Ed. N. Svartholm.Almquist and Wiksell,Förlag AB, Stockholm, 1968.
2. f.i. see O.W. Greenberg, J. Sucher, Phys. Lett. 99B, (1981) 339; H. Harari, N. Seiberg, Weizmann preprint WIS-81/38; H. Fritzsch, G. Mandelbaum, Phys. Lett. 102B, (1981) 319; for recent review see M.E. Peskin, Proc. of the Lepton-Photon Conf. 1981, Ed. W. Pfeil, Bonn 1981.
3. J. Kim et al., Rev. Mod. Phys. 52, (1981) 211.
4. N. Cabibbo, Phys. Rev. Lett.12, (1964) 1324;
M. Gell-Mann, Phys. Rev. Lett.12, (1964) 155;
M. Kobayashi and K. Maskawa, Progr. Theor. Phys. 49, (1973) 652.
5. G. Morpurgo, Physics 2, (1965) 95.
6. G.R. Farrar and D.R. Jackson, Phys. Rev. Lett. 35, (1975) 1416.
7. R.D. Field and R.P. Feynman, Nucl. Phys. B136 (1978) 1.
8. A. Bodek et al., Phys. Rev. D20, (1979) 1471; J.J. Aubert et al., EMC Collaboration, CERN preprint.
9. S. Derenzo, Phys. Rev. 181, (1969) 1854.
10. CDHS Group, Proc. of the Bonn Conf. 1981.
11. See J.g. Branson, Proc. of the Bonn Conf. 1981, p. 288.
12. CELLO Collaboration, to be published.
13. See Ref. [3], p. 279.
14. H. Dahmen, J. Nölle and L. Schülke, Siegen preprint 1978.
15. E.J. Konopinski and H. Mahmond, Phys. Rev. 92, (1958) 1045.
16. See lecture of Professor Ecker.
17. See Ref. [2].
18. PDG, Rev. Mod. Phys. 52, (1980) No. 2; see also K. Kleinknecht, Review London Conf. 1974.
19. A.K. Yaouanc et al., Phys. Rev. D15 (1977) 844; J.F. Donoghue et al., Phys. Rev. D12, (1975) 2875; A.K. Yaouanc et al., Phys. Rev. D9, (1974) 1415.

20. A. Sirlin, Nucl. Phys. B71, (1974) 29.
21. D.H. Wilkinson, Nature 257 (1975) 189.
22. R.E. Shrock and L. Wang, Phys. Rev. Lett. 41, (1978) 1692.
23. See for example H. Fritzsch and P. Minkowski, Phys. Reports 73, (1981) 68.
24. C.G. Callan and D.G. Gross, Phys. Rev. Lett. 22 (1969) 156.
25. A. Bodek et al., Phys. Rev. D20, (1979) 1471.
26. S. Stein et al., Phys. Rev. D12, (1975) 1884.
27. Gargamelle data, Phys. Lett. 84B, (1979) 281.
28. H. Abramowicz et al. (CDHS), CERN-EP/81-163.
29. CDHS, J. Knobloch et al., Proc. Neutrino Conf., Hawaii 1981.
30. CHARM, J.V. Allaby et al., Proc. Neutrino Conf. Hawaii 1981.
31. P. Langacher et al., Neutrino Conf., Bergen 1979, p. 276.
32. J. Blietschau et al., Nucl. Phys. B133, (1978) 205.
33. C.J.Prescott et al., Phys. Lett. 84B, (1979) 524.
34. R.F. Marshak, Riezzudin and C.P. Ryan, "Theory of Weak Interactions", Wiley
35. R.J. Schnitzer, Phys. Rev. D13, (1976) 74.
36. M.K. Gaillard and B.W.Lee, Phys. Rev. Lett. 33, (1974) 108; G. Altarelli and L. Maiani, Phys. Lett. 52B, (1974) 351.
37. G. Altarelli, N. Cabibbo and L. Maiani, Nucl. Phys. B88, (1975) 285.
38. For a recent review see L. Foà, Bonn Conf. 1981, p.775.
39. See talk of F. Pauss at this conference.
40. Branching ratios taken from Ref. [18]. In the case of K^+ the pure leptonic decay $\Gamma(K^+ \to \mu^+ \nu)$ has been omitted.
41. N. Cabibbo and L. Maiani, Phys. Lett. 52B, (1974) 351.
42. B. Guberina, S. Nussinov, R.D. Peccei and R. Rückl, Phys. Lett. 91B, (1980) 116.
43. See e.g. J. Leveille, Michigan preprint UM-HE 81-18 (1981).
44. J.C. Patt and C.H. Woo, Phys. Rev. D3, (1971) 2920.
45. G. Altarelli et al., Nucl. Phys. B88, (1975) 285.
R.L. Kingsley et al., Phys. Rev. D11, (1975) 1919.

46. J. Kirkby, Proc. of the Intern. Symp. on Lepton and Photon Interactions at High Energy, FNAL, Batavia, IL (1979).
47. G. Trilling, Talk given at the Intern. Conf. on High Energy Physics, Madison, Wisconsin (1980).
48. M. Bander, D. Silverman and A. Soni, Phys. Rev. Lett. 44, (1979) 7.
49. H. Fritzsch and P. Minkowski, Phys. Lett. 90B, (1980) 455.
50. F. Wagner, Moriond Conf. 1981, Vol. II, 409, G. Schreier and F. Wagner, to be published.
51. N. Isgur and G. Karl, Phys. Rev. D18, (1978) 4187.

Acta Physica Austriaca, Suppl. XXIV, 203–228 (1982)

WEAK DECAYS OF OPEN AND HIDDEN TOP[+]

by

J.H. KÜHN
Max-Planck-Institut für Physik und Astrophysik
Munich, Fed. Rep. Germany

INTRODUCTION

Within our present standard model of electroweak interaction it is most natural to expect top: a sixth quark which - together with the well established b - completes the third weak isodoublet of quarks. Experimentally no signal for toponium has been found up to 37 GeV. Although this may be disappointing one hand, the perspectives to test new and interesting physics get increasingly exciting with increasing quarkonium mass. A variety of QCD predictions will be tested in a decisive manner. The quarkonium potential, which determines levels, wave functions and transition rates, will be measured over a rather wide range. Short distance properties of QCD will be investigated through the study of hadronic decays, in particular into three gluons. Although far cleaner and more convincing, these investigations will be still quite similar to the corresponding ones for the $b\bar{b}$ systems.

[+]Lectures given at the XXI. Internationale Universitätswochen für Kernpyhsik, Schladming, Austria, February 25-March 6, 1982.

There is however a completely new aspect, which becomes rapidly important above 40 GeV - weak effect on toponium. Already for ψ and Y strong and electromagnetic decays are of comparable strength. On the other hand we expect that weak and electromagnetic interactions will be equally important at energies above 50-100 GeV. We thus predict for toponium an extremely exciting interplay between weak, electromagnetic and strong interactions [1-7] and may easily find situations where their relative strength is even inverted.

Evidently toponium will be very well suited to study the coupling of the neutral current to heavy quarks - for relatively low masses ($\lesssim$ 70 GeV) mainly through γ-Z_o interference, for masses closer to Z_o neutral current decays will be even dominant.

In addition weak decays of a single quark will become increasingly important and above $M_{t\bar{t}} \gtrsim 100$ GeV they dominate all competing modes. The situation becomes even more extreme for the semiweak decay $t \rightarrow Wb$ once the t-mass is above threshold. Then the t-quark will decay so rapidly that it has not sufficient time to form T-mesons together with lighter quarks.

I should stress that already now a detailed discussion of T meson and toponium decays is not of purely academic interest. Within the next five years the energy region accessible in e^+e^- annihilation will be expanded drastically. PETRA will soon increase its energy up to 45 GeV and eventually even to 60 GeV. TRISTAN (60 GeV) and LEP (120-180 GeV) will come into operation. Linear colliders with energies up to 1 TeV are under serious consideration. Thus it is extremely important for the planning of experiments and the design of detectors to have clear predictions within the context of the standard models - despite the fact that we may hope for the unexpected.

The following discussion of weak decays of open and

hidden top will be organized as follows: I will firstly treat in some detail electroweak annihilation decays (Sect. Ia). Then I will briefly comment on parity mixing in toponium (Sect. Ib). Single quark decays as means to determine t-quark lifetimes will be discussed in Sect.Ic. Decay properties of top mesons will be studied in Sect.II. Finally we will throw a glance towards ultrahigh energies - accessible eventually with linear colliders-and present some predictions for decays of superheavy quarks (Sect.III).

I. WEAK EFFECTS ON TOPONIUM

a. Annihilation Decays

Perhaps the most important step after the discovery of a new flavor is to determine whether the new quark fits again into the repeatedly successful scheme of weak isodoublet generations. The quark charge is measured most easily in e^+e^- annihilation through the step in R or through the electronic width of the lowest lying $^3S_1$ state $\Gamma(^3S_1 \rightarrow e^-e^+)$.

The neutral current coupling also requires the interaction of quark and antiquark, and therefore toponium is the right system to investigate. In principle there are three amplitudes which contribute to the decay into a lighter fermion-antifermion pair: The annihilation through a virtual photon (Fig. 1a), through a virtual Z_0 (Fig.1b), and through W exchange (Fig.1c). All these three amplitudes lead to the same final states. Thus they interfere, partly constructively. They are all proportional to the wave function at the origin, and therefore their relative strength is fixed independent of potential model dynamics.

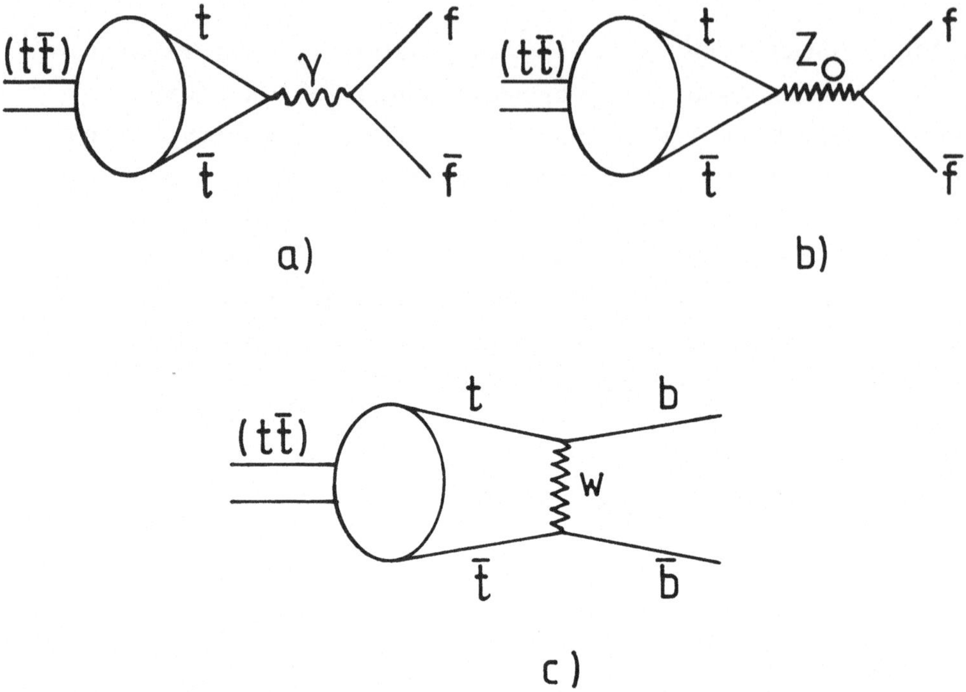

Fig. 1. Feynman diagrams for the decay of toponium into a fermion-antifermion pair. a) annihilation through the virtual photon, b) through the virtual Z_o, c) through W exchange.

If we confine ourselves for the moment to leptons as final states, then only the first two amplitudes contribute (Fig. 2). Since it is only the vector part of the neutral current which couples to the $^3S_1$ state, all interference effects depend on g_v^t, the neutral current vector coupling constant.

For masses below 60 GeV this amplitude is still relatively small and therefore will be observed mainly through its interference with the dominant electromagnetic current [8-11]. This interference term is proportional to $(A_{NC}^{lepton} + V_{NC}^{lepton}) V_{NC}^{t} \times J_{em}^{lepton} \; J_{em}^{t}$. The vector part of the leptonic neutral current leads only to a negligible change of the total cross section and will be neglected for the moment. The remaining parity-violating term leads to a parity-violating asymmetry, which can be studied only if the polarization of initial or final states is determined. The first option can be used with longitudinally polarized beams, the second through weak decays of τ leptons. Let me note in passing that this asymmetry is quite different from the $\mu^+\mu^-$ asymmetry in the continuum, which originates from a $A^e_{nc}\ A^\mu_{nc} \times J^e_{EM}\ J^\mu_{EM}$ interference term and is parity-conserving.

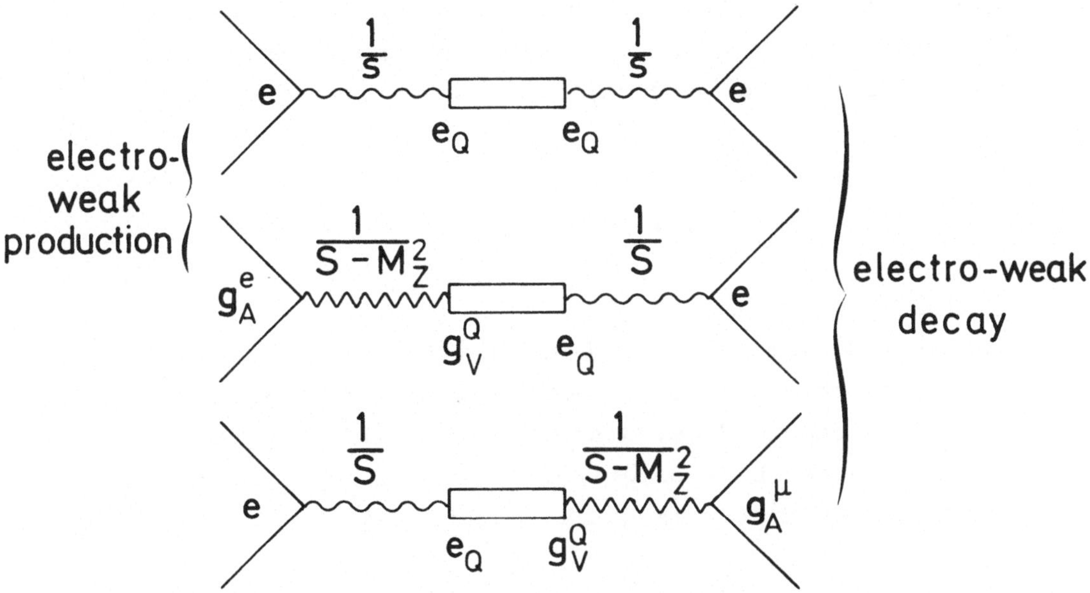

Fig. 2. Amplitudes for electroweak production of toponium and its decays into $\mu^+\mu^-$ where only lowest order neutral current effects are considered.

The production amplitude is given by

$$A = 4\pi\alpha f_V \varepsilon^\mu\ \bar{e}\gamma_\mu(1+r_e R\gamma_5)e$$

where

$$\varepsilon^{\mu} \equiv \text{polarization vector of } 1^{--}$$

$$\varepsilon_{\mu} M^2 f_V \equiv \langle 0 | J_{\mu}^{EM_Q} | 1^{--} \rangle$$

$$r_e \equiv \frac{g_A^e \, g_V^{\mu_Q}}{e \, e_Q}$$

$$R \equiv \frac{M^2}{M_Z^2 - M^2} \quad .$$

For the extreme case $r_e R = +1$ (-1) only right (left) handed electrons couple and the cross section is zero for the "wrong" polarization. In general the parity-violating asymmetry in the production cross section is

$$\delta = \frac{\sigma(+) - \sigma(-)}{\sigma(+) + \sigma(-)} = \frac{2 r_e R}{1 + (r_e R)^2}$$

which is shown in Fig. 3 ($\sin^2\theta_W = 0.23$). Longitudinally polarized beams also lead to an alignment of the resonance spin and this allows to observe a parity-violating $\cos\theta$-term in the leptonic decay distribution, which reverses its sign with the beam polarization. The integrated forward-backward asymmetry is easily calculated:

$$A_{FB} \equiv \frac{\int_0^1 d\cos\theta \frac{d\sigma}{d\cos\theta} - \int_{-1}^0 d\cos\theta \frac{d\sigma}{d\cos\theta}}{\int_{-1}^1 d\cos\theta \frac{d\sigma}{d\cos\theta}} = \frac{3}{2} \frac{r_{\mu} R}{1 + (r_{\mu} R)^2} \quad .$$

Analysis of $\tau^+\tau^-$ Final States

Even for unpolarized beams we may try to analyze the helicity of the μ- and τ-pairs through their weak decays. Particularly asymmetries in τ-decays into leptons and hadrons have been analyzed by Koniuk, Leroux and Isgur [2] and later also by others [3], [5]. The τ-net helicity

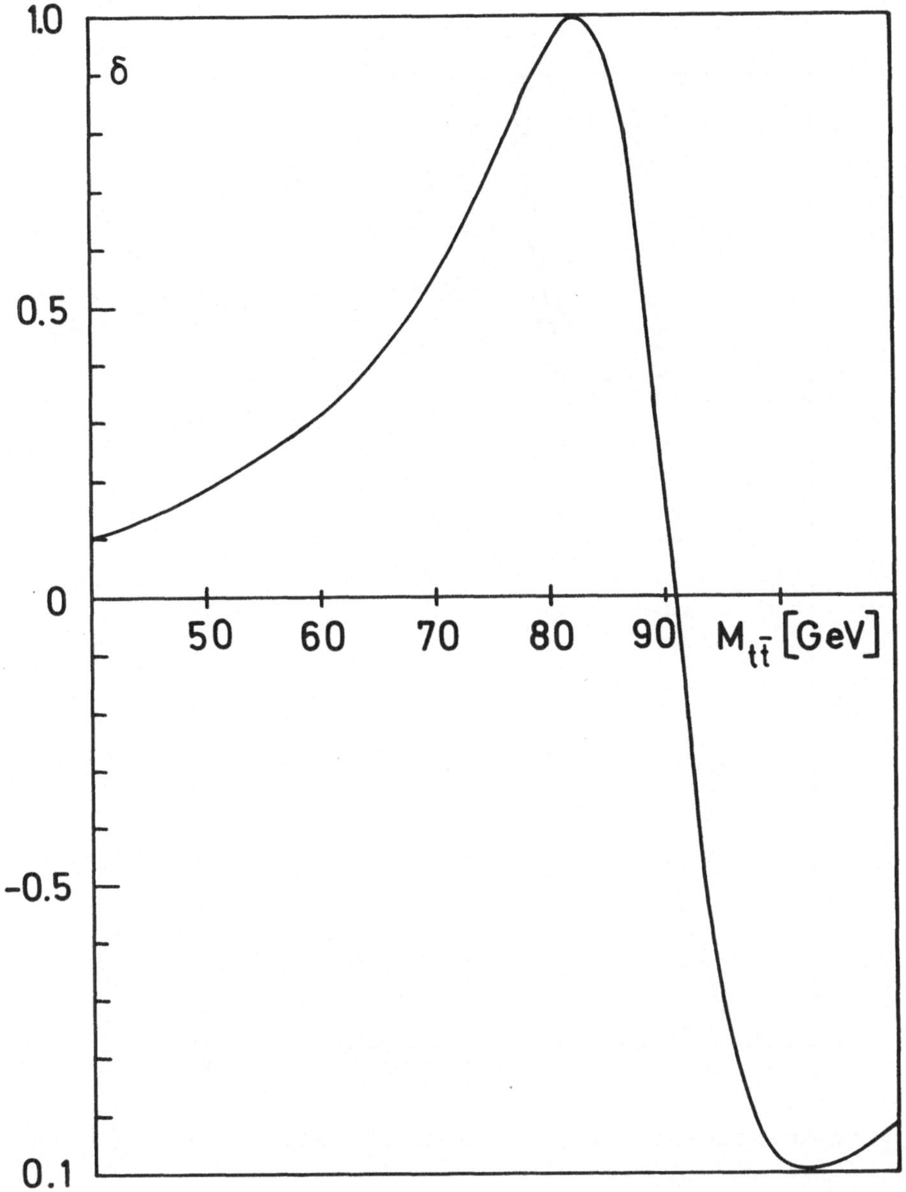

Fig.3. Relative difference between production cross section of ($t\bar{t}$) for right-handed and left-handed beams $\delta = (\sigma_R - \sigma_L)/(\sigma_R + \sigma_L)$ as a function of $M_{t\bar{t}}$.

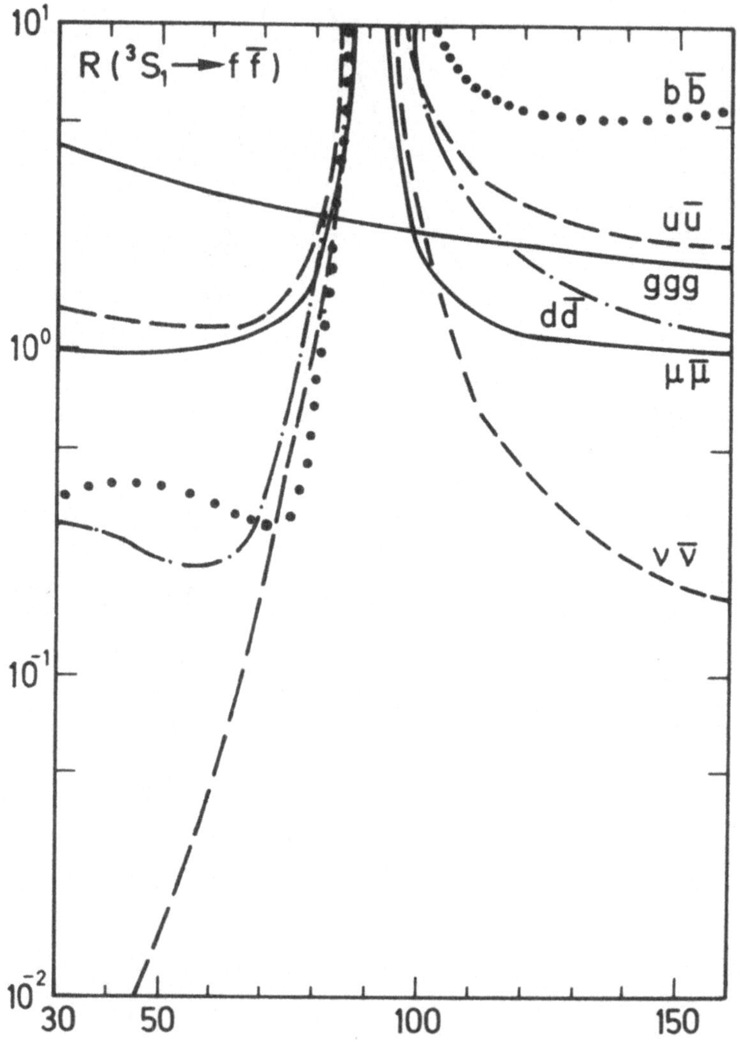

Fig. 4. Relative rates for the annihilation into various $f\bar{f}$ channels:

$$R(f\bar{f}) \equiv \frac{\Gamma(^3S_1 \xrightarrow{\gamma, Z_0, W} f\bar{f})}{\Gamma(^3S_1 \xrightarrow{\gamma} \mu^+\mu^-)} .$$

$$h(\theta) = -2R\left(r_\tau + r_e \frac{2\cos\theta}{1+\cos^2\theta}\right)$$

has an angle-independent contribution from the weak decay, proportional to r_τ, which leads to a distortion of the decay spectra. The antisymmetric contribution proportional to r_e is a result of the net polarization of the 1^{--} due to its electroweak production.

To consider the simplest case, let me discuss the $\tau\to\pi\nu$ decay in more detail. The pion is emitted preferentially in the direction of the τ-spin. $r_\tau \neq 0$ therefore leads to an angle independent net helicity of τ, which in turn implies a θ-independent linear term in the energy distribution of the pion. $r^e \neq 0$ on the other hand leads to a θ-dependent distoıtion of dN/dE_π, which implies an energy dependent forward-backward asymmetry in the pion distribution:

$$\frac{\int_0^1 d\cos\theta \frac{d\sigma}{d\cos\theta dE_\pi} - \int_{-1}^{0} \ldots\ldots}{\int_{-1}^{1} \ldots\ldots} = -\frac{3}{2} r_e R \frac{E_\pi}{E_\tau} .$$

Similar asymmetries have been analyzed also for the other decay modes of τ[9].

Electroweak Annihilation Rates

As mentioned previously, weak effects modify the rates only modestly for toponium masses up to 60-70 GeV. This changes quite rapidly once we approach M_Z. We first find a rather complicated interplay between all three amplitudes of Fig. 1, with partly constructive partly destructive interference as shown in Fig. 4. Between 80 and 110 GeV the annihilation through Z_o dominates completely. The same is of course true for the continuum background of the toponium resonance. Thus the resonance will be indistinguishable from the background if we look for differences in the final state like event shapes, and manifest itself only through the change in the cross section.

Higgs-γ-Decay

One of the most intriguing problems in the context of gauge models for electroweak interactions is to verify the existence of the predicted Higgs boson. Among the various reactions, which might serve this purpose, the decay of toponium into the Higgs boson and a photon is particularly suited, if the Higgs mass lies below $M_{t\bar{t}}$ (Fig. 5).

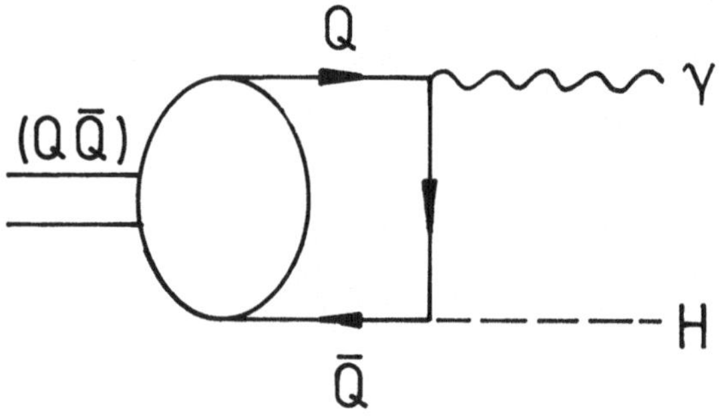

Fig. 5 . Feynman diagram for the Higgs-γ-decay.

A monochromatic photon line of energy $E_\gamma = (M_{t\bar{t}}^2 - M_H^2)/2M_{t\bar{t}}$ would be a characteristic signature for this two-body decay, and the Higgs could be furthermore identified through its characteristic decay modes into the heaviest available flavors $-\tau^+\tau^-$, and charmed mesons for M_H below $B\bar{B}$ threshold and $B\bar{B}$ above $\gtrsim 11$ GeV. The relative rate for this semiweak process should be increasingly important for heavier quarkonia, since the Higgs-$t\bar{t}$ coupling is proportional to the quark mass. The decay is only possible for states with negative charge conjugation. The $^3S_1$ decay has been calculated previously [12],

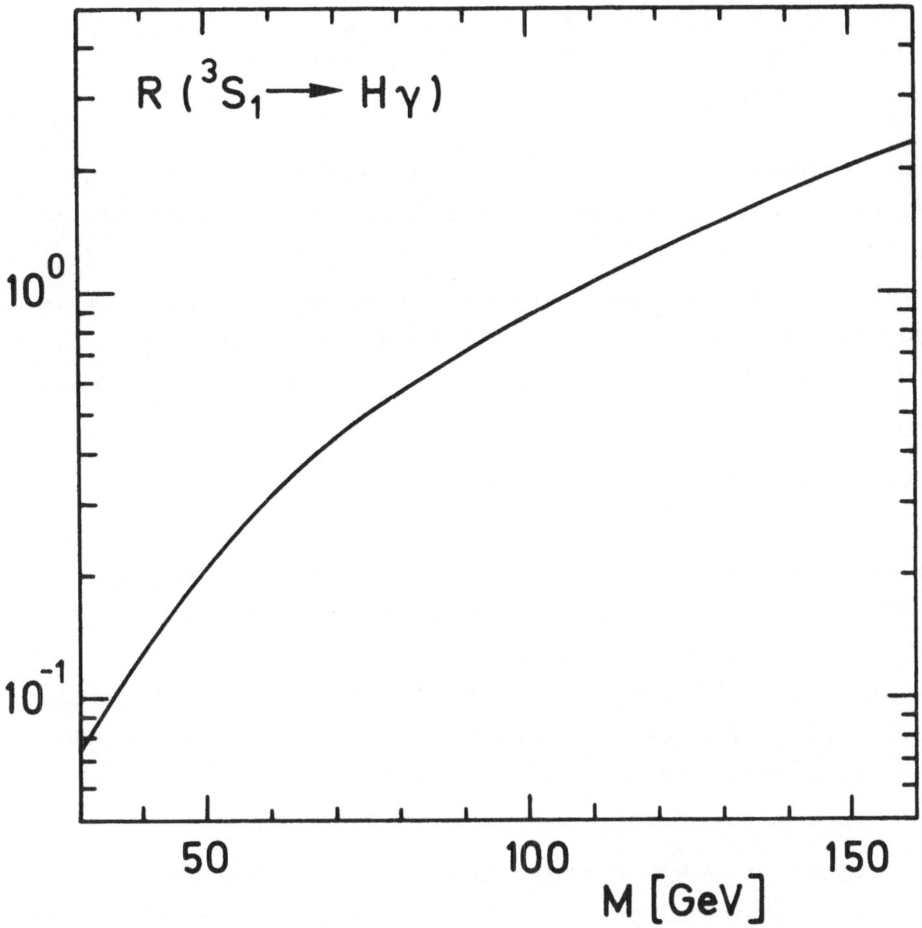

Fig. 6. Relative decay rate $R(H\gamma) \equiv \Gamma(^3S_1 \rightarrow H\gamma)/\Gamma(^3S_1 \overset{\Upsilon}{\rightarrow} \mu^+\mu^-)$.

$$\Gamma(H\gamma) \equiv \frac{\Gamma({}^3S_1 \to H\gamma)}{\Gamma({}^3S_{1\gamma} \to \mu^+\mu^-)} = \frac{G_F}{\sqrt{2}\ 4\pi\alpha}\ (M^2 - M_H^2)\ ,$$

and is shown in Fig. 6. A similar result holds for the ${}^1P_1$ decay [1]. In effect this implies a branching ratio of roughly 2 % for $M_{t\bar{t}}$ = 50 GeV and M_H = 10 GeV, a rate which should not impose serious experimental difficulties. Thus toponium will either lead to quite restrictive limits on the existence of the Higgs or to its discovery.

Ib. Parity Mixing in Toponium

It has been predicted a long time ago that one of the observable effects of neutral currents be the mixing of atomic levels with different parities. Such a parity violation has been observed experimentally some time ago despite the smallness of the relevant energies and mixing angles [13]. For heavy quarkonia the typical masses and momenta are far larger than for atoms and one might therefore expect relatively larger mixing angles.

Quarkonia are eigenstates of parity and charge conjugation, which however may mix through the neutral current. It is in fact the violation of C which could help to detect the mixing: The 2 ${}^3S_1$ state ($J^{PC} = 1^{--}$) mixes with the 1 ${}^3P_1$ state ($J^{PC} = 1^{++}$) and through this the direct decay (2 ${}^3S_1$) → (1 ${}^3S_1$) is allowed.

In the standard quarkonium spectroscopy the levels are eigenstates of J^{PC}. The P and C violating neutral current will lead to transitions between levels of different P and C; the total angular momentum J however has to remain unchanged. Thus one might expect mixing of the lowest lying states ${}^1S_0$ (0^{-+}) and ${}^3P_0$ (0^{++}) or of ${}^3S_1$ (1^{--}), ${}^3P_1$ (1^{++}) and ${}^1P_1$(1^{-+}). Since we are furthermore interested only in CP conserving interactions, the only remaining possibility is the mixing between vector and axial states, and I shall concentrate on the lowest levels $1\ ^3S_1$, $2\ ^3S_1$ and $1\ ^3P_1$.

The mixing angles are defined through

$$|\tilde{1}\rangle = |1\rangle + \theta_1 \, |P\rangle$$

$$|\tilde{2}\rangle = |1\rangle + \theta_2 \, |P\rangle$$

where I used the shorthand notation $1 = 1\ ^3S_1$, $2 = 2\ ^3S_1$, $P = {}^3P_1$ for the unperturbed levels, and the twiddle indicates the new eigenstates. The mixing angle can be calculated for small θ in lowest order perturbation theory,

$$\theta_N = \frac{\langle N|\, H_{NC} \,|P\rangle}{\sqrt{2M_N}\,(M_N - M_P)\,\sqrt{2M_P}} \qquad N = 1,2 \quad .$$

There are two diagrams which contribute - s and t channel exchange of Z_o.

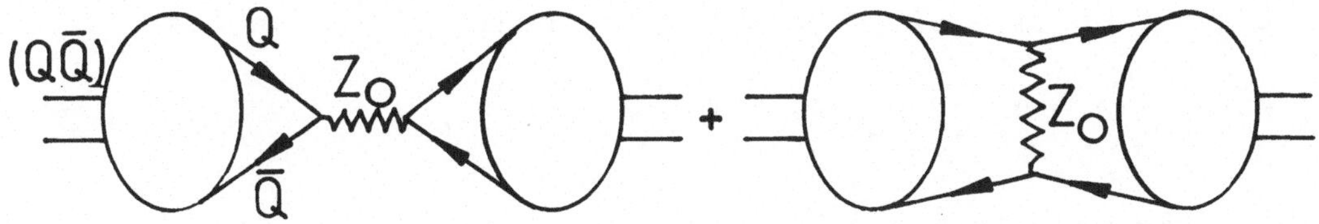

Fig. 7

If one evaluates these amplitudes for quarkonia of masses even up the 200 GeV, M_Z is still far larger than the typical relative momentum, and the t-channel exchange can be approximated by an effective four fermion interaction. The relevant amplitudes are then evaluated in a straight-forward manner,

$$\langle N|H_{NC}|P\rangle = f_{V,N} g_V M_N^2 \left(\frac{1}{M_Z^2 - M_{t\bar{t}}^2} + \frac{1}{3}\frac{1}{M_Z^2}\right) f_A g_A M_P^2 \quad .$$

How could one observe this mixing? There is in fact one signal, namely, the direct decay $2\ ^3S_1 \to 1\ ^3S_1 + \gamma$. It violates charge conjugation and is forbidden if weak interactions are ignored. In contrast to atomic physics, where neutral currents induce parity violation, here a C violation decay is the signal to look for.

The relevant rates can be calculated in a model independent way and expressed through the width $\Gamma(P \to 1+\gamma)$ and $\Gamma(2S \to P+\gamma)$, f_V, f_A, and the parameters of the Weinberg-Salam model. The mixing of P with 1 and 2 contributes, and in fact the two amplitudes interfere always constructively. For superheavy states, the main contribution will come from the 2-P mixing due to their small mass difference in a nearly Coulombic potential.

The transition amplitude $\langle\tilde{2}|H_{EM}|1\rangle$ for the dipole transition is easily evaluated for small mixing angles:

$$\langle\tilde{2}|H_{EM}|\tilde{1}\rangle = \theta_2 \langle P|H_{EM}|1\rangle + \theta_1 \langle 2|H_{EM}|P\rangle \ ,$$

and the rate is given by

$$\Gamma(\tilde{2} \to \tilde{1}\gamma) = (M_2 - M_1)^3 .$$

$$\left(|\theta_2|\sqrt{\frac{\Gamma(P \to 1 + \gamma)}{(M_P - M_1)^3}} + |\theta_1| \sqrt{\frac{\Gamma(2 \to P + \gamma)}{(M_2 - M_P)^3}}\right)^2 .$$

To evaluate this numerically one needs information about the wave functions of all three states under discussion. A realistic estimate would necessarily rely on extensive numerical calculations - and would be untrustable because of the unknown short distance behavior of the potential. However, once the various quarkonium levels and their radiative transitions have been experimentally observed, f_{V1}, f_{V2} and the mass splitting are readily determined, and the only missing quantity f_A can be calculated rather accurately.

To illustrate the order of magnitude, the ratio $\Gamma(2 \to 1 + \gamma)/\Gamma(2 \xrightarrow[\gamma,Z]{} \mu^+\mu^-)$ has been evaluated in a variety of models. The rates are displayed in Fig. 8. They are relatively insensitive to model details and in any case prohibitively small. The reason for this smallness can be traced to the fact that the relevant energy scale for these weak effects on the wave function is not $M_{t\bar{t}}$, but rather the characteristic momentum for the wave function, which, even for $M_{t\bar{t}} \gtrsim 100$ GeV, is still rather small compared to the characteristic energy scale for electroweak interactions.

Ic. Single Quark Decays

For extremely massive quarks not only top mesons decay weakly, but weak decays of a single quark in toponium will compete favorably with annihilation decays of toponium into three gluons or into a fermion-antifermion pair. The rate for this process [1]

$$\Gamma_{SQD} = 2 \cdot 9 \frac{G_F^2 m_t^5}{192\pi^3} f\left(\frac{m_t^2}{M_W^2}, \frac{m_b^2}{m_t^2}\right) ,$$

$$f(\rho,\mu) = 2 \int_0^{(1-\sqrt{\mu})^2} du \left(\frac{1}{1-u\rho}\right)^2 [(1-\mu)^2 + u(1+\mu) - 2u^2] \sqrt{\Delta(1,\mu^2,u^2)}$$

$$= \begin{cases} (1-\mu^2)(1-8\mu+\mu^2) - 12\mu^2 \ln\mu & \text{for } \rho = 0 \\ 2\rho^{-4}[6(\rho+(1-\rho)\ln(1-\rho)) - 3\rho^2 - \rho^3] & \text{for } \mu = 0 \end{cases} ,$$

increases roughly with M^5 and thus dominates for masses beyond 110 GeV (Fig. 9).

These decays are of considerable theoretical interest: The fact that t and b form a weak isodoublet is - strictly speaking - only established, if the strength of their weak coupling is measured to be $f\sqrt{G_F M_W^2/\sqrt{2}}$, where f depends on various mixing angles and is close to 1. A T-lifetime-determination would measure this coupling but appears rather remote ($\tau \approx 10^{-18}$ s for m_t = 25 GeV).

The determination of the branching ratio for SQD's in toponium, however, would also give the desired information and - as we shall see - appears feasible for $M_{t\bar{t}}$ as low as 40 GeV and is straightforward for $M_{t\bar{t}}$ =50 GeV. Inspite of their small branching ratio (Table 1) semileptonic SQD's lead to quite striking signatures which should discriminate them against the "background" from hadronics and current-induced decays. SQD is independent of the wave function and the quantum numbers of the bound state. Thus the branching ratio for SQD's

$$Br_{SQD} = \frac{\Gamma_{SQD}}{\Gamma_{tot}} \approx \frac{\Gamma_{SQD}}{\Gamma_{ANNIHILATION}}$$

increases for radial excitations.

The production cross section times the branching ratio for SQD's

$$\sigma(e^+e^- \to (t\bar{t})) \cdot Br_{SQD} \approx$$

$$\approx \sigma(e^+e^- \to \mu^+\mu^-) \frac{9\pi}{2\alpha^2\sqrt{2\pi}} \frac{\Gamma(t\bar{t}) \to e^+e^-)}{\Delta E_{CM}} \cdot \frac{\Gamma_{SQD}}{\Gamma_{ANN}}$$

is furthermore approximately independent of the wave function and thus will be the same for all states - at least as long as annihilation decays dominate the total width.

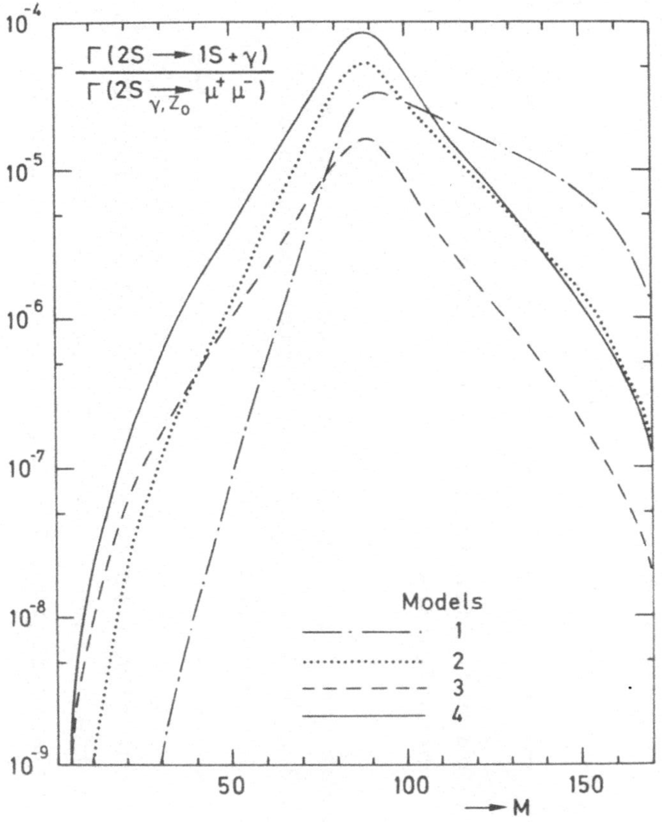

Fig. 8. Rate for the C violating decay $2S \to 1S + \gamma$ for various assumptions on the potential.

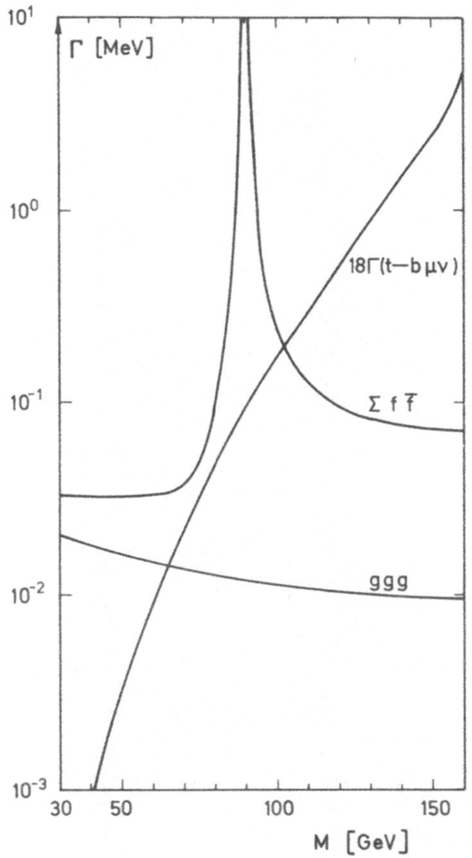

Fig. 9. Comparison of rates for SQD, hadronic decays and fermionic decays as a function of $M_{t\bar{t}}$.

Hard Leptons from SQD [7]

Hard leptons from the decay of toponium

$(t\bar{t}) \rightarrow T + \bar{B} + \ell + X$

or from the subsequent decay of top mesons

$(t\bar{t}) \rightarrow B + \bar{T} + X$
$\quad\quad\quad\quad\quad \hookrightarrow \ell + X$

lead to especially clean signature for SQD's.

Decay Mode / $M_{t\bar{t}}$ GeV	ggg	e^+e^- $+\mu^+\mu^-$ $+\tau^+\tau^-$	$c\bar{c}$	$u\bar{u}$ $+d\bar{d}$ $+s\bar{s}$	$b\bar{b}$	SQD	prompt inclusive $e^{\pm} + \mu^{\pm}$ yield from SQD
50	31	28.8	11.8	16.4	3.9	6.2	2.8
70	18	21.6	8.2	12.8	2.2	30.7	13.6

Table 1. Branching fraction of $(t\bar{t})$ into various channels in %.

As we can see from Fig. 10, their contribution in the region $0.3<\xi< 0.5$ is larger by a factor two than our estimate for the background from $(t\bar{t}) \xrightarrow{\gamma, Z} c\bar{c}$ or $b\bar{b} \rightarrow \ell^{\pm} + X$. Note that this background can be determined experimentally from a measurement of lepton production off resonance. For $M_{t\bar{t}} = 70$ GeV our signal is enhanced by a relative factor 6 and thus is the dominant source of leptons in nearly the whole kinematic region. The lepton background from $b\bar{b}$

and $c\bar{c}$ can be reduced further if one uses the differences in the shapes of events from SQD's and fermionic decays. Suitable cuts in acoplanarity and thrust might allow to measure prompt leptons even for $M_{t\bar{t}}$ as low as 40 GeV.

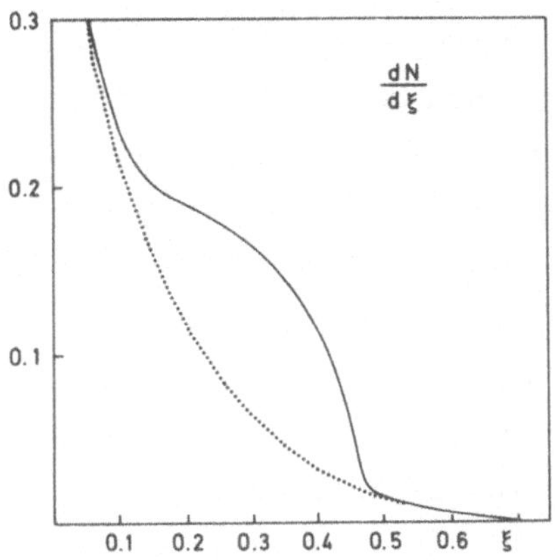

Fig. 10

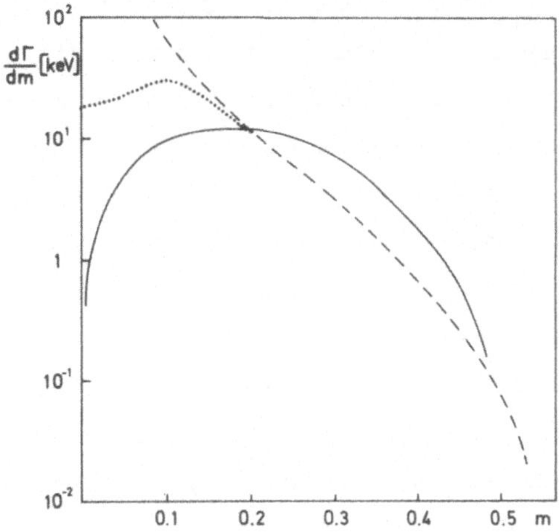

Fig. 11

Fig. 10. Lepton spectrum $dN/d\xi$ as a function of $\xi = 2E/M_{t\bar{t}}$ from the background reaction $e^+e^- \to b\bar{b}$ or $c\bar{c}$ (dotted line) and from SQD plus background (solid line) for $M_{t\bar{t}} = 50$ GeV.

Fig. 11. Minor distribution for SQD's (solid line), four-jet events from hadronic decays (dashed line), four-jet events with the additional requirement that the mass of each two-jet subsystem is larger than 1/9 $M_{t\bar{t}}$ (dotted line).

Since these leptons originate from a parity-violating V-A interaction, their angular distribution is correlated with the spin of the t-quarks and thus with the spin of toponium. In particular polarization of $(t\bar{t})$ due to γ-Z^0 interference will lead to a sizeable forward-backward asymmetry of the decay leptons.

Event Shapes [7]

Nonleptonic SQD's lead to quite characteristic events with six jets in two independent planes. ALthough we do not hope to resolve all of them individually, we expect that they will lead to rather isotropic event shapes and thus can be distinguished from the competing annihilation decays. Current induced decays lead to pencil-like events ($q\bar{q}$) with a small admixture of planar ($q\bar{q}g$) and a negligible contamination by nonplanar events ($q\bar{q}gg$ and $q\bar{q}q\bar{q}$). Hadronic decays lead to dominantly planar three-jet events with a small admixture of nonplanar four-jet events ($gggg$ and $q\bar{q}gg$) which have been studied by Koller, Streng, Walsh and Zerwas [14]. To compare their yield of nonplanar events to that from SQD's in Fig. 11, we use minor m, which vanishes for planar events ($0 \leq m \leq 1/\sqrt{3}$) and thus measures the non-planarity of the final state. We find that for $M_{t\bar{t}} = 50$ GeV the signal for nonleptonic SQD's is clearly visible. Of course there are additional signatures which enhance them even further [7]. We therefore conclude that SQD's will allow to determine the lifetime of the t-quark quite unambiguously. In fact, quite in contrast to the situation in charm decays, the measurement gets easier with decreasing lifetimes, which in our case implies an increasing branching ratio for SQD's.

II. WEAK DECAYS OF T-MESONS

Weak decays are not only increasingly important for toponium, but also for T^* they will be the dominant mode. Let me recall that the magnitude of the hyperfine splitting between singlet and triplet states decreases with increasing mass of even one of the constituents:

$$M_\rho - M_\pi = 640 \text{ MeV}$$
$$M_{K^*} - M_K = 400 \text{ MeV}$$
$$M_{D^*} - M_D = 140 \text{ MeV}$$
$$M_{B^*} - M_B \approx 50 \text{ MeV},$$

as predicted in QCD from the decreasing color magnetic interaction

$$M = C_o + C_1 \frac{\vec{S}_1 \ \vec{S}_2}{m_1 \ m_2}$$

which implies $\Delta M_{HFS} \sim 1/m_1 m_2$. Hadronic decays of D^* into $D + \pi$ are therefore barely possible, and the corresponding decays for heavier mesons are strictly forbidden by phase space.

The magnetic dipole transition, which is already quite important for D^* decays, is thus expected to be the main decay mode for B^* and will hopefully be found soon at Cornell or DESY through its sharp photon line of roughly 50 MeV. The rate for these electromagnetic transitions is given in the nonrelativistic model by

$$\Gamma = \frac{1}{3} \alpha k^3 \left(\frac{Q_1}{m_1} - \frac{Q_2}{m_2}\right)^2 ,$$

a result which is roughly consistent with the measured rates for $\rho \to \pi\gamma$, $\omega \to \pi\gamma$, $K^* \to K\gamma$, $\phi \to \eta\gamma$ and $D^* \to D\gamma$. Thus M1-rates decrease with the quark mass - of course quite in contrast to the weak decay rates which in the spectator model increase with M_t^5. Thus the ratio

$$\frac{\Gamma_W(T^*)}{\Gamma_{EM}(T^*)} = \left(\frac{m_t}{13 \text{ GeV}}\right)^8$$

exceeds unity by a large factor for $m_t \gtrsim 18$ GeV [6]. How could we test this idea? Obviously weak decays of T^* and

T lead to rather similar lepton spectra and event shapes. Thus the cascade of T^* to t with subsequent weak T decay will lead to events which are rather similar to those from direct weak decay of T^*, in particular since the absence of a rather soft (10 MeV) photon line will be difficult to prove. There is however some hope through the study of angular distributions of decay leptons. Weak decays of T will obviously lead to isotropic distributions in the T-rest frame. Leptons from weak decays of T^* however will be correlated with the spin of the T^*:

$$dN \sim (1 + \cos\theta)\, d\cos\theta$$

$$\cos\theta = \frac{\vec{p}_e \cdot \vec{S}_{T^*}}{|\vec{p}_e|}$$

and this leads to nontrivial lepton angular distributions. Let me give some examples:

(a) e^+e^- annihilation with longitudinally polarized beams close to $T\bar{T}$ threshold: at the quark level one of course expects that the quark spin is alligned with the beam direction. A t-quark with $S_Z = + 1/2$ is expected to convert into $T^*(S_Z = 1)$, $T^*(S_Z = 0)$ and $T(S = 0)$ with relative weight 2:1:1. One thus expects a sizeable forward-backward asymmetry

$$\frac{N_F(e^+) - N_B(e^+)}{N_F(e^+) + N_B(e^+)} = \frac{1}{4}$$

from the weak decays of $T^*(S_Z = 1)$.

(b) Even with unpolarized beams one helicity of the in-electrons is preferred, due to γ-Z interference, and for energies close to the $T\bar{T}$ threshold this net polarization δ is the same as the one already shown in Fig. 3. The resulting forward-backward asymmetry is in this case of course far smaller but still more than 10 % above 65 GeV.

(c) Another possibility is offered through the study of lepton-lepton angular correlations in simultaneous leptonic decays of t and $\bar{t}$. These are of the form

$$d N \sim [\frac{1}{3}(1 + \frac{1}{3} \cos\theta_{+-}) + \frac{3}{4}]d\cos\theta_{+-}$$

$$\cos \theta_{+-} \equiv (\vec{p}_+ \cdot \vec{p}_-)/(|\vec{p}_+||\vec{p}_-|) \quad ,$$

but of course the rate is diminished by the squared leptonic branching ratio.

III. SUPERHEAVY QUARKS

A quark with $m_t > m_W$ may decay semi-weakly into W and its lighter isodoublet partner. The decay width for this reaction increases rapidly with M (Fig. 12) and exceeds 1 fermi^{-1} above $m_t \gtrsim$ 100 GeV. For such a large decay width our notion of mesons breaks down. The heavy quark decays, even before it has time ($\gtrsim$ 1 fermi/c) to form a meson with spatial extension $\gtrsim$1 fermi together with a light quark out of the sea (Fig. 13). As a clear signature for such a situation one expects that the angular distributions of all decay products (W, leptons) follow the predictions of the free quark model. Remember that these distributions are quite different if the quarks firstly convert into scalar or vector mesons.

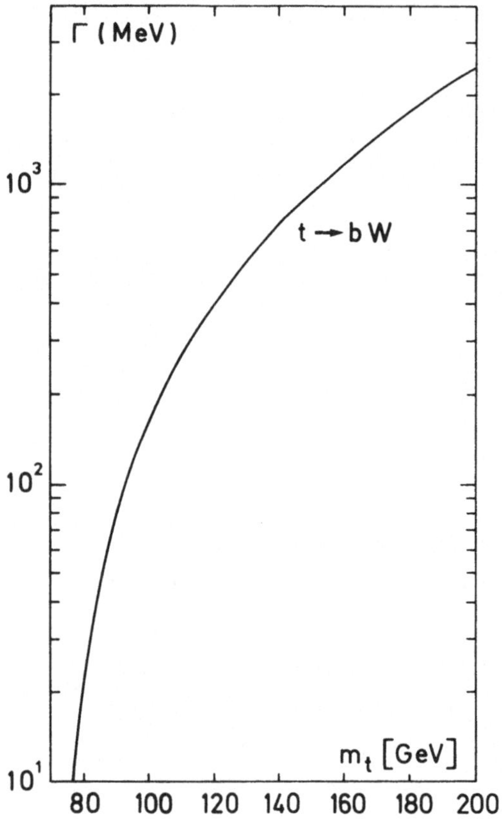

Fig. 12. Rate for the semiweak decay of t into Wb.

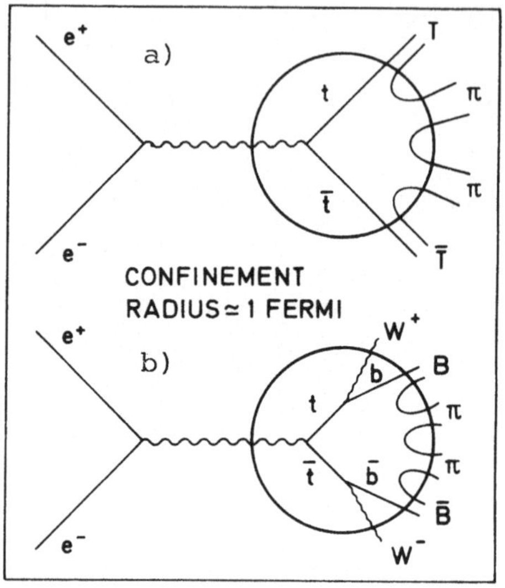

Fig. 13. Open top production in e^+e^- annihilation.

a) The weak decay width of t is far less than m_π. T-mesons are produced and decay far away from the production point.

b) The weak decay width of t is far smaller than m_π. The t-quark decays within the confinement region. No T-mesons can be produced.

Similarly the whole concept of quarkonia breaks down for $m_t \gtrsim 200$ GeV. Then the decay width of its constituents of roughly 1 GeV is comparable to the mass difference between the ground state and the radial excitations. This corresponds to the situation where the t-quark lifetime is smaller than the characteristic time of revolution. We thus arrive finally at a situation where the t-quark decays before it can participate in strong interactions.

ACKNOWLEDGEMENTS

I want to thank H.Mitter and F.Widder for the invitation to Schladming and their kind hospitality.

REFERENCES

1. For a review and detailed discussion see J.H. Kühn, "Weak Interactions of Quarkonia", lectures given at the XX. Cracow School of Theoretical Physics, May 1980, Acta Physica Polonica B12, (1981) 347.
2. J. Ellis and M.K. Gaillard, "Physics with very High Energy e^+e^- Colliding Beams", preprint CERN 76-18 (1976).
3. J. Bjorken, in Proc. of 1977 Hamburg Conference .
4. K. Fujikawa, Prog. Theor. Phys. 61, (1979) 1186.
5. G. Goggi and G. Penso, Nucl. Phys. B165, (1980) 429.
6. I.I. Bigi and H. Krasemann, Z. Phys. C7, (1981) 127.
7. J.H. Kühn and K.H. Streng, preprint MPI-PAE/PTh 62/81 (to be published in Nucl. Phys. B).
8. I.I. Bigi, J.H. Kühn and H. Schneider, preprint MPI-PAE/PTh (1978) 28/78.
9. R. Koniuk, R. Leroux and N. Isgur, Phys. Rev. D17, (1978) 2915.
10. R. Budny, Phys. Rev. D20, (1979) 2763.

11. L.M. Seghal and P. Zerwas, Nucl. Phys. B183, (1981) 417.
12. F.A. Wilczek, Phys. Rev. Lett. 39, (1977) 1304.
13. For a recent review see E.N. Fortson and L. Wilets, in Advances in Atomic and Molecular Physics, ed. B. Bederson and D.R. Bates, Academic Press, New York 1980, Vol. 16, p. 319 ff.
14. K. Koller, K.H. Streng, T.F. Walsh and P. Zerwas, SLAC preprint (1981), in preparation.

Acta Physica Austriaca, Suppl. XXIV, 229–273 (1982)

ELECTROWEAK PROCESSES BEYOND THE TREE APPROXIMATION+

by

G. ALTARELLI
Istituto di Fisica, Universitá di Roma I, Roma, Italy
Istituto Nazionale di Fisica Nucleare, Sezione di Roma,Italy

I. INTRODUCTION

This set of three lectures is intended to push further at a more advanced level the study of electroweak processes which was started in the parallel courses of Drs. Ecker and Wagner. Much work has been devoted recently to a careful analysis of the higher order corrections to the tree diagram relations on which the experimental tests of the standard model of electroweak interactions are based. The subject seems by now mature for a clear and reasonably compact presentation of methods and results. I shall devote the first part of these lectures to expose in a simple and systematic way how the most important terms in the corrections to the electroweak processes and parameters ($\sin^2\theta_W$, M_W, M_Z, and so on) arise and can be evaluated. This will lead us into a discussion of leptonic and semileptonic electroweak processes in the formalism of operator expansion and renormalization group methods. The connection with the theory

+Lectures given at the XXI. Internationale Universitätswochen für Kernphysik, Schladming,Austria,February 25 - March 6,1982 of non-leptonic weak processes is naturally introduced by the close similarity of approach and by the structure of the operator mixings among some semileptonic and non-leptonic amplitudes. In the second part of these lectures I shall therefore review the theoretical status of inclusive decay rates of heavy flavored particles. This is in fact a subject of high current interest and one that allows to present an outlook of methods and results of importance for the theory of all non-leptonic processes.

PART ONE

1.1 General Framework.

The main experimental tests of the standard model [1] of the electroweak interactions have to do with the determination of the neutral current couplings to the fundamental fermions. In particular the value of $\sin^2\theta_W$ determined from different experiments should coincide. Within experimental errors this is in fact found to be true from measurements on the ratio of neutral to charged current deep inelastic cross sections of neutrinos and antineutrinos on matter, from the left- right asymmetry in polarized electron-nucleon inelastic scattering, from neutrino and antineutrino elastic scattering on electrons, from the forward-backward asymmetry in $e^+e^- \to \mu^+\mu^-$, and, at least in principle, many more such comparisons can be made.
The analysis of each experiment is usually performed by using theoretical relations derived in the Born approximation. In this lowest order approach, for each experiment i, the resulting expression for $(\sin^2\theta_W)_{Born}$ is related to a given function R_i of the measured quantities. If all corrections to the Born approximation are negligible and the theory is correct, a single value should be obtained from each experiment:

$$R_i = (\sin^2\theta_W)_{Born} \qquad (1)$$

(independent of i). Inclusion of higher order effects change this relation into:

$$R_i = \sin^2\theta_W(\lambda) + \delta_i(\lambda) \tag{2}$$

where $\sin^2\theta_W(\lambda)$ is a theoretical quantity to be defined at the scale λ in a given renormalization scheme, and $\delta_i(\lambda)$ is a process-dependent correction to the Born approximation. To lowest order $\delta_i(\lambda)$ may contain terms

$$\delta_i(\lambda) \sim o\left(\frac{\alpha}{\pi} \ln \frac{M^2}{\lambda^2}\right) + o\left(\frac{\alpha}{\pi} \ln \frac{M^2}{\mu_i^2}\right) + o\left(\frac{\alpha}{\pi} \ln \frac{m^2}{\mu_i^2}\right) + o\left(\frac{\alpha}{\pi} \frac{1}{\sin^2\theta_W}\right) + o\left(\frac{\alpha}{\pi}\right) + \ldots \tag{3}$$

where μ_i is the energy scale of the process under consideration and M is the M_W or M_Z mass. It is convenient to choose for λ, i.e. the reference theoretical scale, $\lambda \simeq M$ which simplifies many formulae and corresponds to focus on $\sin^2\theta_W(M)$, which is of interest also for comparison with GUT predictions. Terms of order $\frac{\alpha}{\pi} \ln \frac{m^2}{m_i^2}$ are usually taken into account already by the experimental team[2] and can be in any case included among terms of order $\frac{\alpha}{\pi}$. It is then simple to see that $\delta_i(M)$ is expected to be of order

$$|\delta_i(M)| \simeq 0.02 \tag{4}$$

which makes this effect (and its sign) very important for a precise comparison of experiments with theory and with GUT predictions. For $\mu_i^2 \simeq 0.3\ \text{GeV}^2$ we obtain the following estimate of the relative importance of the various terms in the expansion of $\delta_i(\lambda)$:

$$\alpha \ln \frac{M^2}{\mu^2} \qquad (\alpha \ln \frac{M^2}{\mu^2})^2 \qquad \frac{\alpha}{\sin^2\theta_W} \equiv \alpha_W$$

$$(\sim 10)\alpha \qquad (\sim 0.75)\alpha \qquad (\sim 5)\alpha \tag{5}$$

The most important contributions are seen to arise from the leading logs and from the α_W terms. It is remarkable that a simple and universal treatment of the leading logs can be given [3,4,5] with the $(\frac{\alpha}{\pi} \ln \frac{M^2}{\mu^2})^2$ terms obtained in most cases at no extra cost, and that also the α_W terms can be extimated by a single step for all processes of interest[4]. This is what we shall describe in the following. A much more intricate job, which depends on the detailed definition of each quantity, is the evaluation of the remaining corrections of order α. These detailed calculations exist in the literature[6 7 8] for the most relevant processes. As expected they contribute little to the final answer, so that learning about the origin and the structure of the main terms is confirmed to be a significant enterprise.

At energy scales $\mu^2 << M^2$ the exact SU(3) × SU(2) × U(1) gauge theory reduces to SU(3) × U(1) plus a number of additional non-renormalizable effective interactions, typically in the form of four fermion operators. These operators specify an effective hamiltonian of interest for weak processes. The coefficient functions of this short distance expansion of the exact theory are corrected by terms of order α and/or α_s which are computable by renormalization group techniques. The computation is expecially simple at the leading logarithmic accuracy. The bare effective hamiltonian which rules the tree diagram approximation can be written as

$$H_o = \sum_a c_a^{(o)} O_a \tag{6}$$

where O_a are local operators and $c_a^{(o)}$ are c-number coefficients. The corrected hamiltonian then takes the form:

$$H(\mu) = \sum_a c_a O_a(\mu) \tag{6'}$$

and the coefficient functions c_a satisfy in general the renormalization group equation

$$\left(-\frac{\partial}{\partial t} + \beta(\alpha)\frac{\partial}{\partial \alpha} + \beta_s(\alpha_s)\frac{\partial}{\partial \alpha_s}\right) c_a + c_b \Gamma_{ba} = 0 , \tag{7}$$

where $t = \ln \frac{M^2}{\mu^2}$, $\beta(\alpha)$ and $\beta_s(\alpha_s)$ are the beta functions which determine the running of α and α_s respectively:

$$\frac{d\alpha(t)}{dt} = \beta[\alpha(t)] , \qquad \frac{d\alpha_s(t)}{dt} = \beta_s[\alpha_s(t)] \tag{8}$$

with

$$\beta(\alpha) = b_Q\alpha^2 + O(\alpha^3) , \qquad b_Q = \sum_f{}' \frac{Q_f^2}{3\pi} ,$$

$$\beta_s(\alpha_s) = -b\alpha_s^2 + O(\alpha_s^3) , \qquad b = \frac{33-2f}{12\pi} , \tag{9}$$

where Q_f are the fermion charges and f is the number of excited flavours (over which the sum $\sum_f{}'$ runs). The general solution of the renormalization group equation (7) reads:

$$c_a = c_a(t=0, \vec{\alpha}(t)) \, \bar{T}\exp \int_0^t \Gamma[\vec{\alpha}(t')]dt' , \tag{10}$$

where $\vec{\alpha} \equiv (\alpha, \alpha_s)$, Γ is the anomalous dimension matrix, and $\bar{T}$ orders t in decreasing order from left to right. In the one loop approximation

$$\Gamma(\vec{\alpha}) = \alpha\gamma + \alpha_s\gamma_s \; ; \tag{9'}$$

γ and γ are evaluated in a standard way from the one-loop renormalization of the relevant operators:

$$(O_a)_{ren} = [\delta_{ab} + (\alpha\gamma + \alpha_s \gamma_s)_{ab} \ln \Lambda^2 + \ldots](O_b)_{bare} , \qquad (10')$$

with Λ being the ultraviolet cut-off.
In determining the set of operators that can get mixed at the one-loop level it is useful to remember that SU(3) × U(1) conserves the number $(f-\bar{f})$ of quarks and leptons of each flavour and separately the number of neutrinos and antineutrinos of each kind, because neither gluons nor γ's interact with ν_i. Also a restricted parity operator P' is conserved which is identical with parity P except that it does not act on ν and $\bar{\nu}$.

1.2 The Leading Log Approximation.

In this section we consider the leading log terms for various processes in succession.
a) Muon decay. Muon decay $\mu^- \rightarrow e^- \nu_\mu \bar{\nu}_e$ is important because it provides a definition of the Fermi coupling G_F.
H_o is given by:

$$H_o = \frac{G_F^o}{\sqrt{2}} \bar{\nu}_\mu \gamma_\lambda (1-\gamma_5) \mu \; \bar{e} \gamma_\lambda (1-\gamma_5) \nu_e . \qquad (11)$$

By Fierz rearrangement we can recast H_o in the form:

$$H_o = \frac{G_F^o}{\sqrt{2}} \bar{\nu}_\mu \gamma_\lambda (1-\gamma_5) \; \nu_e \; \bar{e} \; \gamma_\lambda (1-\gamma_5) \mu . \qquad (12)$$

This is useful because the inert neutrinos are separated into one current. Thus effectively under SU(3) × U(1)H_o behaves as a one-current operator. The anomalous dimension in one loop is of course determined by photon emission only $(\gamma_5 = o)$. Moreover it is simple to realize that γ is also zero because the anomalous dimension is fixed by the massless theory and both the vector and axial vector currents $\bar{e}\gamma_\lambda\mu$ and $\bar{e}\gamma_\lambda\gamma_5\mu$ are conserved in this limit. Note that the flavour change from e to μ does not spoil the argument because it in no way alters the structure of the relevant diagrams in the massless theory. The infrared divergences which appear in the vertex and self-energy diagrams can be easily separated from the ultraviolet divergences of interest for computing γ. Thus as a result there are no leading log terms in μ decay. G_F can be defined by the relation:

$$r^{-1} = \frac{G_F^2\, m_\mu^5}{192\pi^3}\left(1 - \frac{8m_e^2}{m_\mu^2}\right)\left[1 + \frac{\alpha}{2\pi}\left(\frac{25}{4} - \pi^2\right) + o(\alpha^2)\right] , \tag{13}$$

where the order α correction is from finite terms due to γ exchange and emission and all other contributions from W and Z exchange are included in G_F.

b) <u>Semileptonic processes mediated by W:</u>

The following argument holds in general for any semileptonic process mediated by charged currents[9], but for definiteness we consider $\nu_\mu + d \rightarrow \mu^- + u$. Then

$$H_o = \frac{G_F^o}{\sqrt{2}}\, \bar{\mu}\gamma_\lambda(1-\gamma_5)\nu_\mu\; \bar{u}\gamma_\lambda(1-\gamma_5)d \ . \tag{14}$$

All virtual photon or gluon virtual one-loop diagrams within each of the two separate currents lead to zero anomalous dimension as before. The only non-vanishing contributions to γ arise from photon emission from the muon with reabsorption by either of the two quarks. Note that the ordinary scaling violations of order $\alpha_s \ln |q^2|$ are a different problem. What is relevant here is that there are no $\alpha_s \ln M^2$ terms but the only large logarithms are of the form $\alpha \ln M^2$. Thus the corrected hamiltonian is given by:

$$H(\mu) = \frac{G_F}{\sqrt{2}}\left[\frac{\alpha(M)}{\alpha(\mu)}\right]^{\gamma/b_Q} \bar{\mu}\gamma_\lambda(1-\gamma_5)\nu_\mu\bar{u}\gamma_\lambda(1-\gamma_5)d \ . \tag{15}$$

A simple explicit computation leads to:

$$\gamma = \frac{3}{4\pi} Q_u = \frac{1}{2\pi} \tag{16}$$

where $Q_u = 2/3$ is the charge of the u quark. Then

$$F \equiv \left[\frac{\alpha(M)}{\alpha(\mu)}\right]^{\gamma/b_Q} \sim \left[1 + \frac{\alpha(\mu)}{2\pi} \ln \frac{M^2}{\mu^2} + \ldots\right] . \tag{17}$$

At $\mu^2 \sim 2\ \mathrm{GeV}^2$, $(F-1) \simeq 1\%$.

c) <u>ν initiated neutral current processes.</u>

Consider

$$\nu_\mu + f \rightarrow \nu_\mu + f \tag{18}$$

where we consider ν_μ because this is the most common case in practice. The bare hamiltonian is given by:

$$H_o = \frac{G_F^o}{\sqrt{2}} \rho\, \bar{\nu}_\mu \gamma_\lambda (1-\gamma_5) \nu_\mu \quad [J_\lambda^Z + \frac{2}{\rho} \bar{\mu} \gamma_\lambda (1-\gamma_5) \mu] \tag{19}$$

where

$$\rho = \frac{M_W^2}{M_Z^2\, c^2\, (M)} , \tag{20}$$

$$J_\lambda^Z = \sum_f \bar{f}\, \gamma_\lambda [\tau_3^f (1-\gamma_5) - 4Q_f\, s^2(M)] f , \quad (\tau_3 = \pm 1) \tag{21}$$

with

$$s^2(M) \equiv \sin^2\theta_W(M) ,$$
$$c^2(M) \equiv \cos^2\theta_W(M) . \tag{22}$$

We introduce $s^2(M)$ and $c^2(M)$ in H_o because when $\mu \to M$ all leading log corrections disappear and the corrected hamiltonian reduces to H_o. In fact we just want to find the difference between $H(\mu)$ and H_o. The presence of the second term in eq.(19) arises from Fierz rearrangement of the charged current contribution.
Since the ν_μ current is completely inert it would seem that the anomalous dimension γ is zero in this case also. But this conclusion would be false here because of the penguin diagram in fig. 1.

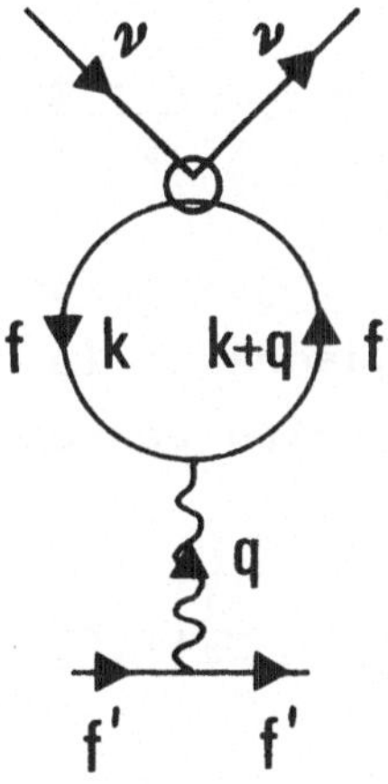

Fig. 1

Starting from the operators (A' is a flavour-diagonal colour-singlet matrix)

$$O^i = \bar{\nu}_\mu \gamma_\lambda (1-\gamma_5) \nu_\mu \bar{f} \gamma_\lambda A^i f \quad , \qquad (P'=+1)$$

$$O_5^i = \bar{\nu}_\mu \gamma_\lambda (1-\gamma_5) \nu_\mu \bar{f} \gamma_\lambda \gamma_5 A^i f \quad , \qquad (P'=-1) \qquad (23)$$

the penguin diagram mixes O^i with O^Q (i.e. the one with $A=Q_f$). Note that gluon exchange starting four the colour singlet operators O^i and O_5^i vanishes because of $\mathrm{Tr}\, t^A = 0$ where t^A is the colour matrix at the gluon coupling. Thus only photon exchange is relevant. O_5^i does not mix because the $\varepsilon_{\mu\nu\lambda\sigma}$ tensor arising from tracing the γ_5 vanishes when contracted with all tensors that can arise from the loop integration, because only the photon momentum q_μ is available. What is important to stress is that, contrary to superficial appearance, the penguin diagram is one-particle irreducible. In fact after removal, by charge renormalization, of the quadratic divergence, the logarithmic divergence is proportional to q^2. This factor of q^2 erases the pole in the photon propagator. The computation leading to this result can be reproduced very briefly. The quantity of interest from the loop is given by

$$I_{\mu\nu} = \frac{ie^2}{(2\pi)^4} \int d^4k \; \mathrm{Tr} \left[\gamma_\mu \frac{\not{k}}{k^2} \gamma_\nu \frac{\not{k} + \not{q}}{(k+q)^2}\right] . \tag{24}$$

We are interested in the logarithmically divergent terms, which by dimension are quadratic in q. We then take the second derivative of $I_{\mu\nu}$ with respect to q_α and q_β at q=0:

$$D_{\alpha\beta} = \frac{1}{2} \frac{\partial^2}{\partial q_\alpha \partial q_\beta}\bigg|_{q=0} \frac{k+q}{(k+q)^2} = \frac{-1}{k^4}(\gamma_\alpha k_\beta + \gamma_\beta k_\alpha + g_{\alpha\beta} k^2) + \frac{4k_\alpha k_\beta \not{k}}{k^6} . \tag{25}$$

For the symmetric integration in the Euclidean region one can use the relations

$$k_\alpha k_\beta = \frac{1}{4} g_{\alpha\beta} k^2 ,$$

$$k_\alpha k_\beta k_\lambda k_\mu = \frac{k^4}{24} [g_{\alpha\beta} g_{\lambda\mu} + g_{\alpha\lambda} g_{\beta\mu} + g_{\alpha\mu} g_{\beta\lambda}] , \tag{26}$$

and

$$\int \frac{d^4k}{k^4} = i\,\pi^2 \ln \Lambda^2 . \tag{27}$$

Then it is a matter of simple algebra to obtain

$$\int d^4k\ \mathrm{Tr}[\gamma_\mu \frac{\not k}{k^2}\gamma_\nu\ D_{\alpha\beta}]\ q^\alpha q^\beta = -\frac{4}{3}\ i\pi^2\ \ln \Lambda^2 (q_\mu q_\nu - q_{\mu\nu} q^2)\ . \tag{28}$$

The $q_\mu q_\nu$ term is zero when contracted with the conserved f' current at the bottom (via the photon propagator $-g_{\nu\rho}/q^2$). Collecting all factors one obtains as a result

$$-\frac{\alpha}{3\pi}\ \ln \Lambda^2\quad \bar{\nu}_\mu \gamma_\lambda (1-\gamma_5)\nu_\mu\ \ f'\gamma_\lambda\ f'\ . \tag{29}$$

This amounts to say that

$$O^i \rightarrow (1-\alpha b_Q \ln \frac{M^2}{\mu^2})\ \frac{\mathrm{Tr}(A^i Q)}{\mathrm{Tr}\ Q^2}\quad O^Q \rightarrow [\frac{\alpha(M)}{\alpha(\mu)}]^{-1}\ \frac{\mathrm{Tr}(A^i Q)}{\mathrm{Tr}\ Q^2}\ O^Q \tag{30}$$

(recalling eq. (9) for b_Q and that $\ln \Lambda^2$ is to be traded with $\ln M^2/\mu^2$).
The bare hamiltonian in eq. (19) can then be rewritten as

$$H_o = \frac{G_F^o}{2\sqrt{2}}\ \rho[\frac{4b_z}{b_Q} - \frac{2}{\rho}\ \frac{1}{3\pi b_Q} - 4s^2(M)]O^Q + H_\perp \tag{31}$$

where $H_\perp$ contains operators orthogonal to O^Q and

$$b_z = \sum_f{}'\ \frac{1}{4}\ (Q_f \tau_3^f\ )\frac{1}{3\pi}\ . \tag{31'}$$

Upon renormalization according to eq. (30) this goes into:

$$H = \frac{G_F}{2\sqrt{2}}\ \rho[\frac{4b_z}{b_Q} - \frac{2}{\rho}\ \frac{1}{3\pi b_Q} - 4s^2(M)]\ \frac{\alpha(\mu)}{\alpha(M)}\ O^Q + H \tag{32}$$

or, by adding and subtracting the bare term in O^Q:

$$H = \frac{G_F\rho}{2\sqrt{2}}\ [\frac{4b_z}{b_Q} - \frac{2}{\rho}\ \ \frac{1}{3\pi b_Q} - 4s^2(M)]\ \ \frac{\alpha(\mu)-\alpha(M)}{\alpha(M)}\ O^Q + H_o\ . \tag{33}$$

But then we immediately see that H can be rewritten in the same form as H_o with the replacement of $s^2(M)$ with $s^2(\mu)$ given by:

$$s^2(\mu) = s^2(M) \frac{\alpha(\mu)}{\alpha M)} - \frac{\alpha(\mu)-\alpha(M)}{\alpha(M)} \frac{1}{b_Q} (b_z - \frac{1}{6\pi\rho}) . \qquad (34)$$

Note that in terms of $\alpha_W = \frac{\alpha}{s^2}$ this equation can be written as

$$\frac{\partial}{\partial \ln \mu^2} \frac{1}{\alpha_W(\mu^2)} = \frac{1}{b_Q} (b_z - \frac{1}{6\pi\rho}) \frac{\partial}{\partial \ln \mu^2} \frac{1}{\alpha(\mu)} = -(b_z - \frac{1}{6\pi\rho}) .$$

$$(35)$$

The b_z term is the contribution of fermions to the $SU(2)_W$ beta function, that is the form to which that beta function reduces in the limit of freezing the heavy gauge bosons. The extra term in $1/\rho$ is not present in the ordinary evolution equation for α_W.
In conclusion the whole leading log effect in ν initiated neutral current processes amounts to a redefinition of $\sin^2\theta_W$ according to eq. (34).

d) <u>Electron initiated neutral current processes.</u>

We aim in particular at treating $e_{R,L}+N\rightarrow e_{R,L}+X$ (the SLAC experiment), but we shall also consider the forward-back-ward asymmetry in $e^++e^- \rightarrow \mu^++\mu^-$. The relevant bare hamiltonian is

$$H_o = \frac{G_F^o \rho}{2\sqrt{2}} \{\frac{1}{2} J_\lambda^Z J_\lambda^Z + \frac{2}{\rho}[\bar{u}\gamma_\lambda(1-\gamma_5)d\ \bar{d}\gamma_\lambda(1-\gamma_5)u + u,d\leftrightarrow c,s + u,d\leftrightarrow t,b]\}$$

$$(36)$$

where J_λ^Z is given by eq. (21) and Cabibbo-like mixings were omitted for simplicity in the charged current contribution.

Of interest to us are the terms:

$$H_o = \frac{G_F^o \rho}{2\sqrt{2}}\{\bar{e}\gamma_\lambda\gamma_5 e[V_u \bar{u}\gamma_\lambda u + V_d \bar{d}\gamma_\lambda d + \ldots] +$$

$$+ \bar{e}\gamma_\lambda e[A_u \bar{u}\gamma_\lambda\gamma_5 u + A_d \bar{d}\gamma_\lambda\gamma_5 d + \ldots] +$$

$$+ \bar{e}\gamma_\lambda\gamma_5 e[C_\mu \bar{\mu}\gamma_\lambda\gamma_5 \mu + \ldots]\} \,. \tag{37}$$

The SLAC experiment is determined by $V_{u,d}$ and $A_{u,d}$ while C_μ is the relevant quantity for the asymmetry in $e^+e^- \to \mu^+ \mu^-$. In Born approximation

$$V_q^o = \tau_3 - 4Qs^2; \quad A_q^o = \tau_3(1-4s^2); \quad C_\mu^o = 1 \,. \tag{38}$$

The difference with the ν case is that charged leptons are not inert under SU(3) × U(1). Thus semileptonic and non-leptonic operators are mixed through the diagram in fig.2 .

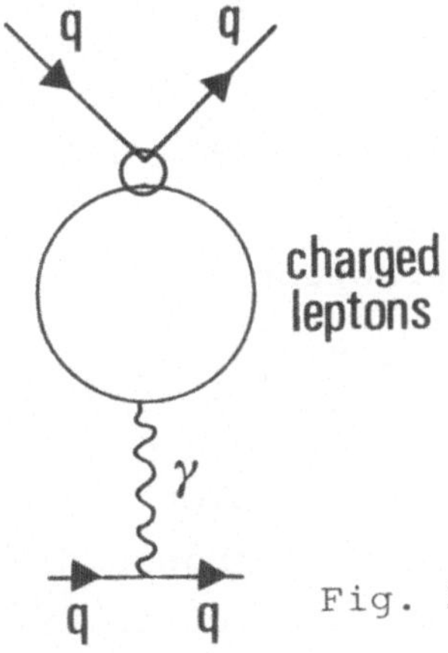

Fig. 2

The one-loop anomalous dimension contains both γ and γ_s in eq. (9) and we are faced with a problem with two running couplings. To simplify matters one can restrict to first order in α and all orders in α_s (that is one gives up the evaluation of the small terms of order $(\alpha \ln M^2)^2$). In this approximation the general solution in eq. (10) for the coefficient functions in the leading log approximation reduces to:

$$c_i = c_k^o[U_s(t,0) + \alpha \int_o^t dt' U_s(t,t')\gamma U_s(t',0)]_{ki} \qquad (39)$$

with

$$U_s(t,t') = \exp\{\gamma_s \int_{t'}^{t} \alpha(t'')dt''\} . \qquad (40)$$

Recall that the only non-vanishing matrix elements of γ_s are between non-leptonic channels. Since we are interested in semileptonic or leptonic terms in the hamiltonian we can set U=1 when acting on O_i. The effective hamiltonian then takes the form:

$$H \sim c_k^o O_k(\mu) + \alpha c_k^o \; [\int_o^t dt' U_s(t,t')]_{ki}\gamma_{ij}O_j(\mu) . \qquad (41)$$

We stress the following points:

1) $[\int_o^t dt' \; U_s(t,t')]_{ki} \neq \delta_{ki}$ t only for k,i both nonleptonic .
2) γ_{ij} can at most change one lepton pair into one quark pair. Thus a purely leptonic process, as $e^+ + e^- \to \mu^+ + \mu^-$, is not affected by strong interactions:

$$H_{Leptonic} \simeq H_o + \alpha t \; c_i^o \; \gamma_{ij} \; O_j^{Leptonic}(\mu) . \qquad (42)$$

3) As the mixing though γ_{ij} of semileptonic with non-leptonic operators arises through a penguin diagram which requires the charged leptons in a vector $(\bar{e}\gamma_\mu e)$ current it follows that only $A_{u,d}$ are affected by strong interactions (recall eq. (37)).

With these points in mind as a first step one studies the matrix γ_{ij} by setting $\gamma_s = 0$. The photon exchange diagrams are of three types shown in fig. 3.

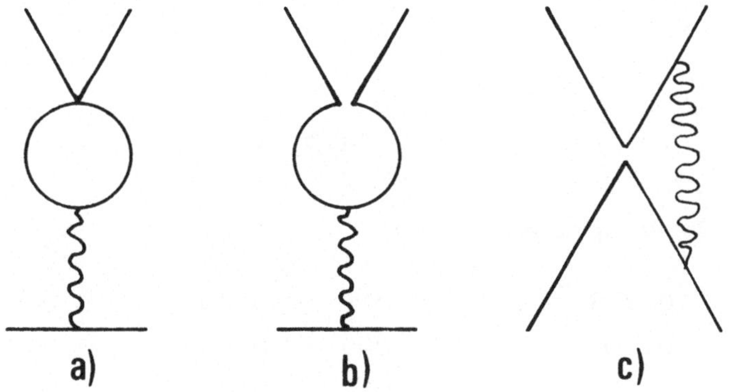

Fig. 3

Diagram (a) leads to redefinition of $\sin^2\theta_W$ according to

$$s^2(\mu) = s^2(M) + \alpha \ln \frac{M^2}{\mu^2} [b_2 - b_Q \, s^2(M)] \ . \tag{43}$$

(More precisely, each fermion is to be weighted by $\log \frac{M^2}{m^2}$ where $m = \max(\mu, m_f)$.) This relation between $s^2(\mu)$ and $s^2(M)$ is the same, at first order in α, as in the ν case except for the term in $1/\rho$ from charged currents whose origin was discussed previously.

Diagrams (b) and (c) lead to the contributions shown in table 1.

Table 1

	Tree Approx.	δa	δb
Vq	$\tau_3 - 4Qs^2$	$-\frac{1}{3}\, Q(1-4s^2)$	$\frac{3}{2}Q\tau_3(1-4s^2)$
Aq	$\tau_3(1-4s^2)$	$-\frac{1}{3}Q(1-4Q\tau_3 s^2)-\frac{2}{3}\underset{\sim}{Q}$	$\frac{3}{2}Q(\tau_3-4Qs^2)$
Cμ	1	0	$-\frac{3}{2}(1-4s^2)^2$

The entries under $\delta_{a,b}$ are to be multiplied by $\frac{\alpha}{\pi} \ln \frac{M^2}{\mu^2}$. $\underset{\sim}{Q} = -1/3,\ 2/3$ for u and d quarks respectively. Note that $(1-4s^2)$ is a small quantity, so that in particular the correction to C_μ is very small. Numerical values of the corrections, evaluated for $s_o^2 = 0.22$, are shown in table 2 also for the α_W and strong interaction terms to be discussed later.

Table 2

	Tree Approx.	δa	$\delta b+\delta c$	α_W	α_s
Vu	0.413	-0.018	0.002	0.006	0
Vd	-0.707	0.009	0.001	0.006	0
Au	0.120	-0.027	0.010	0.006	0.000
Ad	-0.120	0.027	0.000	0.006	0.003
C_μ	1	0	0.000	-0.006	0

The next step is to evaluate γ_S, whose contributions can only affect Au,d, as already mentioned. The problem is equivalent to studying the QCD leading log corrections to the flavour conserving parity violating non-leptonic hamiltonian. This problem was solved long ago[10]. Here it suffices to mention that the final result is an invisible correction to Au and a small correction to Ad, as shown in table 2, in the column α_s, as evaluated from $\Lambda \simeq$ 400 MeV.

1.3 Correction of Order α_W.

In this section we extimate the corrections of order α_W which are the most important ones after the leading logs. These terms are those which are non-vanishing in the limit

$$\alpha \to 0 , \qquad \sin^2\theta_W \to 0 ,$$

$$\alpha_W = \frac{\alpha}{\sin^2\theta_W} \quad \text{fixed} \quad . \tag{44}$$

In this limit the photon coincides with the U(1) gauge boson and decouples from all particles. The Z boson coincides with W_3 and $M_Z = M_W$. Neglecting Higgs couplings to fermions the theory has an exact global SU(2). The bare hamiltonian is given by:

$$H_o = \frac{G_F^o}{\sqrt{2}} \vec{J}_\lambda \vec{J}_\lambda \tag{45}$$

with

$$\vec{J}_\lambda = \sum_f \bar{f}_L \gamma_\lambda \vec{\tau} f_L \qquad (f_L = \frac{1-\gamma_5}{2} f) \quad . \tag{46}$$

The corrected hamiltonian is obtained from double $\vec{W}$ exchange. All terms proportional to fermion masses or external momenta are down by powers m/M and can be neglected. Because of the global SU(2) symmetry the result will be of the form

$$H = \frac{G_F^o}{\sqrt{2}} [(1+\varepsilon_1) \vec{J}_\lambda \vec{J}_\lambda + \varepsilon_2 J_\lambda^o J_\lambda^o] \tag{47}$$

where J_λ^o is an SU(2) singlet current:

$$J_\lambda^o = \sum_f \bar{f}_L \gamma_\mu f_L \quad ; \tag{48}$$

ε_1 is not to be computed because it is absorbed in the definition of G_F:

$$G_F = G_F^o (1+\varepsilon_1) \quad ; \tag{49}$$

ε_2 is finite and can be computed from the box diagrams in fig. 4 in the limit of zero external momenta.

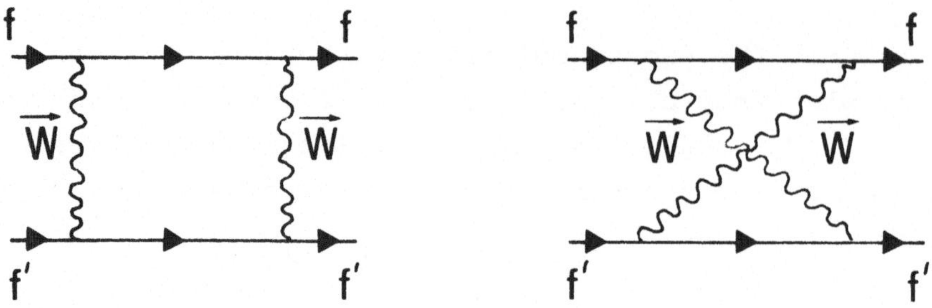

Fig. 4

We can simply sketch the derivation. The quantity of interest is

$$\delta H = \frac{g^4}{(2\pi)^4}\int d^4k\,\gamma_\mu \frac{\not{k}}{k^2}\gamma_\nu a_+ \frac{\tau_i}{2}\frac{\tau_j}{2} \times [\gamma_\mu \frac{-\not{k}}{k^2}\gamma_\nu a_+ \frac{\tau_i}{2}\frac{\tau_j}{2} - (\mu, i \leftrightarrow \nu, j)]\frac{1}{(k^2-M^2)^2} \tag{50}$$

where $a_+ = \frac{1-\gamma_5}{2}$ and the relative minus sign arises from the anticommutation relations of fermions. One writes $\tau_i\tau_j$ as $\frac{1}{2}[\tau_i,\tau_j]+\delta_{ij}$ and only the δ_{ij} term is to be kept for ε_2. A factor of 3 arises from $\delta_{ij} \times \delta_{ij}$. A simple algebra leads to

$$\gamma_\mu\gamma_\rho\gamma_\nu a_+ \times \gamma_\mu\gamma_\rho\gamma_\nu a_+ = 16\gamma_\mu a_+ \times \gamma_\mu a_+ \quad ,$$

$$\gamma_\mu\gamma_\rho\gamma_\nu a_+ \times \gamma_\nu\gamma_\rho\gamma_\mu a_+ = 4\gamma_\mu a_+ \times \gamma_\mu a_+ \quad . \tag{51}$$

By recalling eq. (26) and taking into account that

$$\int \frac{d^4k}{k^2(k^2-M^2)^2} = \frac{-i\pi^2}{M^2} \quad , \tag{52}$$

$$\bar{f}\gamma_\mu a_+ f\; \bar{f}'\gamma_\mu a_+ f' \rightarrow \frac{1}{2} J^o_\lambda J^o_\lambda \quad , \tag{53}$$

one obtains the result ($\alpha_W = \frac{g^2}{4\pi}$, $G_F = \frac{g^2}{8M^2}$)

$$-i\,\frac{G_F}{\sqrt{2}}\left(-\frac{9}{16}\frac{\alpha_W}{\pi}\right) J^o_\lambda J^o_\lambda \tag{54}$$

or

$$\varepsilon_2 = -\frac{9}{16}\frac{\alpha_W}{\pi} \quad . \tag{55}$$

The implications of this result will be clear from the examples in the following section. In particular the entries of table 2 under the heading "α_W" are immediately obtained from eqs. (37,53,55).

1.4 Applications.

In this section we collect some examples of applications to physical processes of interest.

a) $\sin^2\theta_W$ from $R = \frac{\bar{\nu}_\mu e \to \bar{\nu}_\mu e}{\nu_\mu e \to \nu_\mu e}$. This is in principle the cleanest method . In terms of the couplings a_L and a_R for left-handed and right-handed electrons,

$$H = \frac{G_F}{\sqrt{2}}\, \bar{\nu}_\mu \gamma_\lambda (1-\gamma_5)\nu_\mu [a_L \bar{e}_L \gamma_\lambda e_L + a_R \bar{e}_R \gamma_\lambda e_R] \ , \tag{56}$$

one has

$$R = \frac{\sigma^{\bar{\nu}}}{\sigma^{\nu}} = \frac{\frac{1}{3}a_L^2 + a_R^2}{a_L^2 + \frac{1}{3}a_R^2} \quad . \tag{57}$$

In the tree approximation $a_L^o = -1+2s_o^2$, $a_R^o = 2s_o^2$,and

$$R_o = \frac{16s_o^4 - 4s_o^2 + 1}{16s_o^4 - 12s_o^2 + 3} \quad . \tag{58}$$

In leading logs one simply reinterprets s_o^2 as $s^2(\mu)$, related to $s^2(M)$ by eq. (34). The α_W corrective terms modify a_L:

$$a_L = a_L^o - \varepsilon_2 \ ,$$

$$a_R = a_R^o \quad . \tag{59}$$

Under this latter correction $R \to R_o + \delta R$ and δR is approximately given by

$$\delta R = \frac{8}{9} \frac{a_R^{o\,2}}{(a_L^{o2} + \frac{1}{3} a_R^{o2})^2} \, 2|\varepsilon_2||a_L^o| \; . \tag{60}$$

For $s^2 \sim 0.22$ one obtains

$$R = 0.788 + 1.35|\varepsilon_2| \simeq 0.788 + 0.008 \; . \tag{61}$$

This corresponds to a shift upward of $s^2(\mu)$ by 0.005. The data are at present not sufficiently precise to allow a determination of $s^2(\mu)$ from these processes at a comparable accuracy.

b) $\sin^2\theta_W$ from $R_\nu = \frac{\sigma^\nu_{NC}}{\sigma^\nu_{CC}} = \frac{\nu+N\to\nu+X}{\nu+N\to\mu^-+X}$. The ratio of neutral to charged current cross sections in deep inelastic neutrino scattering, together with the same quantity from anti-neutrinos, provide one of the most important methods of determining $\sin^2\theta_W$. For purposes of illustration we limit ourselves to the valence approximation, although this is certainly inadequate for a realistic procedure. In this approximation the relevant terms in the bare hamiltonian are

$$H_o = \frac{G_F^o}{\sqrt{2}} \bar{\nu}_\mu \gamma_\lambda (1-\gamma_5)\nu_\mu \{u_L \bar{u}\gamma_\lambda(1-\gamma_5)u + d_L \; \bar{d}\gamma_\lambda(1-\gamma_5)d + u_R \bar{u}\gamma_\lambda(1+\gamma_5)u + $$
$$+ d_R \; \bar{d}\gamma_\lambda(1+\gamma_5)d + \ldots\} \; . \tag{62}$$

In terms of the couplings $u_{L,R}$ and $d_{L,R}$ one has:

$$\sigma^\nu_{NC} \sim \int_0^1 dx \; x(u(x)+d(x))[u_L^2+d_L^2+\tfrac{1}{3}(u_R^2+d_R^2)] \; ,$$
$$\sigma^\nu_{CC} \sim \int_0^1 dx \; x(u(x)+d(x)) \; , \tag{63}$$

and

$$R_\nu = u_L^2 + d_L^2 + \frac{1}{3}(u_R^2 + d_R^2) \tag{64}$$

In tree approximation

$$u_L^o = \frac{1}{2} - \frac{2}{3} s_o^2 \qquad u_R^o = -\frac{2}{3}s_o^2$$

$$d_L^o = -\frac{1}{2} + \frac{1}{3} s_o^2 \qquad d_R^o = \frac{1}{3}s_o^2 \tag{65}$$

$$R_\nu = As_o^4 + Bs_o^2 + C \qquad (A = \frac{20}{27}\ ;\ B = -1;\ C = \frac{1}{2})\ . \tag{66}$$

The leading logarithms produce two effects:

1) change s_o^2 into $s^2(\mu)$ related by eq. (34) to $s^2(M)$,
2) rescale the charged current cross section by a factor F^2 where F is given by eq. (17).

Thus in this approximation

$$R_\nu = \frac{1}{F^2}\,[As^4(\mu) + Bs^2(\mu) + C] \tag{67}$$

which implies to first order in F-1:

$$s^2(\mu) - s_o^2 = \frac{2R_\nu(F-1)}{2As_o^2+B}\ ; \tag{68}$$

when the terms of order α_W are also considered we have:

$$u_L = u_L^o + \frac{1}{2}\,\varepsilon_2 \qquad d_L = d_L^o + \frac{1}{2}\,\varepsilon_2$$

$$u_R = u_R^o \qquad d_R = d_R^o\ , \tag{69}$$

so that

$$u_L^2 + d_L^2 = u_L^{o^2} + d_L^{o^2} + \varepsilon_2(u_L^o + d_L^o) \simeq u_L^{o^2} + d_L^{o^2} - \frac{1}{3}\,s_o^2\varepsilon_2\ . \tag{70}$$

Thus one has

$$s^2(\mu) - s_o^2 = \frac{2R_\nu(F-1)+\frac{1}{3}s_o^2\varepsilon_2}{2As_o^2+B} \quad . \tag{71}$$

Note that while the valence approximation is bad for s_o^2, it is presumably adequate for the correction in eq. (71). For $R_\nu \sim 0.31$, $s_o^2 = 0.230$ and $\mu = 1.4$ GeV one obtains:

$$s^2(1.4) \simeq 0.221 \quad ,$$
$$s^2(M) \simeq 0.217 \quad . \tag{72}$$

Results of exact calculations[8], also including the effect of experimental cuts, are shown in table 3.

Table 3

EXP	s_o^2	Energy cut GeV	$s^2(M)$ ($\overline{MS}$)
CHARM	0.220±0.015	2	0.210
BEBC	0.217±0.045	15	0.206
CDHS	0.230±0.013	10	0.219

When finite corrections in α are also considered the definition of $\sin^2\theta_W(M)$ must be further refined by specifying the renormalization scheme. This is why the label $\overline{MS}$ (modified minimal subtraction) appears in table 3. The previous table shows that the simple approximate evaluation described in these lectures indeed provides a reasonably accurate description of the effect.

c) $\sin^2\theta_W$ from the electron-deuteron left-right asymmetry.

The electron-deuteron asymmetry is given by:

$$A = \frac{d\sigma_R - d\sigma_L}{d\sigma_R + d\sigma_L} = -Q^2 \frac{G_F}{4\sqrt{2}\pi\alpha} \left[a(x,Q^2) + \frac{1-(1-y)^2}{1+(1-y)^2} b(x,Q^2)\right] . \tag{73}$$

Neglecting scaling violations, charm densities, and setting $s=\bar{s}$ for the strange quark content of the proton one has:

$$a(x) = \frac{9}{5}\left[\frac{2V_u-V_d}{3} - \frac{4}{5}\,\frac{s+\bar{s}}{D}\,\frac{2V_d+V_u}{3}\right] , \tag{74}$$

$$b(x) = \frac{9}{5}\left[\frac{2A_u-A_d}{3}\quad\frac{q-\bar{q}}{D}\right] , \tag{75}$$

with

$$q = u + d + s ,$$

$$D = q + \bar{q} - \frac{3}{5}(s+\bar{s}) . \tag{76}$$

Values of $(2V_u-V_d)$, $(2V_d+V_u)$, $(2A_u-A_d)$ can directly be derived from tables 1,2. We have:

$$\frac{2V_u-V_d}{3} = 1 - \frac{20}{9}\,s^2(\mu) + 0.001 + 0.002 ,$$

$$\frac{2V_u+V_d}{3} = -\frac{1}{3} + 0.001 + 0.006 ,$$

$$\frac{2A_u-A_d}{3} = 1 - 4s^2(\mu) + 0.007 + 0.002 - 0.001 , \tag{77}$$

where in the first column we find the tree diagram expression with $s^2(\mu)$ replacing s_o^2, and the correction from $\delta_b+\delta_c$, from α_W and α_s occur in that order in the following columns; $s^2(\mu)$ and $s^2(M)$ are to be related by eq. (43).
From the analysis at SLAC (done with neglect of the strange sea: $s=\bar{s}=0$, an approximation to be improved before it really makes sense to apply the higher order corrections) it was found that

$$s^2(\mu) = 0.224 \pm 0.020 \qquad (\mu^2 = 2\ \text{GeV}^2) . \tag{78}$$

From this we obtain

$$s^2(M) \simeq 0.217 \ . \tag{79}$$

Note that although $s^2(\mu)$ is different than in the ν case, we end up by chance to precisely the same value of $s^2(M)$ because the relation between $s^2(\mu)$ and $s^2(M)$ is different in the two cases. Starting from the same value of $s^2(\mu)$ the value $s^2(M) \simeq 0.216$ was obtained in a more complete analysis of higher order corrections in ref. [8]. This again shows that our simplified analysis indeed contains the bulk of the effect.

1.5 Masses of Weak Bosons

The masses of weak bosons are naturally defined from the position of the pole in the corresponding propagator. In the leading logarithmic approximation G_F (whose evolution has no leading logs as we saw in the section on μ decay, and can then be taken as a fixed quantity) is related to the $W^{\pm}$ pole position and to the residue at the pole:

$$M_W^2 = \frac{\pi \ \alpha(M)}{\sqrt{2} G_F s^2(M)} \ . \tag{80}$$

In the tree approximation one replaces in this formula the values of α and s^2 as measured in low energy experiments:

$$(M_W^2)_{Bare} = \frac{\pi \ \alpha \ (0)}{\sqrt{2} G_f s^2(\mu)} \simeq [\frac{37.28}{s(\mu)}]^2 \ . \tag{81}$$

The relation between the two quantities is thus given by:

$$M_W^2 = (M_W^2)_{Bare} \frac{\alpha(M)}{\alpha(0)} \frac{s^2(\mu)}{s^2(M)} \ . \tag{82}$$

Beyond the leading logarithmic approximation further corrective terms of order α (and α_W) appear, which depend on every detail of the definition of M_W^2, α and s^2. We have already seen how the ratio $s^2(\mu)/s^2(M)$ can be evaluated for ν induced and electron induced neutral current processes. As for $\frac{\alpha(M)}{\alpha(O)}$ the leading log approximation for this ratio is given by:

$$\frac{\alpha(M)}{\alpha(O)} = \frac{1}{1-\frac{\alpha(O)}{3\pi}\sum_f Q_f^2 \ln\frac{M^2}{m_f^2}} = \frac{1}{1-\frac{\alpha(O)}{3\pi}\, t_Q} \tag{83}$$

with

$$|\alpha(O)|^{-1} = 137.036 \; . \tag{84}$$

In table 4 the separate contributions to t_Q of the different fermions are shown for $M \simeq 90$ GeV.

Table 4

	e	μ	u,d	s	c,τ	b	t	TOTAL
m_f	5.10^{-4}	0.105	0.140	0.250	1.8	5	20	
Q_f^2	1	1	$\frac{5}{3}$	$\frac{1}{3}$	$\frac{7}{3}$	$\frac{1}{3}$	$\frac{4}{3}$	
t_{Q_f}	24.20	13.51	21.55	3.92	18.26	1.93	4,01	87.84

Note that the masses of u,d and of course, of t are to some extent uncertain. For example for $m_{u,d} \simeq 0.3$ GeV one would obtain $t_{Qu,d}(0.3) \simeq 19.01$, and for $m_{u,d} \simeq 0.01$ GeV $t_{Qu,d}(0.01) \simeq 30.35$. One unit of t_Q corresponds to $\simeq 24$ MeV on M_W. Thus the value of M_W is uncertain by ~ 250 MeV because of the ambiguity on quark masses. As a result from the values in table 4 one obtains:

$$[\alpha(M)]^{-1} \simeq 127.8 \ . \tag{85}$$

Exactly the same procedure is followed for M_Z and leads to

$$M_Z^2 = \frac{M_W^2}{c^2(M)} = (M_Z^2)_{Bare} \frac{\alpha(M)}{\alpha(0)} \frac{s^2(\mu)}{s^2(M)} \frac{c^2(\mu)}{c^2(M)} \ . \tag{86}$$

Recalling that from $\sigma^{\nu}_{NC}/\sigma^{\nu}_{CC}$ we derived $s^2(\mu) \simeq 0.221$ and $s^2(M) \simeq 0.217$, we obtain $(M_W)_{Bare} \simeq 72.30$ GeV, $(M_Z)_{Bare} \simeq$ $\simeq 89.85$ GeV, and

$$M_W = (M_W)_{Bare} \cdot 1.045 \simeq 82.86 \text{ GeV} \ ,$$

$$M_Z = (M_Z)_{Bare} \cdot 1.042 \simeq 93.64 \text{ GeV} \ . \tag{87}$$

From the SLAC asymmetry the values $s^2(\mu) \simeq 0.224$ and $s^2(M) \simeq 0.217$ were obtained. Thus $M_{W,Z}$ are unaltered, and the difference is only in the bare values.
Results of more complete calculations are also available. As an illustration we quote in table 5 the results of [3] where $s^2(\mu)$ was defined from $\overset{(-)}{\nu}_\mu e \to \overset{(-)}{\nu}_\mu e$ at $\mu \simeq 0.3$ GeV.

Table 5

$s^2(0.3)$	$(M_W)_{Bare}$	$\delta M_{\text{Leading Logs}}$	δM_{finite}	M_W
0.22	79.48	3.50	-0.70	82.28
0.23	77.74	3.27	-0.65	80.36
0.24	76.10	3.07	-0.61	78.56

Table 5

$s^2(0.3)$	$(M_Z)_{Bare}$	$\delta M^{Leading}_{Logs}$	δM_{finite}	M_Z
0.22	90.00	3.83	-0.40	93.43
0.23	88.59	3.64	-0.36	92.06
0.24	87.29	3.46	-0.33	90.40

Once more the leading logarithmic approximation is rather good. Actually the present experimental uncertainty on $s^2(M)$ and the theoretical uncertainty on the values of the quark masses are of the order (or larger) of δM_{finite}.

1.6 Widths of Weak Bosons.

Results on the widths of weak bosons can similarly be derived. We consider for example the purely leptonic channels $Z \to \nu\bar{\nu}$ and $W^+ \to \mu^+ \nu_\mu$. In the tree approximation one easily derives for one neutrino type:

$$\Gamma(Z \to \nu\,\bar{\nu})_{Bare} = \frac{1}{24} \frac{\alpha(0)}{s^2(\mu)c^2(\mu)} (M_Z)_{Bare} \quad . \tag{88}$$

Then in leading logs:

$$\Gamma(Z \to \nu\bar{\nu}) = \Gamma(Z \to \nu\bar{\nu})_{Bare} \left(\frac{M_Z}{(M_Z)_{Bare}}\right)^3 \quad . \tag{89}$$

The width of Z into no visible particles can be taken as a measure of the number N_ν of neutrino types. Then one obtains:

$$N_\nu = 1.12 \frac{\sum_i \Gamma(Z \to \nu_i \bar{\nu}_i)}{159 \text{ MeV}} \quad . \tag{90}$$

That is the correction amounts to one neutrino type out of seven.

Similarly one obtains:

$$\Gamma(W\to\mu\nu) = \Gamma(W\to\mu\nu)_{Bare}\left(\frac{M_W}{(M_W)_{Bare}}\right)^3 \simeq 1.11\ \Gamma(W\to\mu\nu)_{Bare} \quad .(91)$$

PART TWO

2.1 Inclusive Decay Rates of Heavy Flavours.

We limit our discussion to inclusive decay rates which are the simplest ones from the theoretical point of view. Without an understanding of inclusive rates it seems hopeless to derive firm predictions on exclusive channels. The restriction to heavy flavours has similar motivations, because one expects that a partonic description should be more and more adequate as the mass increases. The question of whether charm particles are heavy enough-probably not- will be considered in detail in the following.

2.2 Semileptonic Width.

This is by far the simplest rate. When the momentum of the lepton pair is sufficently large with respect to the hadron binding energy a parton description suggests that the rate for semileptonic decay is reduced to that of the corresponding quark decay: $Q\to q+\ell+\nu_\ell$. Deviations from the parton model associated with interference with spectator and annihilation are absent or suppressed in this case, so that quark decay should be appropriate. Then the semileptonic width is given by:

$$\Gamma_{SL}(Q)=\sum_q |U_{Qq}|^2 I\left[\frac{m_q}{m_Q}, \frac{m_\ell}{m_Q}, 0\right] \frac{G_F^2 m_Q^5}{192\pi^3}\left\{1-\frac{2}{3}\frac{\alpha_s(m_Q)}{\pi}\left(\pi^2-\frac{25}{4}\right)f\left(\frac{m_q}{m_Q}\right)+..\right\}, \tag{92}$$

where U_{Qq} is the entry $Q \to q$ in the quark matrix, $I(x,y,z)$ is the three body phase space factor, and the last bracket describes the QCD leading correction for real and virtual gluon emission. The function $f(x)$[11] decreases monotonically from $f(0)=1$ down to $f(1) \simeq 0.41$.

If one knows the semileptonic branching ratio B_{SL} for a given heavy flavoured hadron, then the semileptonic width estimated from eq. (92) leads to a prediction of the total lifetime:

$$\tau = \frac{B_{SL}}{\Gamma_{SL}} \quad . \tag{93}$$

For example the average $\bar{B}_{SL}$ of D^o and D^+ is known with good accuracy from experiments at the ψ'' resonance:

$$\bar{B}_{SL}(D^{+,o}) = (8.2 \pm 1.2)\ \% \ . \tag{94}$$

As the dominant transition $c \to s$ is isoscalar, Γ_{SL} is expected to be about the same for D^+ and D^o. Thus for the average lifetime of D mesons one expects:

$$\bar{\tau}(D^{+,o}) \simeq \frac{1}{\cos^2\alpha}\ (\sim 5.10^{-13}\mathrm{sec}) \tag{95}$$

where $\cos\alpha$ is the mixing angle for the transition $c \to s_c$ where s_c is the Cabibbo rotated strange quark (in the notation of Maiani for mixing angles[12]). From present data one can set a bound $\cos^2\alpha \gtrsim 0.6$. The uncertainty on the figure 5 in eq. (95) arises from the charm mass value (the dependence being with m_c^5) and on the value of α_s and amounts to one or two units. The experimental values are at present [13]

$$\tau(D^+) \simeq (6.9\ {}^{+3}_{-2})\ 10^{-13}\mathrm{sec}$$

$$\tau(D^o) \simeq (2.4\ {}^{+0,7}_{-0,5})\ 10^{-13}\mathrm{sec}$$

$$\tau(F^+) \simeq (2.8\ {}^{+1,2}_{-1,0})\ 10^{-13}\mathrm{sec}$$

$$\tau(A_c) \simeq (2.5 \,^{+1,5}_{-1,0})\ 10^{-13}\,\mathrm{sec}\ . \tag{96}$$

The agreement for the average lifetime of D^+ and D^o is good within the experimental and theoretical uncertainties.

2.3 Non-Leptonic Amplitudes.

To proceed further toward a computation of B_{SL} and of non-leptonic inclusive rates we shall divide the discussion into two conceptually different points: first the set up of an effective hamiltonian for nonleptonic weak amplitudes and then the attempts at evaluating its matrix elements. We shall see that the first part can be carried through on rather solid ground, while the second step is more tentative. Let us consider a non-leptonic weak process induced by charged currents. In lowest order in the weak coupling the transition matrix element is given by the time-ordered product of the two weak charged currents folded with the W propagator:

$$H_{FI} \simeq g_W^2 \int d^4x D_W(x^2, M_W^2) \langle F|T[J^\mu(x), \tau_\mu^+(0)]|I\rangle \tag{97}$$

where M_W, g_W, and D_W are the W boson mass, coupling and propagator, respectively. For flavour-changing amplitudes the leading contributions in the limit $M_W \to \infty$ arise from the four-fermion operators of dimension six in the short-distance operator expansion for the T-product[14]. In the particular example of charm-changing processes the relevant terms are of the form

$$H_{FI}^{\Delta c=1} = \frac{G_F}{\sqrt{2}} \{c_+(t,\alpha_s)\langle F|O_+(o)|I\rangle + c_-(t,\alpha_s)\langle F|O_-(o)|I\rangle + \ldots\} \tag{98}$$

where

$$t = \ln \frac{M_W^2}{\mu^2} \quad . \tag{99}$$

Here μ is a reference mass scale, $\alpha_s=\alpha_s(\mu)$ being the renormalized QCD coupling at the scale μ, and

$$O_{\pm} = \frac{1}{2}[(\bar{s}'c)_L(\bar{u}d')_L \pm (\bar{u}c)_L(\bar{s}'d')_L] =$$

$$= \frac{N\pm 1}{2N}(\bar{s}'c)_L(\bar{u}d')_L \pm \sum_A (\bar{s}'t^A c)_L(\bar{u}t^A d')_L \quad . \tag{100}$$

The shorthand notations for left-handed (right-handed) currents,

$$(\bar{q}_1 q_2)_{L,R} = \bar{q}_1\gamma_\mu(1\mp\gamma_5)q_2 \ ,$$

$$(\bar{q}_1 t^A q_2)_{L,R} = \bar{q}_1\gamma_\mu(1\mp\gamma_5)t^A q_2 \quad , \tag{101}$$

were used here; s' and d' are the Cabibbo- like quark mixtures coupled to the c and u quarks respectively, in the standard electroweak theory [1,15]; t^A are the SU(N) colour matrices in the quark (fundamental) representation, with the normalization

$$\mathrm{Tr}(t^A t^B) = \frac{1}{2}\,\delta^{AB} \ . \tag{102}$$

The second equality in eq. (100) is obtained through Fierz rearrangement of $(\bar{u}c)$ $(\bar{s}'d')$, according to the identity

$$(\bar{q}_1 q_2)_L(\bar{q}_3 q_4)_L = \frac{1}{N}(\bar{q}_1 q_4)_L(\bar{q}_3 q_2)_L + 2\sum_A (\bar{q}_1 t^A q_4)(\bar{q}_3 t^A q_2) \ . \tag{103}$$

We also recall the additional relation

$$\sum_A(\bar{q}_1 t^A q_2)_L(\bar{q}_3 t^A q_4)_L = \left(\frac{W^2-1}{2N^2}\right)(\bar{q}_1 q_4)_L(\bar{q}_3 q_2)_L - \frac{1}{N}\sum_A(\bar{q}_1 t^A q_4)_L(\bar{q}_3 t^A q_2)_L \tag{104}$$

In eq. (98) the dots stand for non-leading terms from operators of higher dimension and also for "penguin" operators [16] of dimension six, which are present in some channels for non-degenerate quark masses, and will be considered later on. In the free field limit

$$c_+ = c_- = 1 \qquad \text{(free fields)}$$

and

$$H^{\Delta c=1} \sim \frac{G_F}{\sqrt{2}} (\bar{s}'c)_L(\bar{u}d')_L \,. \tag{105}$$

In the leading logarithmic approximation O_+ and O_- are multiplicatively renormalizable, because anomalous dimensions are determined by the massless theory, and $O_\pm$ have definite and different transformation properties under the $SU(N)_{flavour}$ symmetry of the massless theory. c_+ and c_- deviate from the free field value by a known amount:[17]

$$c_\pm = \left[\frac{\alpha_s(\mu^2)}{\alpha_s(M_W^2)}\right]^{\gamma_\pm} , \tag{106}$$

$$\gamma_+ = \frac{-9(N-1)}{N(11N-2f)} \underset{f=4}{\overset{N=3}{=}} -\frac{6}{25} ,$$

$$(\gamma_+^2 = \frac{1}{\gamma_-})$$

$$\gamma_- = \frac{9(N+1)}{N(11N-2f)} \underset{f=4}{\overset{N=3}{=}} \frac{12}{25} , \tag{107}$$

where N and f are the numbers of colour and of flavours respectively. Recently a calculation of the effective hamiltonian at the next to the leading log accuracy was completed[18].

$$c_\pm \simeq \left[\frac{\alpha_s(\mu^2)}{\alpha_s(M_W^2)}\right]^{\gamma_\pm} \left\{1 + \frac{\alpha_s(\mu^2)-\alpha_s(M_W^2)}{\pi} \rho_\pm + \ldots\right\} , \tag{108}$$

where $\rho_\pm$ is independent of the renormalization scheme for the operators $O_\pm$, but only depends on the definition of α_s. Numerically, in the $\overline{MS}$ definition of α_s, for N=3 and f=4,6 one obtains:

$$c_\pm = \begin{cases} \left[\dfrac{\alpha_s(\mu^2)}{\alpha_s(M^2)}\right]^{\begin{smallmatrix}-0.24\\+0.48\end{smallmatrix}} \left\{1 + \dfrac{\alpha_s(\mu^2)-\alpha_s(M_W^2)}{\pi}\begin{pmatrix}-0.469\\+1.36\end{pmatrix}\right\}, & f=4 \\[2ex] \left[\dfrac{\alpha_s(\mu^2)}{\alpha_s(M_W^2)}\right]^{\begin{smallmatrix}-0.286\\+0.572\end{smallmatrix}} \left\{1 + \dfrac{\alpha_s(\mu^2)-\alpha_s(M_W^2)}{\pi}\begin{pmatrix}-0.574\\+1.65\end{pmatrix}\right\}, & f=6 \end{cases} \quad (109)$$

It is important to observe that the next to the leading correction is of "normal" size, i.e. $\rho_\pm \sim O(1)$, and follows exactly the same pattern of enhancement and suppression of the leading term. Actually most of the correction could be reabsorbed in a redefinition of α_s (by scaling up $\Lambda_{\overline{MS}}$ by a factor of about 1.8). As a consequence the above results considerably reinforce the leading log predictions and put the status of the effective hamiltonian on a much more solid basis.

It is well known that for strange particle decays the structure of the effective hamiltonian works in favour of the $\Delta I=\frac{1}{2}$ rule[17]. In fact for strangeness changing amplitudes O_- is pure $\Delta I=\frac{1}{2}$ and c_- is enhanced by gluon effects, while O_+ also contains $\Delta I=\frac{3}{2}$ and c_+ is suppressed. However $\mu \sim m_s$, where m_s is the strange quark mass, is too low to rely massless perturbation theory in the descent from M_W down to m_s. Below the charm mass m_c various effects connected with the no longer negligible c mass appear, such as penguin diagrams and separate anomalous dimensions for operators in different $SU(3)_f$ representations[16][19]. On the other hand the presence of still heavier quarks like b and t is of not great relevance and can be effectively taken into account by adjusting the number f of the excited flavours between 4 and 6. Penguin diagrams shown in fig. 5 are zero for

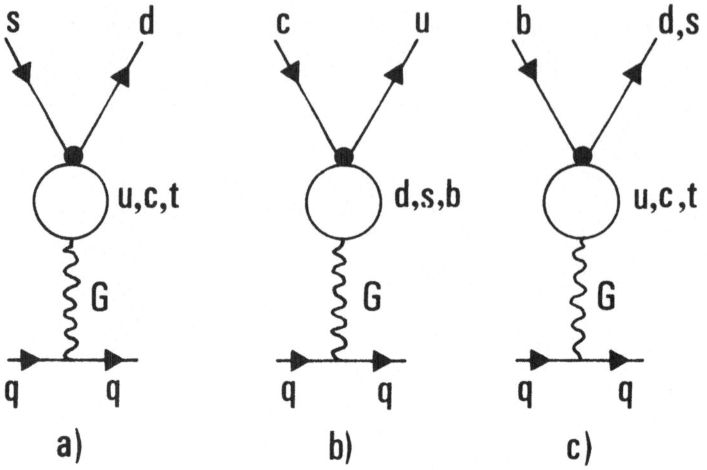

Fig. 5

equal quark masses because of GIM. Their coefficient is rather small: $(c_+ + c_-)\frac{\alpha_s}{12\pi}\ln\frac{m_c^2}{m_s^2}$. (The argument of the log is the heavy quark mass in the loop divided by the energy scale for the process. The value of the K_L-K_S mass difference shows that the presence of the t quark is unimportant.) But the claim, originally made in [16], is that the matrix element is particularly large, due to a low energy effect connected with the smallness of the current mass of u,d quarks. In conclusion the $\Delta I=\frac{1}{2}$ rule observed in K and hyperon decays corresponds to a factor $\frac{A_{1/2}}{A_{3/2}} \simeq 20$ in amplitude. The QCD effect in the coefficients $\tau_\pm$ can account for the square root of this factor. The remaining factor is presumably due to low energy effects in matrix elements, also including perhaps those of penguin diagrams. This view is to some extent supported by the fact that the $\Delta I=\frac{1}{2}$ rule is less pronounced in Ω^- decays, where low energy effects are presumably less pronounced.

For c decays the penguin diagrams, in the limit $m_b \gg m_d$, $m_s \sim 0$, are given by:

$$H \sim \frac{G_F}{\sqrt{2}}(c_+ + c_-)U_{cb}U_{ub}\frac{\alpha_s(m_c)}{12\pi} \ln \frac{m_b^2}{m_c^2}(\bar{c}t^A u)_L(\bar{q}t^A q)_{L+R} . \quad (110)$$

The existing bounds on mixing angles give $|U_{cb}U_{ub}| \lesssim \theta_c/4$. Thus, as the matrix elements are not expected to be large, penguin diagrams are certainly irrelevant for charm decays. At most they can contribute in Cabibbo suppressed channels, but their contribution is expected to be small even at that level. For b decays, in the limit $m_t >> m_u$, $m_c \sim 0$, one similarly obtains:

$$H \simeq \frac{G_F}{\sqrt{2}}(c_+ + c_-)U_{tb}U_{ts}\frac{\alpha_s(m_b)}{12\pi} \ln \frac{m_t^2}{m_b^2}(\bar{b}t^A s)_L(\bar{q}t^A q)_{L+R} . \quad (111)$$

Lacking any information on the mixing angles one can only say that this contribution is presumably small, but could be not entirely negligible.

As a first conclusion we have seen that the structure of the effective non-leptonic hamiltonian is known in the form:

$$H \sim \frac{G_F}{\sqrt{2}} [c_+O_+ + c_-O_- + \text{penguins}] , \quad (112)$$

and that penguin diagrams are negligible in the $|\Delta c|=1$ sector where the hamiltonian reduces to:

$$H \simeq \frac{G_F}{\sqrt{2}} [\frac{c_+ + c_-}{2} (\bar{c}s')_L(\bar{d}'u)_L + \frac{c_+ - c_-}{2} (\bar{c}u)_L(\bar{d}'s')_L] . \quad (113)$$

Rather confusing is the status of matrix elements, in particular the relative importance of the quark decay with spectator(s) and the annihilation mechanisms which we are now going to discuss.

Initially it was assumed, according to parton ideas, that the c quark decay was the main mechanism, with the other constituents in the hadron acting as spectators.

This obviously leads to approximately the same B_{SL} or lifetimes for all charmed hadrons. Since the discovery of a longer lifetime for D^+ than for D^o (and approximately also F^+ and Λ_c) it was clear that quark decay cannot be the only effect. After the first data by DELCO followed by new evidence from MARK II, now there is also an increasing amount of data from emulsions and bubble chambers that led to the lifetimes quoted in eqs. (96). In particular the best number for the ratio of D^+ and D^o lifetimes appears now to be:

$$\frac{\tau(D^+)}{\tau(D^o)} \simeq 2.5 \begin{array}{l} +0.8 \\ -0.6 \end{array} . \qquad (114)$$

This more recent value is much closer to 1 than previously reported, and more understandable theoretically as we shall see.

Although quark decay cannot be, at least for charm, the only effect, however it certainly is there. It is thus important to summarize what would happen for the decay of a free heavy quark. We shall later discuss the possible role of the spectators.

In the leading log approximation the total non-leptonic width of a heavy quark Q arises from $Q \to q + q_\alpha + \bar{q}_\beta$ and is given by[20]:

$$\Gamma_{NL} \simeq \frac{G_F^2 m_Q^5}{192\pi^3} \sum_{q, q_\alpha, \bar{q}_\beta} |U_{Qq}|^2 \, |U_{q_\alpha q_\beta}|^2 \, I[\frac{m_q}{m_Q}, \frac{m_{q\alpha}}{m_Q}, \frac{m_{q\beta}}{m_Q}] \, \frac{2c_+^2 + c_-^2}{3} , \qquad (115)$$

where the last factor is larger than 1 as follows from eqs. (106), (107) for $c_\pm$. It arises as follows. By writing H in eq. (113) in terms of one single ordering of quark fields, obtained by Fierz rearrangement of the second term according to eq. (104), one gets:

$$H \simeq \frac{G_F}{\sqrt{2}} \{ (\tfrac{2}{3}c_+ + \tfrac{1}{3}c_-)(\bar{c}s')_L + (c_+ - c_-)(\bar{c}t^A s')_L (\bar{d}'t^A u)_L \} . \tag{116}$$

The incoherent sum of the rates into color singlet and color octet $\bar{d}'u$ final states leads to

$$(\tfrac{2}{3}c_+ + \tfrac{1}{3}c_-)^2 + \tfrac{2}{9}(c_+ - c_-)^2 = \frac{2c_+^2 + c_-^2}{2} \tag{117}$$

where the factor of 2/g arises from the ratio of TrI × TrI=9 for the singlet-singlet contribution and $\sum_{AB}[Tr(t^A t^B)]^2 = 2$ for the octet-octet contribution (recall eq. (102)). The overall normalization is chosen as to reproduce the free field result for $c_+ = c_- = 1$.

Ignoring phase space distorsions and finite corrections of order α_s this leads to:

$$B_{SL} \simeq \frac{1}{2 + 2c_+^2 + c_-^2} \sim \begin{matrix} 20\ \%\ \text{Free fields} \\ 13 - 16\ \%\ \text{Leading Logs} \end{matrix} . \tag{118}$$

Phase space corrections are however of importance, because the three body phase space is sensitive to masses. In particular non-leptonic amplitudes could be somewhat suppressed with respect to semileptonic modes, because of quark masses. It has been observed[21] that by setting $m_s \sim 500$ MeV and $m_{u,d} \sim 300$ MeV the effect is quite large and could bring $B_{SL}(D^+)$ back in the 20% range. On the other hand the effect is quite small for $m_s \sim 300$ MeV and $m_{u,d} \sim 100$ MeV, which looks more reasonable in view of the final state being made up of π and K, and because no analogue effect is seen in τ semileptonic decay. Thus I think that this effect is presumably marginal.

Corrections of order α_s to the non-leptonic rate have also been computed recently[18]. In the $\overline{MS}$ prescription for α_s they amount to the replacement:

$$\frac{2c_+^2+c_-^2}{3} \longrightarrow \frac{2c_+^2+c_-^2}{3}\{1+\frac{2}{3}\frac{\alpha_s(m_c)}{\pi}(\frac{31}{4}-\pi^2+\frac{19}{4}\frac{c_-^2-c_+^2}{2c_+^2+c_-^2})+$$

$$+2\frac{\alpha_s(m_c)-\alpha_s(M_W)}{\pi}\frac{2c_+^2\rho_+ + c_-^2\rho_-}{2c_+^2+c_-^2}\} \qquad (119)$$

where $\rho_\pm$ are defined in eq. (108). Numerically for $\Lambda \simeq 0.25$ GeV the curly bracket is 1.21. The size of the correction is normal and its sign is once more in the same direction of the leading log effect. The prediction of a depletion of B_{SL} for quark decay below the free field value of 20% is thus confirmed by the order α_s corrections which lower the semi-leptonic width and increase the nonleptonic rate. We shall see that the annihilation mechanism cannot affect much the D^+ rate, and that spectator effects which for D^+ could in principle be relevant are not expected to be large. Thus the prediction of a small B_{SL}, obtained from the study of the decay properties of a heavy quark, is most probably relevant for D^+ decay.

Now comes naturally the question of possible spectator effects. An interesting point was put forward in [22]. At the Cabbibo allowed level the c quark decays according to $c \to su\bar{d}$. The D^+ meson is unique in that the spectator $\bar{d}$ finds an identical antiquark in the final state of c decay. Note parenthetically that also for Λ_c then would be a u spectator identical with a u quark from c decay. Assuming for a moment that the D^+ wave function was a δ-function at the origin, then the two $\bar{d}$'s would be emitted in the same space-time point. The final state would be the same as produced from the vacuum by an interaction where $\bar{c}$ is replaced by $\bar{d}$: for example $(\bar{c}s)_L(\bar{d}u)_L \to (\bar{d}s)_L(\bar{d}u)_L$, because the D^+ is color singlet and spin singlet, which means that color and helicity are just right for the replacement of an incoming c with an outgoing $\bar{d}$. Thus the analogue of O_- would vanish and the non-leptonic rate of D^+ would be reduced by a factor

$$\frac{4c_+^2}{2c_+^2+c_-^2} \sim \frac{1}{2} \quad .$$

Now, of course this is an upper bound because the wave function is not a δ at the origin and the overlap between spectator and active antiquark must be far smaller. On the other hand we experimentally know that this effect cannot be dominant. In fact we know that the average B_{SL} for D^+ and D^o is small, and that the D^+ lifetime appears as normal while the D^o lifetime looks shorter. Thus we conclude that the interference effect between spectator and active quarks cannot be a prominent effect, although it may be of help in providing some difference in the right direction for the D^+ and D^o lifetimes.

We now consider the annihilation mechanisms. First we recall that by this name we mean exchange of a W between constituents of a given hadron, either in the s or t channels. The name annihilation is strictly speaking only appropriate for an s-channel exchange.

The first point to recall is that annihilation without gluon emission cannot work, for two main reasons. The first is the helicity suppression factor . In the massless limit a pseudoscalar meson cannot go into a spin-one $q\,\bar{q}$ pair as produced by a V-A interaction. Even for an s quark the suppression factor $\frac{m_s^2}{m_c^2} \lesssim \frac{1}{10}$ is too small. Also colour factors are completely unfavourable. The D^o decay needs a $c\bar{u}$ pair in a colour singlet combination, which means an amplitude proportional to $X_- = \frac{2}{3}c_+ - \frac{1}{3}c_-$ (from eq. (113)) while F^+ and D^+ need $c\bar{s}$ or $c\bar{d}$ in a colour singlet state, with amplitude $X_+ = \frac{2}{3}c_+ + \frac{1}{3}c_-$. Since $\frac{X_+^2}{X_-^2} \sim 20$ this would make the F^+ lifetime much shorter than that of D^o and destroy the Cabibbo suppression pattern for D^+. Thus annihilation without gluon emission cannot work. However it is there and can provide a sizeable rate for $F^+ \to \tau^+ \nu_\tau$ and contribute an important fraction of the D^+ Cabibbo suppressed modes.

With single gluon emission the helicity suppression factor is removed and the colour factors look all right. In fact, in the D^o case the amplitude for a $c\bar{u}$ pair in a colour octet is proportional to $(c_+ + c_-)$ while in D^+ and F^+ the amplitude corresponding to colour octet $c\bar{s}$ and $c\bar{d}$ is proportional to $(c_+ - c_-)$. Not only the pattern $\tau(D^+) > \tau(F^+) > \tau(D^o)$ is consistent with the data but other signatures for the relevance of the annihilation diagrams are observed. For example final states with no $s\bar{s}$ quarks in F^+ decay were seen, as well as dominance of isospin $\frac{1}{2}$ amplitudes in D^o decays and rather abundant Cabibbo suppressed modes in D^+ decays[23].

If on qualitative grounds the annihilation diagrams with gluon emission are a convincing candidate for explaining the shorther lifetime of D^o, F^+ and Λ_c with respect to D^+ the quantitative estimate of the relative importance of annihilation versus quark decay for charm is a much more difficult problem [24]. A crude non-relativistic estimate relates $\frac{\Gamma_{D^o}}{\Gamma_{D^+}}$ to f_D, the analogue of f_π for D mesons, (via $|\psi(o)|^2$):

$$\frac{\Gamma_{D^o}}{\Gamma_{D^+}} \sim 1 + 0.7 \left(\frac{f_D}{m_u}\right)^2 \quad . \tag{120}$$

Since all attempts at an evaluation of f_D lead to $f_D \sim 200\text{-}300$ MeV at most, it seems difficult to explain $\tau(D^+)/\tau(D^o) \gtrsim 3$ by this mechanism. It is thus satisfactory that the most recent data show a tendency to a smaller ratio than previously claimed.

Note that the annihilation mechanism should not affect the semileptonic rate in a sizeable way, because these modes are forbidden in the one-gluon limit (the leptons being colourless) and the emission of two gluons is presumably more suppressed both by α_s and by phase space.

For merely indicative purposes I collect a tentative guess of the quantitative importance of the various effects that we mentioned in our discussion for B_{SL} of D^o and D^+, which takes into account the empirical facts that $\bar{B}_{SL} \simeq (8.1 \pm 1.2)$ % and $2 \lesssim \frac{\tau(D^+)}{\tau(D^o)} \lesssim 5$:

		B^o_{SL}	B^+_{SL}
	NAIVE	20%	20%
Q DECAY	QCD	(8÷13)%	(8÷13)%
+ SPECTATOR INTERFERENCE		(8÷13)%	(11÷16)%
+ MASS EFFECTS		(10÷15)%	(13÷18)%
+ ANNIHILATION		(4÷5)%	(12÷17)%
		Average	8÷11 %

For b flavoured particles violations of the parton model are expected to go down. Mass effects are negligible (except for two or more τ or c in the final state). Interference with spectators tends to go rapidly down, because the overlap between spectator and active quarks is reduced when the phase space is increased. Finally the annihilation mode is decreasing with a power of m_Q. Which power it is is not clear.

ACKNOWLEDGMENTS

I am very grateful to the Organizing Committee and to Prof.Mitter in particular, for the kind invitation and hospitality. I am also indebted to L.Maiani for clarifying discussions.

REFERENCES

1. S. Weinberg, Phys. Rev. Letters 19 (1967) 1264; A. Salam, in "Elementary Particle Physics", ed. by N. Svartholm (Almquist and Wiksell, Stockholm, 1968), p. 367.
   For reviews see E.S. Abers and B.W. Lee; Phys. Reports 9, 1 (1973); J.C. Taylor, "Gauge Theories of Weak Interactions" (Cambridge Univ. Press, 1976); G. Altarelli, "Phenomenology of Flavordynamics" in "QFD, QCD and Unified Theories", ed. by K.T. Mahanthappa, J. Randa, Plenum Pub. Corp. New York, 1980.
2. See also: A. De Ruyula, R. Petronzio, A. Savoy-Navarro, Nucl. Phys. B154 (1979) 394.
3. F. Antonelli, L. Maiani, Nucl. Phys. B186 (1981) 269.
4. S. Bellucci, M. Lusignoli, L. Maiani, Nucl. Phys. B189 (1981) 329.
5. See also: W.J. Marciano and A. Sirlin, Phys. Rev. D22 (1980) 2695.
   G.C. Ross, C.H. Llewellyn Smith and J.F. Wheater, Nucl. Phys. B177 (1981) 263.
   S. Dawson, J.S. Hagelin and L. Hall, Harvard preprint HUTD-80/A090.
   D.Yu. Bardin, P.Ch. Christova and O.M. Fedorenko, Nucl.

Phys. B175 (1980) 435.
S.Sakakibara, Dortmund preprint DO/THI/80.
J. Kubo and S. Sakakibara, Dortmund preprint DO-TH 80/25.
E.A. Paschos and W. Wirbel, Dortmund preprint DO-TH 81/4.
I. Antoniadis, C. Roiesnel and C. Kounnas. Ecole Polytechnique preprint, LPTENS 81/4.

6. J. Kiskis, Phys. Rev. D8 (1973) 2129.
R. Barlow and S. Wolfram, Phys. Rev. D20 (1979) 2198.
D.Yu. Bardin and O.M. Fedorenko, Yad Fiz. 30 (1979) 811.
A. Sirlin and W.J. Marciano, New York University preprint NYO/TR3/81.
D.Yu. Bardin, O.M. Fedorenko and N.M. Shumeiko, Yad.Fiz. 32 (1980) 782, and Dubna preprint E2-80-503.
P. Salomons and Y. Ueda, Phys. Rev. D11 (1975) 2606.
N. Byers, R. Ruckl and A. Yano, Physica 96A (1979) 163.
M. Green and M. Veltman, Nucl. Phys. B169 (1980) 137 (E: B175 (1980) 547).
K. Aoki, Z.Hioki, R. Kawabe, M. Konuma and T. Muta, Kyoto preprint RIFP-416.
G. Passarino and M. Veltman, Nucl. Phys. B160 (1979) 151.
M. Consoli, Nucl. Phys. B160 (1979) 208.

7. F. Antonelli, G. Corbo, M. Consoli, O. Pellegrino, Nucl. Phys. B183 (1981) 195.
8. C.H.Llewellyn Smith, J.F. Wheater, Oxford Preprint 68/81.
9. A.Sirlin, New York University Preprint, NYU/TR8/81.
10. G. Altarelli, R.K. Ellis, L. Maiani, R. Petronzio, Nucl. Phys. B88 (1975) 215.
11. N. Cabibbo, L. Maiani, Phys. Lett. 79B (1978) 109.
N. Suzuki, Nucl. Phys. B145 (1978) 420.
N. Cabibbo, G. Corbo, L. Maiani, Nucl. Phys. B155 (1979) 93.
12. M. Kobayashi and K. Muskawa, Progr. Theor. Phys. 49 (1973) 652.
S. Pakvasa and H. Sugawara, Phys. Rev. D14 (1976) 305.
L. Maiani, Phys. Lett. 62B (1976) 183.

13. C. Fisher, private communication; see also the Proceedings of the Moriond Workshop on Heavy Flavours, ed. by J. Tran Thanh Van, 1982.
14. K. Wilson, Phys. Rev. 179 (1979) 1499.
15. S.L. Glashow, J. Iliopoulos, L. Maiani, Phys. Rev. D2 (1970) 1285.
16. M.A. Shifman, A.I. Vainshtein and V.I. Zakharov, Nucl. Phys. B120 (1977) 316; JEPT (Sov. Phys.) 45 (1977) 670; see also C.T. Hill and G.G. Ross, Phys. Lett 94B (1980) 234.
N.B. Wise and E. Witten, Phys. Rev. D20 (1979) 1216.
17. B.W. Lee and M.K. Gaillard, Phys. Rev. Lett. 33 (1974) 108.
G. Altarelli and L. Maiani, Phys. Lett. 52B (1974) 351.
18. G. Altarelli G. Curci, G. Martinelli, S. Petrarca, Phys. Lett. 99B (1981) 141; Nucl. Phys. B187 (1981) 461.
19. For recent analysis see:
F.J. Gilman and M.B. Wise, Phys. Rev. D20 (1979) 2392.
B. Guberina and R.D. Peccei, Nucl. Phys. B 163 (1980) 289.
F. Buccella, M. Lusignoli, L. Maiani and A. Pugliese, Nucl. Phys. B152 (1979) 461.
20. B.W. Lee, M.K. Gaillard and G. Rosner, Rev. Mod. Phys. 47 (1975) 277.
G. Altarelli, N. Cabibbo and L. Maiani, Nucl. Phys. B88 (1975) 285; Phys. Lett. 57B (1975) 277.
S.R. Kingsley, S.Treiman, F. Wilczek and A. Zee, Phys. Rev. D11 (1975) 1914.
J. Ellis, M.K. Gaillard and D. Nanopoulos, Nucl. Phys. B100 (1975) 313.
21. J.L. Cortes et al., PAR-LPTHE 80/31 (1980).
U. Baur, H. Fritzsch, Univ. München Preprint (1981).
22. B. Guberina, S. Nussinov, R.D. Peccei and R. Rückl, Phys. Lett. 85B (1979) 111; see also R.D. Peccei, R. Rückl MPI-PAE/PTH 75/81/1981.
23. H.H. Trilling, Rapporteur's talk at Int. Conf. on High

Energy Physics, Madison 1980.

24. M. Bander, D. Silverman and A. Soni, Phys. Rev. Lett 44 (1980) 7.

H. Fritzsch and P. Minkowsky, Phys. Lett. 90B (1980) 455; Nucl. Phys. B171 (1980) 413.

W. Bernreuther, O. Nachtmann and B. Stech, Z. Phys. C4 (1980) 257.

Acta Physica Austriaca, Suppl. XXIV, 275–306 (1982)

NONSTANDARD WEAK BOSONS[+]

by

J.J. SAKURAI[*]
Max-Planck-Institut für Physik und Astrophysik
Munich, Fed. Rep. Germany

All the low-energy successes of the standard electroweak gauge model are shown to follow in a more phenomenological model based on global SU(2) broken by γ-W^o mixing. Weinberg's mass predictions need not be valid. Connections with recent composite models of W and Z are also discussed.

INTRODUCTION

In the past several years the low-energy predictions of the standard electroweak gauge model of Glashow [1], Salam, Ward [2] and Weinberg [3] have been brilliantly

[*]Senior Scientist Awardee (Alexander von Humboldt-Stiftung) on leave from the University of California, Los Angeles.

[+]Lectures given at the XXI. Internationale Universitätswochen für Kernphysik, Schladming, Austria, February 25 - March 6, 1982.

confirmed by various neutral-current experiments [4]. It is yet to be tested whether the model is correct at values of q^2 where the electromagnetic and weak forces are expected to compete in strength. The most central and specific of the yet-to-be-verified high q^2 predictions are, of course, the fifteen-year-old mass predictions of Weinberg[3] which, when sharpened by the electroweak radiative corrections [5], lead numerically to

$$m_W = (82.0 \pm 2.4)\ \text{GeV}\ ,$$

$$m_Z = (93.0 \pm 2.0)\ \text{GeV}\ . \qquad (1)$$

These predictions now (meaning within several months?) being tested at the CERN $\bar{p}$ collider are most remarkable in that the quantitative details of electroweak physics at q^2 of order 8,000 GeV^2 (timelike) are inferred on the basis of our knowledge of weak interaction phenomena at q^2 of order 1-100 GeV^2 (spacelike).

Despite the extraordinary nature of these mass predictions the theoretical physics community appears to be in general agreement with the view that the W and Z bosons will be found at the mass values predicted by (1). The majority opinion - presumably shared also by the Royal Swedish Academy of Sciences - is best summarized by the following remark of Glashow [6] at the LEP Summer Study (September 1978):

> "Since the low-energy limit of unified theory is so well confirmed, few can doubt the truth of its central prediction: the existence of $W^{\pm}$ at $\sim$ 80 GeV and of Z^0 at $\sim$ 90 GeV."

In contrast to this majority view found in countless review papers and even in some text books, I would like to present in this lecture a minority view:
The extraordinary successes of the standard electroweak

model in accounting for low q^2 phenomena do not immediately imply the correctness of the model at high q^2. Specifically I would like to show that as yet there is no direct experimental evidence for weak-electromagnetic unification in the usual sense, that W and Z need not be gauge bosons, and that the Weinberg mass relations may still fail.

The line of investigation I am reporting here started more than four years ago by Bjorken [7] in a rather unpopular invited talk, "Alternatives to Gauge Theories", presented at the Ben Lee Memorial Conference, and was developed further by P.Q. Hung and myself [8]. Subsequently other heretics like Dombey [9] contributed to the propagation of these ideas. Furthermore in the past year or so, our point of view has come to attract an increasing amount of attention in connection with composite models of W and Z, as will be discussed towards the end of this lecture.

γ-W^o Mixing

Let us suppose that the history of neutral-current physics were different. Imagine that the low-energy phenomenology of the neutral-current interactions were first established experimentally, and at that time the theorists were not sophisticated enough to think about the gauge-theory ideology. How might the subject have been developed?

Given the existence of neutral currents as well as charged currents with large parity violation, the most natural thing is to assume (a) pure V-A to ensure maximal parity violation, and (b) weak isospin invariance or global SU(2) to accommodate the charged and neutral currents. So we may begin with

$$L_{eff} = (4G/\sqrt{2})\ \underline{J}_{\lambda} \cdot \underline{J}_{\lambda} \qquad (2)$$

where $\underline{J}_\lambda$ stands for a triplet of weak isospin currents built up of left-handed fermions. The form (2) is quite natural; indeed, almost immediately after the V-A proposal and long before the birth of the standard electroweak model, it was actually written down by Bludman [10] and shortly afterwards by Z'eldovich [11].

A closer look at the experimental data reveals that, although (2) is satisfactory for the charged-current interactions, its neutral-current predictions do not quite agree. For one thing the observed neutral currents are known not to be of the pure V-A form. To fit the neutral-current data it is found necessary to modify the interactions of the third current in such a way that the entire Lagrangian may look like

$$L_{eff} = (4G/\sqrt{2}) \sum_{\alpha=1,2} J_\lambda^\alpha J_\lambda^\alpha + (4G/\sqrt{2}) (J_\lambda^3 - \text{"}\sin^2\theta_W\text{"} J_\lambda^{em})^2 \tag{3}$$

where at this stage "$\sin^2\theta_W$" is some constant experimentally determined to be about 0.23. All the successes of the standard model achieved so far follow just as we let

$$J_\lambda^3 \rightarrow J_\lambda^3 - \text{"}\sin^2\theta_W\text{"} J_\lambda^{em} \tag{4}$$

in the SU(2) symmetric Lagrangian (2). In addition, if we assume that the weak interactions are mediated by a $W^{\pm,o}$ triplet coupled to weak isospin, the modification (4) implies that the W^o couplings have to be altered slightly as we switch on the electromagnetic couplings which are known to break SU(2). We are led to a particular way of breaking global SU(2), namely γ-W^o mixing represented by

$$L_{\gamma W} = -(1/4)\, \lambda\, (F_{\mu\nu} W_{\mu\nu}^3 + W_{\mu\nu}^3 F_{\mu\nu}) \,. \tag{5}$$

With this the γ-W^o junction goes like q^2, in conformity with electromagnetic gauge invariance.

For definiteness let us now consider simple low-order diagrams for elastic νq scattering. If there were no γ-W^o mixing, then we would have just W^o exchange; on the quark side the W^o is coupled to the third component of weak isospin formed out of u and d. See Fig. 1(a). But because of (5) this is not the whole story. We must also consider Fig. 1(b) where the quark emits a photon that converts itself into a W^o which, in turn, can interact with the neutrino. It is important to keep track of the q^2 dependence here. The $1/q^2$ behavior due to photon exchange just cancels the q^2 behavior coming from the γ-W^o junction. As a result, in the limit where $m_W{}^2$ is much larger than q^2, we have just a current-current coupling of the kind represented by the last term of (3).

The q^2 dependence of the effective $\mu\bar{\nu}\nu$ vertex that appears in Fig. 1(b) is what we expect from a neutrino with a finite charge radius; because the neutrino is chargeless, its electromagnetic (Dirac) form factor must go like q^2 for small q^2. Historically Bjorken [7] first noted that, as far as low-energy neutrino-induced neutral current processes are concerned, all the consequences of the standard model follow just by postulating, in addition to global SU(2), a charge radius for the neutrino with magnitude

$$\langle r^2 \rangle = (6G/\sqrt{2})\, \sin^2\theta_W/\pi\alpha \quad . \qquad (6)$$

However, to fit the electron-deuteron parity violation data at SLAC, the energy dependence of electron-positron annihilation into muon pairs at PETRA, etc., new weak-interaction contributions to the charge radii of e, μ, u and d must also be postulated. Furthermore the amount of the contribution needed in each case is just what we expect if the charge radius arises in a "universal manner" from γ-W^o mixing.

The effective γ-W^o interaction (6) is not a priori weak; to treat its effect quantitatively to all orders, the usual approach based on low-order diagrams is not so convenient. It is much better to use the propagator matrix formalism, as we will see in a moment.

The problem of gauge-invariant mixing between a spin-one object and the photon was discussed extensively in the sixties with reference to vector meson dominance. I even wrote a book on this and related subjects. However, the most transparent (and correct!) treatment of this mixing problem is found in a 1961 paper of Kobzarev, Okun' and Pomeranchuk [13], based on the propagator matrix formalism, whose existence I became aware of only four years ago. I now rewrite their paper in a manner suitable for our situation.

Suppose we have just one particle with no mixing. The propagator in momentum space is

$$D(q^2) = 1/(q^2 + m^2) \quad , \tag{7}$$

and the mass is determined by looking at the zero of the inverse propagator

$$q^2 + m^2 = 0 \; . \tag{8}$$

If we have two particles with mixing, the propagator is a 2 x 2 matrix, denoted by $\Delta(q^2)$, and the masses are determined by the roots of the equation

$$\det\,(\Delta^{-1})\Big|_{q^2 = -m_i^2} = 0 (i = 1,2) . \tag{9}$$

We now specialize to the γ-W^o complex. Our basic Lagrangian is written as

$$L = \frac{1}{4} \phi^T_{\mu\nu} K \phi_{\mu\nu} - \phi^T_\mu M^2 \phi_\mu + \phi^T_\mu J_\mu \tag{10}$$

where

$$\phi_{\mu\nu} = \begin{bmatrix} F_{\mu\nu} \\ W^o_{\mu\nu} \end{bmatrix} , \qquad \phi_\mu = \begin{bmatrix} A_\mu \\ W^o_\mu \end{bmatrix} , \qquad J_\mu = \begin{bmatrix} eJ^{em}_\mu \\ gJ^3_\mu \end{bmatrix} . \tag{11}$$

There is a mixing term present only in the kinetic part, not in the mass part ("current mixing" in the unfortunate terminology of Kroll, Lee and Zumino [14]):

$$K = \begin{bmatrix} & \lambda \\ \lambda & 1 \end{bmatrix} , \qquad M^2 = \begin{bmatrix} 0 & 0 \\ 0 & m^2_W \end{bmatrix} . \tag{12}$$

Notice that the diagonal elements in the M^2 matrix are 0 and m^2_W; we start with a W^o mass identical to the charged W mass, in agreement with the idea of global SU(2) before mixing.

The Lagrangian (10) can easily be shown to lead to the inverse propagator matrix

$$\Delta^{-1}(q^2) = \begin{bmatrix} q^2 & \lambda q^2 \\ \lambda q^2 & q^2 + m^2_W \end{bmatrix} . \tag{13}$$

The pho on and the Z mass are to be obtained by looking for the zeros of det (Δ^{-1}). First, we see that

$$\det (\Delta^{-1}(q^2)) \Big|_{q^2 = 0} = 0 \tag{14}$$

is automatically satisfied. An initially massless photon remains massless in the presence of the gauge invariant mixing term (5), a hardly surprising result. More interesting is the second root obtained by solving

$$\det (\Delta^{-1}(q^2)) \Big|_{q^2 = -m^2_Z} = 0 \tag{15}$$

with m_Z^2 assumed to be nonvanishing where, after mixing, we have renamed the neutral boson

$$W^o \xrightarrow{\text{mixing}} Z \ . \tag{16}$$

Specifically (15) leads to

$$\begin{vmatrix} -m_Z^2 & -\lambda m_Z^2 \\ -\lambda m_Z^2 & m_W^2 - m_Z^2 \end{vmatrix} = 0 \ . \tag{17}$$

In this way we obtain a very important mass formula [8]

$$m_Z^2 = m_W^2/(1-\lambda^2) \ . \tag{18}$$

Note that, as the W^o changes into Z, its mass is raised by a factor of $1/\sqrt{(1-\lambda^2)}$.

We can also compute the current-current interaction that follows from (10) by writing $(1/2)J^T\Delta J$ in terms of the Z boson propagator. A straightforward computation gives [8]

$$L_{eff} = \frac{1}{2}\{e^2 J_\lambda^{em} \frac{1}{q^2} J_\lambda^{em} + \frac{g^2}{1-\lambda^2} [J_\lambda^3 - \frac{e}{g}\lambda J_\lambda^{em}] \frac{1}{q^2+m_Z^2} [J_\lambda^3 - \frac{e}{g}\lambda J_\lambda^{em}]\}. \tag{19}$$

The first term is due to photon exchange, the second due to Z exchange. The low q^2 limit of (19) is of vital interest. The dimensionless constant for the neutral-current interations, according to (19), is larger than g^2 by a factor of $1/(1-\lambda^2)$, but m_Z^2 is also heavier by the same factor [see (18)]. As a result, in the $q^2 \to 0$ limit, the neutral-current strength is unaltered from its global SU(2) value; it is still characterized by the same G (a la Fermi) as in the charged-current interactions. Explictly we have for the neutral-current part of (19)

$$\mathcal{L}^{NC}(q^2 \cong 0) = (4G/\sqrt{2})[J_\lambda^3 - (e\,\lambda/g)J_\lambda^{em}]^2 \tag{20}$$

where we have used the charged-current relation

$$G/\sqrt{2} = g^2/8m_W^2 \quad . \tag{21}$$

What the experimentalists call "ρ", the neutral-to-charged current strength ratio, is seen to be unity. The only modification is in the structure of the current, which is precisely of the form (3) provided we identify

$$e\lambda/g = \text{"}\sin^2\theta_W\text{"} \quad . \tag{22}$$

So our very simple considerations based on global SU(2) broken by γ-W^o mixing give rise to exactly the same low q^2 limit - both in form (the J^3 - "$\sin^2\theta_W$"J^{em} combination) and in strength ("ρ" = 1) - as the standard electroweak gauge model [8].

Weak-Boson Mass Formula

Even though our low q^2 predictions are identical to those of the standard model, there can be substantial differences at high q^2. To see this let us go back to the weak-boson mass relation (18) and rewrite it using (22):

$$m_Z^2 = m_W^2/[1-(g/e)^2\text{"}\sin^4\theta_W\text{"}]. \tag{23}$$

Furthermore we can eliminate g in favor of Fermi's G and m_W^2 [see(21)]. The final result can then be written as [15]

$$m_Z^2 = m_W^2/[1-(m_W/37.4 \text{ GeV})^2\text{"}\sin^4\theta_W\text{"}] \quad . \tag{24}$$

Notice that we cannot obtain m_W and m_Z separately even if $\sin^2\theta_W$ is known. However, the boson mass formula (24) is nontrivial in that, knowing one of them (say, m_W), we can infer the value of the other (m_Z). See Fig. 2.

In this model what the experimentalists call "ρ" is unity; yet m_Z^2 is, in general, not equal to $m_W^2/\cos^2\theta_W$. From the mere fact that "ρ" is unity experimentally [4] (within a few %), we cannot conclude that the Z mass is equal to $1/\cos\theta_W$ times the W mass. After all, "ρ" = 1 is nothing more than the statement that global SU(2) survives at $q^2 = 0$ as far as the absolute strength of the neutral-current interactions is concerned.

There is another important point that follows from our formalism. We first note that the mixing parameter λ cannot exceed unity in magnitude. Otherwise after diagonalization we would not have the usual particle interpretations for Z. A practical consequence of this is that the denominator of (24) cannot become non-negative, which leads to an absolute upper bound for the $W^{\pm}$ mass:

$$m_W < 37.4 \text{ GeV}/\text{"}\sin^2\theta_W\text{"} \cong 170 \text{ GeV for "}\sin^2\theta_W\text{"} \cong 0.22, \quad (25)$$

a result first obtained by Bjorken [7] from somewhat more general considerations. The inequality (25) is quite remarkable. From the fact that "$\sin^2\theta_W$" is no smaller than 0.22, we have deduced that the range of the charged-current interactions cannot be shorter than

$$(1/170) \text{ GeV}^{-1} \cong 1.2 \times 10^{-16} \text{ cm}. \quad (26)$$

A multi-boson generalization of (25) will appear later.

Unlike the $W^{\pm}$ mass there is no analogous bound for the Z mass. Indeed, as the W mass approaches its upper bound, the Z mass increases indefinitely; see (24) and Fig. 2. However, there may be other reasons for believing that the Z mass cannot be ridiculously high. For one thing a high Z mass would imply a large dimensionless constant for the neutral-current interactions, which makes it difficult to understand the successes of lowest order calculations for weak interaction processes. As Llewellyn

Smith [16] argues, it is consistent to use single Z exchange to describe the low-energy interactions only if m_Z is substantially less than the center-of-mass energy $\sqrt{s_o}$ at which unitarity violation takes place. By requiring m_Z to be less than half of $\sqrt{s_o}$, he predicts

$$m_W < 140 \text{ GeV}, \quad m_Z < 280 \text{ GeV}. \tag{27}$$

The Unification Condition

Compare our mass formula (24) with Weinberg's results [3]

$$m_W = 37.4 \text{ GeV}/\sin\theta_W ,$$

$$m_Z = m_W/\cos\theta_W . \tag{28}$$

Clearly Weinberg has more predictions. This is not surprising. In our approach based on γ-W^o mixing, in addition to e and G there are two independent parameters e/g and λ (or equivalently e/g and "$\sin\theta_W$"). In contrast in the standard electroweak model is only one adjustable parameter e/g, which is equal to $\sin\theta_W$.

Weinberg's relations (28), represented by a point in Fig. 2, are seen to be a special case of our formula (24); they are obtainable by imposing a particular relation between the coupling constant ratio e/g and the mixing coefficient λ, viz.

$$e/g = \lambda = \sin\theta_W . \tag{29}$$

We wish to call this the unification condition. Let us write it as

$$e = g \sin\theta_W . \tag{30}$$

This relation, first written down by Glashow [1] in 1961, connects measurable quantities in the electromagnetic and weak interactions. The two quantities, g and $\sin\theta_W$, that appear on the right-hand side of (30) can be deduced, respectively, by measuring the decay width of $W^- \to e^-\bar{\nu}_e$ [or, more practically, by knowing the W mass and using (21)] and by studying the neutral-current structure at low energies, as has been done already. So what we usually regard as purely weak interaction quantities are sufficient to determine Millikan's electronic charge that appears on the left-hand side of (30), hence weak-electromagnetic unification! But this unification condition has not yet been tested experimentally because the W mass has not been determined, and as a consequence g is still an unknown quantity. The fact that eight different ways to determine $\sin^2\theta_W$ agree more or less with each other does not throw light on the question of whether or not the unification condition (29) is correct. From the empirical observation that the neutral currents can be written as a linear combination of J_λ^3 and J_λ^{em} we cannot conclude that the weak and electromagnectic forces are unified.

The main message I am trying to convey is the following. The extraordinary successes of the standard electroweak gauge model in accounting for low-energy neutral-current data also follow in a wider framework based on global SU(2) broken by γ-W^o mixing. These successes by no means guarantee Weinberg's mass relation and are, in fact, logically independent of the unification idea for which the 1979 Nobel Prize was given. So we must still spend \$ 2×10^8 or so to see whether Weinberg is really right.

Let us examine a little more deeply the physical meaning of the unification condition. We go to high q^2 regions and look at the behavior of the effective interaction (19). Let us rewrite it this time using J_λ^3 and J_λ^Y (the weak hypercharge current) defined by

$$J_\lambda^Y = J_\lambda^{em} - J_\lambda^3 . \tag{31}$$

For $q^2 >> m_W^2, m_Z^2$, the effective interaction (19) reduces to [8]

$$L_{eff} \cong \frac{1}{2q^2} \{\frac{e^2}{1-\lambda^2} J_\lambda^Y J_\lambda^Y + (\frac{e^2+g^2-2eg}{1-\lambda^2}) J_\lambda^3 J_\lambda^3 + [\frac{2e(e-\lambda g)}{1-\lambda^2}] J_\lambda^Y J_\lambda^3\}. \tag{32}$$

If there is no constraint between e/g and λ, SU(2) is still violated even at high values of q^2. For SU(2) to become exact we must have

$$L_{eff} \cong \frac{1}{2q^2} (g^2 J_\lambda^3 J_\lambda^3 + \text{const } J_\lambda^Y J_\lambda^Y). \tag{33}$$

Here the coupling coefficient of $J_\lambda^3 J_\lambda^3$ is g^2, the same as the charged-current constant, and the $J_\lambda^Y J_\lambda^3$ term that violates SU(2) is absent. The second term of (33) can be regarded as being due to the exchange of a singlet B boson coupled to weak hypercharge.

Comparing the SU(2) expected behavior (33) with the general behavior (32), we infer that SU(2) becomes exact at high q^2 if e/g is set equal to λ, which is precisely the unification condition (29). So the unification condition ensures a kind of "asymptotic restoration" of SU(2) [7], [8]; the interactions of the $W^{\pm,o}$ triplet again become weak-isospin invariant while the photon, in Bjorken's words, is "transmogrified" into the U(1) B boson coupled to pure weak hypercharge. Needless to say, this restoration idea refers to a short-distance behavior yet to be tested experimentally. It is by no means guaranteed by the extraordinary successes of the standard model in accounting for low q^2 phenomenology.

It has been argued that the imposition of the unification condition is almost sufficient to convert the phenomenological γ-W^o mixing model into the conventional electroweak gauge model. (I say here "almost" because the Higgs sector, or more generally the weak-boson-mass generation mechanism, is yet to be specified.) There are two points worth mentioning in this connection. First, suppose we compute

$$\nu_e + \bar{\nu}_e \to W^+ + W^- \tag{34}$$

due to electron exchange (in the t channel) and Z exchange (in the s channel) within the framework of our general γ-W^0 mixing model. When λ is assumed to be unrelated to e/g, the amplitude for the production of longitudinally polarized $W^\pm$'s is seen to violate tree unitarity. However, when we impose the unification condition, we obtain a high energy behavior as good as that which follows from the conventional electroweak gauge model [8], [17].

Next, I would like to report on an interesting observation made by Feynman, who, in 1978-79, gave lectures on weak interactions at Caltech. Rather than digesting other people's papers, he wanted to reconstruct the standard electroweak model in his own original way. To this end, instead of the usual SU(2) × U(1) gauge interactions of $W^{\pm,0}$ and B, he considers Yang-Mills type $W^{\pm,0}$ interactions, both among themselves and with quarks and leptons, and the photon (<u>not</u> B) interactions that take into account the fact that the $W^\pm$, as well as the quarks and leptons, are electrically charged. When he attempts to impose the Yang-Mills gauge invariance and the electromagnetic gauge invariance, he realizes that something is lacking for the two sets of invariance to be preserved simultaneously. The needed term turns out to be precisely the γ-W^0 mixing interaction <u>with</u> the unification $\lambda = e/g$. Feynman never published this observation; so I published it in my 1979 Haiwaii lectures [18] which interested people may consult for details.

Multiboson Models

We have considered so far a broken SU(2) model based on just one W triplet. We may still keep the ideal of global SU(2) before mixing with the photon but allow the possibility that many weak-boson triplets - or even a continuum of weak quanta - may participate in the weak interactions. Suppose there are N triplets altogether all coupled to weak isospin. The strength of the ith triplet to $\underline{J}_\lambda$ is characterized by g_i, and the neutral member of the triplet mixes with the photon with coupling constant λ_i. We can derive the analog of (19), which turns out to be [19], [20]

$$L_{eff} = \frac{1}{2}\left[\frac{e^2}{q^2}(J_\lambda^{em})^2 + \sum_{i=1}^{N} \frac{f_i^2(J_\lambda^3 - b_i J_\lambda^{em})^2}{q^2+m_{Zi}^2}\right] \tag{35}$$

with

$$\sum_{i=1}^{N}(f_i^2/m_{Zi}^2) = \sum_{i=1}^{N}(g_i^2/m_{Wi}^2) = 8G/\sqrt{2} ,$$

$$\sum_{i=1}^{N}(f_i^2 b_i/m_{Zi}^2) = e \sum_{i=1}^{N}(\lambda_i g_i/m_{Wi}^2) = (8G/\sqrt{2})\text{"}\sin^2\theta_W\text{"} \ . \tag{36}$$

The low q^2 limit of the neutral-current part of (35) is

$$L^{NC}(q^2 \simeq 0) = (4G/\sqrt{2})\ \{(J_\lambda^3 - \text{"}\sin^2\theta_W\text{"}J_\lambda^{em})^2 + C(J_\lambda^{em})^2\} \tag{37}$$

where the coefficient C is given by

$$C = \frac{1}{2}e^2 \sum_{i,j}\left[\frac{(\lambda_i g_i - \lambda_j g_j)^2}{m_{Wi}^2 m_{Wj}^2}\right] / [\sum_k (g_k^2/m_k^2)]^2 \geq 0 \ . \tag{38}$$

Comparing (37) with (20), we observe that this generalized model does give low q^2 results different from those of the standard model by the presence of an extra $(J_\lambda^{em})^2$ term, usually referred to as the C term [21].

Physically speaking, the coefficient C is a measure of deviations from the single Z hypothesis. Gounaris and Schildknecht [22] have derived a sum rule for this coefficient in terms of weighted cross sections in e^+e^- collisions:

$$C = \frac{1}{16}\,\frac{\int ds\ \sigma(e^+e^-\to\gamma, Z_1, Z_2, \ldots Z_N\to\mathrm{all})/s - \int ds\ \sigma(e^+e^-\to\gamma, Z\to\mathrm{all})/s}{[1 + (1-4\sin^2\theta_W)^2]\int ds\sigma(e^+e^-\to\ , Z\to\mathrm{all})/s} \tag{39}$$

where $\int ds\sigma(e^+e^-\to\ , Z\to\mathrm{all})/s$ is to be evaluated in a single Z model; the value of this integral can be shown to be independent of the Z mass once "$\sin^2\theta_W$" is given.

At present only e^+e^- data at low energies are available; so we cannot evaluate (39) directly. Furthermore we cannot detect the presence of the C term in neutrino-induced reactions because the neutrino is chargeless, nor in the SLAC parity experiment because $(J_\lambda^{em})^2$ is purely parity-conserving. Fortunately the C term could reveal its presence in

$$e^+ + e^- \to e^+ + e^-,\ \mu^+ + \mu^- . \tag{40}$$

The current PETRA limit reported elsewhere at this winter school appears to be around

$$C < 0.025. \quad (95\%\ \mathrm{CL}) \tag{41}$$

The weak-boson mass formula (24) has a multiboson generalization. First, we define the average boson masses by

$$1/\bar{m}_W^2 \equiv \{\sum_i (g_i^2/m_{Wi}^4)\}/\{\sum_k (g_k^2/m_{Wk}^2)\}$$

$$1/\bar{m}_Z^2 \equiv \{\sum_i (f_i^2/m_{Zi}^4)\}/\{\sum_k (f_k^2/m_{Zk}^2)\} \ . \qquad (42)$$

Then the multiboson analog of (24) can be derived to be [19]

$$\bar{m}_Z^2 = \frac{\bar{m}_W^2}{1-(\bar{m}_W/37.4 \text{ GeV})^2 \text{"}\sin^4\theta_W\text{"}} \ . \qquad (43)$$

Again we note the positivity requirement

$$\bar{m}_W < 170 \text{ GeV } (\text{"}\sin^2\theta_W\text{"} = 0.22). \qquad (44)$$

We see that the main activities in the charged-current channel must take place at energies below 170 GeV even in multiboson models.

In a single Z model we have a sum rule for the colliding beam cross section R,

$$\bar{R} = \int_{(Z \text{ region})} ds R/s \equiv (3/4\ \pi\alpha^2) \int ds\sigma\ (e^+e^- \to Z \to \text{all}) =$$

$$= (3/8\ \alpha^2)[(1-4\sin^2\theta_W)^2+1]Gm_Z^2/\sqrt{2}, \qquad (45)$$

which easily follows from integrating the usual one-level resonance formula. If m_Z assumes the standard model value, the right-hand side of (45) is a very big number, $\sim$500. Using Schwarz-inequality type arguments, we can derive an analogous expression in the multiboson case, which is now an inequality [20], [22]

$$\bar{R} \equiv (3/4\ \pi\alpha^2) \sum_i \int_{(Z_i \text{ region})} ds\ \sigma(e^+e^- \to Z_i \to \text{all}) \geq (3/8\ \alpha^2)[(1-4\sin^2\theta_W)^2+1]G\bar{m}_Z^2/\sqrt{2}. \tag{46}$$

This is great! If we build an e^+e^- colliding beam apparatus that covers $\sqrt{s} \sim \bar{m}_Z$, a large cross section is guaranteed. But the bad news is that $\bar{m}_Z$ need not necessarily lie in the energy range covered by LEP, SLC or some other favorite project of yours. It can, for instance, be at 160 GeV. Clearly it is safer to plan an e^+e^- collider with as high an energy as the funding agencies - or ultimately the taxpayers - allow.

Composite W and Z

In the past few years a number of authors [23], [24], [25] have speculated on the possibility that not only the quarks and leptons but also the W and Z bosons are bound states of some more fundamental objects - preons, rishons, haplons, etc. In such composite models the weak interactions we observe - nuclear beta decay, inelastic neutrino scattering, etc. - are phenomenological manifestations of a new kind of superstrong confinement dynamics mediated by hypergluons in much the same way as the hadronic interactions we used to study in the sixties - $\rho \to \pi\pi$, Kp scattering, etc. - are now regarded as phenomenological manifestations of QCD, the fundamental dynamics of strong interactions. From this point of view the weak forces we observe are like Van der Waal forces.

A historical analogy may be of some pedagogical value here. In 1960 I thought that the particles later identified as ρ, ω and ϕ were the gauge particles of strong interactions [26]. (I was trying to construct a gauge theory of strong interactions at a time when the concept of colored quarks was yet to be born!) By 1970 most people came to accept the view that ρ, ω and ϕ are not "elementary" gauge particles but quark-antiquark bound states. In 1980, right after the originators of the standard electroweak gauge model were canonized in Stockholm, the physics community almost unanimously believed that W and Z were the gauge particles of weak interactions. By 1980 most people will perhaps be led to the view that W and Z are composites of some more fundamental objects, just as ρ and ω are. The gauge theory of strong interactions based on ρ, etc., is now considered to be obsolete. Will the gauge theory of weak interactions based on W and Z become obsolete by 1990?

Numerous workers have constructed composite models. At this moment we don't yet know which, if any, of the various composite models proposed so far is likely to survive. Here I mention just two of them, a model proposed by Abbott and Farhi [24] and another model proposed by Fritzsch and Mandelbaum [25]. I regard these two models as representative examples of recent unorthodox approaches to electroweak physics rather than as candidates of the "ultimate theory".

The fundamental constituents of the Abbott-Farhi model are a non-Hermitian scalar doublet and a fermion doublet, both confined. In this model a familiar left-handed quark or lepton is a composite object made up of an "elementary" fermion and a scalar boson. The weak bosons W and Z, which are more relevant to my talk here, are supposed to emerge as bound states of scalars

$$W_\mu^+ = \phi^+ D_\mu \phi^o - \phi^o D_\mu \phi^+, \text{ etc.},$$

where D_μ stands for the covariant derivative appropriate for the new underlying gauge theory.

The second model I consider is QHD ("Quantum Haplo

Dynamics") of Fritzsch and Mandelbaum [25]. (An anonymous Greek physicist pointed out to me that "haplo" means "naive" as well as "single".) Here the fundamental constituents are a fermion doublet α, β with $Q=-1/2$ and $Q= +\frac{1}{2}$, and two scalar singlets x, y with $Q=-1/6$ and $1/2$, respectively. The leptons and quarks are bound states of spin 1/2 fermions and spinless bosons:

$$\nu_e = (\bar{\alpha}\ \bar{y}), \qquad u = (\bar{\alpha}\ \bar{x}) ,$$

$$e^- = (\bar{\beta}\ \bar{y}), \qquad d = (\bar{\beta}\ \bar{x}) . \tag{47}$$

In contrast to the Abbott-Farhi model, the weak bosons are more like the ρ meson in that they are fermion-antifermion bound states:

$$W^+ = (\bar{\alpha}\ \beta), \quad W^- = (\bar{\beta}\ \alpha), \quad W^0 = [(\bar{\alpha}\ \alpha) - (\bar{\beta}\ \beta)]/\sqrt{2} . \tag{48}$$

In both models the W interactions with the quarks and leptons are "residual" interactions of the basic confinement dynamics. Fig. 3 shows how the vertex

$$\nu_e \leftrightarrow W^+ + e^- \tag{49}$$

may look in terms of the fundamental constituents of the Fritzsch-Mandelbaum model. In the absence of the electromagnetic couplings the phenomenological W interactions with the left-handed quarks and leptons can be constructed to satisfy global SU(2) â la Bludman [10].

In these models, even though the $W^{\pm,0}$ are composite, the photon, like the gluon and hypergluon, is elementary. From a certain point of view this is highly satisfactory. The "elementary" gauge bosons of the world all remain exactly massless. There is no such thing as a spontaneously broken local gauge symmetry-except in condensed matter physics. The ugly Higgs world is banished away.

The observed neutral-current structure at low q^2 can be obtained by taking advantage of the γ-W^o mixing mechanism discussed earlier. This mechanism now looks completely analogous to γ-ρ^o mixing. See Fig. 4. There is no reason to expect that the constant λ that characterizes the strength of Fig. 4(b) is related to the coupling constant ratio e:g where g characterizes the strength of the phenomenological $\nu_e \leftrightarrow W^+e^-$ vertex appearing in Fig. 3. In models where the photon is elementary but the weak bosons are composite, the "fundamental" electromagnetic forces and the "phenomenological" weak forces are not unified.

Because the unification condition, in general, fails in this class of composite models, the Weinberg mass relations are predicted to be invalid. The weaker mass relation (24) may still hold provided excited weak bosons and/or a weak continuum make negligible contributions to low q^2 physics.

It is, however, possible that the unification condition (29) may be approximately satisfied for dynamical reasons. For example, the underlying confinement dynamics may respect SU(2) at asymptotically high values of q^2, which, as we saw earlier, leads to the unification condition. The unification condition may also be realized if the W boson "imitates" the vector-meson dominance idea of the sixties [12], [14], [27]. If the isovector electromagnetic form factor of the neutrino, etc., is completely dominated by a single W^o pole, we can easily show that λ must be equal to e/g [28].

If the unification condition is a consequence of some dynamical approximation, the Weinberg mass relations are most likely to be satisfied not exactly but only approximately, say, within ±20%. In contrast, in the standard electroweak gauge model, the Weinberg mass predictions are, in principle, as accurate as the g-2 prediction for the muon.

Strength of the γ-W^o Junction

I have repeatedly emphasized that the γ-W^o junction is very much like the γ-ρ junction. But empirically the γ-W^o constant λ^2 is much larger than the analogous γ-ρ^o constant. From $\rho^o \to e^+e^-$ and ρ dominance in photoproduction, etc., the γ-ρ^o constant normalized in the same manner can be inferred to be [12], [29]

$$\lambda^2_{\gamma\rho} \cong 1/300, \tag{50}$$

which is of order 1/137. In contrast, for the γ-W^o junction we have from (22) and the empirical value of $\sin^2\theta_W$

$$\lambda^2 = \text{"}\sin^4\theta_W\text{"}/(e/g)^2 \cong 0.23\ (m_W/79\ \text{GeV})^2. \tag{51}$$

How are we going to account for this big difference?

I now show that a value of λ^2 as large as (51) is not at all unreasonable in composite models characterized by a large mass scale. My argument is based on the concept of Q^2 duality [30], [31], which I briefly review.

Consider a series of vector mesons $V_1, V_2, \ldots$ which can be regarded as quark-antiquark bound states $(Q\bar{Q})$. The Q^2 duality hypothesis states that the vector meson peaks we see in the cross section ratio R for electron-positron annihilations into hadrons, when suitably averaged, closely approximate the free quark-pair cross section ratio for

$$e^+ + e^- \to Q + \bar{Q}. \tag{52}$$

Quantitatively it enables us to relate the strength of

the $\gamma \to V_i$ junction to the vector meson spacing and the constituent charge. This pypothesis can be justified using a finite-energy (or rather finite $-Q^2$) sum rule [31] and also has been tested in the theoretical laboratories of nonrelativistic potential models [32] and two-dimensional QCD [33] with remarkable successes.

When applied to our situation, the Q^2 duality hypothesis leads to [34]

$$\lambda^2 = (\alpha/3\pi) N_H \, e_H^2/m_W^2 \, P(m_W^2) . \tag{53}$$

Here e_H stands for the constituent charge, N_H denotes the number of hypercolors in the theory, and $P(m_W^2)$ is the density of vector boson spectrum, i.e., the number of vector bosons per unit mass squared interval. For definiteness we may use the Fritzsch-Mandelbaum model. The constituent charge is given by

$$e_H^2 = \{\frac{1}{\sqrt{2}} [\frac{1}{2} - (- \frac{1}{2})]\}^2 = \frac{1}{2} . \tag{54}$$

The hypercolor factor N_H is unspecified, but we may take it to be 3 just as in QCD. (There may be an additional factor of 3 if α and β have ordinary colors also.) In the absence of detailed dynamics the hardest thing to guess is $P(m_W^2)$. It is undoubtedly related to the mass scale of QHD. An analogy with QCD may be helpful here. The scale of QCD is set by the famous parameter Λ_C, estimated to be of order 0.1 GeV. With this value of Λ_C we know empirically that $P(m^2)$ in hadron spectroscopy turns out to be of order 1 GeV^{-2}, a typical Regge slope for hadrons; more specifically, $P(m^2)$ for the I=1 vector mesons is computed to be 0.8 GeV^{-2} from $\rho(0.78)$ and $\rho'(1.60)$. In the QHD case the parameter Λ_H, analogous to Λ_C, is conjectured by Fritzsch and Mandelbaum to be of order 0.1 TeV. If QHD is like QCD with 1 GeV replaced by 1 TeV, a reasonable guess for $P(m_W^2)$ will be in the neighborhood of 1 TeV^{-2}.

From (51) and (53) we can deduce $P(m_W^2)$ needed to fit the observed value of $\sin^2\theta_W$. This leads to [34]

$$P(m_W^2) = 0.81\ (79\ GeV/m_W)^4\ TeV^{-2}\ , \tag{55}$$

in excellent accord with our conjecture that $P(m_W^2)$ is of the order 1 TeV^{-2}. So the observed large value of λ^2, or of "$\sin^2\theta_W$", is no mystery in QHD characterized by a large ($\sim$1 TeV) mass scale.

We see that W and Z are anomalously light on the mass scale of QHD:

$$m_{W,Z}^2 \sim 0.01\ TeV^2 << 1\ TeV^2. \tag{56}$$

From this point of view W and Z are like the pion of QCD. The importance of having a large mass scale in QHD can also be inferred by estimating the "size" of W and Z from the bound-state wave function needed to explain the observed large value of λ^2 [35].

Because the weak boson spacing is predicted to be of order 1 TeV^2, "excited" W and Z are likely to play insignificant roles in electroweak physics of the eighties and even of the nineties. The mass formula (24) based on the single pole hypothesis is probably an excellent approximation to the more general mass formula (43). Furthermore the C coefficient that appears in multiboson models is probably negligible.

Even though I have estimated the strength of γ-W^o mixing in the particular context of the Fritzsch-Mandelbaum model, I believe that the main conclusion reached is largely independent of the specific model used. To make λ^2 sizable, the important thing is that the mass squared difference between the photon and the W boson (before mixing) is very much smaller than the characteristic squared mass of the dynamical mechanism responsible for the W boson as a bound state. On the scale of QHD the W boson looks nearly massless, almost like the photon. It then follows, as in any quantum-mechanical system with a small energy-level difference, that γ and W^o must be strongly mixed.

Conclusion

I now summarize the main results.

(a) The striking successes of the standard electroweak gauge model in accounting for low q^2 neutral-current data can be reproduced equally well in a more phenomenological model based on global SU(2) broken by γ-W^o mixing.

(b) The boson masses, m_W and m_Z, need not satisfy the Weinberg mass relations. Instead we expect the weaker mass formula (24). There is an upper bound for m_W, in the neighborhood of 170 GeV.

(c) The validity of the Weinberg mass relations or, more generally, the correctness of the weak-electromagnetic unification idea, rests on asymptotic restoration of SU(2), a yet-to-be-tested short-distance behavior.

(d) Multiboson or continuum generalizations also lead to $m_W < 170$ GeV in the charged-current channel.

(e) Models with composite W and Z (but elementary γ) must rely on the γ-W^o mixing mechanism to reproduce the observed low q^2 behavior of neutral-current phenomenology.

(f) It is not unreasonable to obtain the observed "large" γ-W^o mixing constant in composite model characterized by a TeV mass scale.

If the standard electroweak gauge model fails at high timelike values of q^2, we will undoubtedly enter one of the most exciting eras in the history of particle physics.

REFERENCES

1. S.L. Glashow, Nucl.Phys. 22, (1961) 579.
2. A. Salam and J.C. Ward, Phys. Lett. 13 (1964) 168; A. Salam, Elementary Particle Theory, ed. N. Svartholm (Almquist and Wiksell, Stockholm, 1968), p. 367.
3. S. Weinberg, Phys. Rev. Lett. 19, (1967) 1264.
4. For a review see e.g. J.E. Kim, P. Langacker, M. Levine and H.H. Williams, Rev. Mod. Phys. 53, (1981) 211; P.Q.Hung and J.J. Sakurai, Ann. Revs. Nucl. Sci. 31, (1981) 375.
5. W.J. Marciano and A. Sirlin, Nucl. Phys. B189, (1981) 442; C.H. Llewellyn Smith and J.F. Wheater, Phys.Lett. 105B, (1981) 486.
6. S.L. Glashow, Proc. LEP Summer Study, CERN Yellow Report 79-01 (1979).
7. J.D. Bjorken, Proc. Ben Lee Memorial International Conference on Parity Nonconservation, Weak Neutral Currents and Gauge Theories, ed. D.B. Cline and F.E. Mills (Hardwood Academic Pub., London, 1978), p. 701; J.D. Bjorken, Proc. XIII Rencontre de Moriond, ed. Tran Than Van (Editions Frontieres Dreux, France, 1978), Vol. II, p. 491; J.D. Bjorken, Phys. Rev. D19, (1979) 335.
8. P.Q. Hung and J.J. Sakurai, Nucl. Phys. B143 , (1978) 81.
9. N. Dombey, Proc. 1981 Banff Summer Institute on Particles and Fields (to be published).
10. S.A. Bludman, Nuov. Cim 9, (1958) 443.
11. Y.B. Zel'dovich, JETP 9, (1959) 682.
12. J.J. Sakurai, Currents and Mesons (University of Chicago Press, 1969).
13. I. Yu. Kobzarev, L.B. Okun' and I. Ya. Pomeranchuk, JETP 14, (1962) 355.

14. N.M. Kroll, T.D. Lee and B. Zumino, Phys. Rev. 157, (1967) 1376.
15. The boson mass formula (24) first appeared in this form in Ref. 8. However, it is implicit also in Bjorken's work (Ref. 7).
16. C.H. Llewellyn Smith, Proc. 1981 Banff Summer Institute on Particles and Fields (to be published).
17 This result is anticipated in view of the well-known theorem of C.H.Llewellyn Smith, Phys. Lett. 46B, (1973) 233, and J.M. Cornwall, D. Levin and G. Tiktopoulos, Phys. Rev. D10, (1974) 1145, stating that tree unitarity enforces gauge invariant vertices.
18. J.J. Sakurai, Proc. Eight Hawaii Topical Conference on Particle Physics (University of Hawaii Press, Honolulu, 1981).
19. E.H. de Groot and D. Schildknecht, Zeit. Phys. C10, (1981) 55.
20. N. Wright, UCLA preprint, UCLA/81/TEP/14.
21. The importance of the C term in broken SU(2) models was first stressed by J.D. Bjorken (Ref. 7).
22. G.J. Gounaris and D. Schildknecht, Zeit. Phys. C12, (1982) 57.
23. H. Terazawa, Y. Chikashige and K. Akama, Phys. Rev. D15, (1977) 480;
H. Harari, Phys. Lett. 86B, (1979) 83;
M.A. Shupe, Phys. Lett. 86B, (1979) 87;
O.Greenberg and J. Sucher, Phys. Lett. 99B (1981) 339;
R. Barbieri, A. Masiero and R. Mohapatra, Phys. Lett. 105B, (1981) 369.
24. L. Abbott and E. Farhi, Phys. Lett. 101B, (1981) 69.
25. H. Fritzsch and G. Mandelbaum, Phys. Lett. 102B, (1981) 319.
26. J.J. Sakurai, Ann. Phys. 11, (1960) 1.
27. M. Gell-Mann and F. Zachariasen, Phys. Rev. 124, (1961) 953.
28. This line of thinking has recently been advocated particularly by R. Kögerler and D. Schildknecht, CERN preprint

TH 3231-CERN (1982).

29. In the vector dominance notation of Ref. 12, $\lambda_{\gamma\rho}$ here is to be identified with e/f_{ρ}.
30. A. Bramon, E. Etim and M. Greco, Phys. Lett. 41B, (1972) 609.
31. J.J. Sakurai, Phys. Lett. 46B, (1973) 207.
32. K. Ishikawa and J.J. Sakurai, Zeit. Phys. C1, (1979) 117; J.S. Bell and R.A. Bertlmann, Zeit. Phys. C4, (1980) 11.
33. J.A. Bradley, C.S. Langensiepen and G. Shaw, University of Manchester preprint, M/C TH 81/09.
34. P. Chen and J.J. Sakurai, Phys. Lett. 110B, (1982) 481.
35. H. Fritzsch and G. Mandelbaum, Phys. Lett. 109B, (1982) 224.

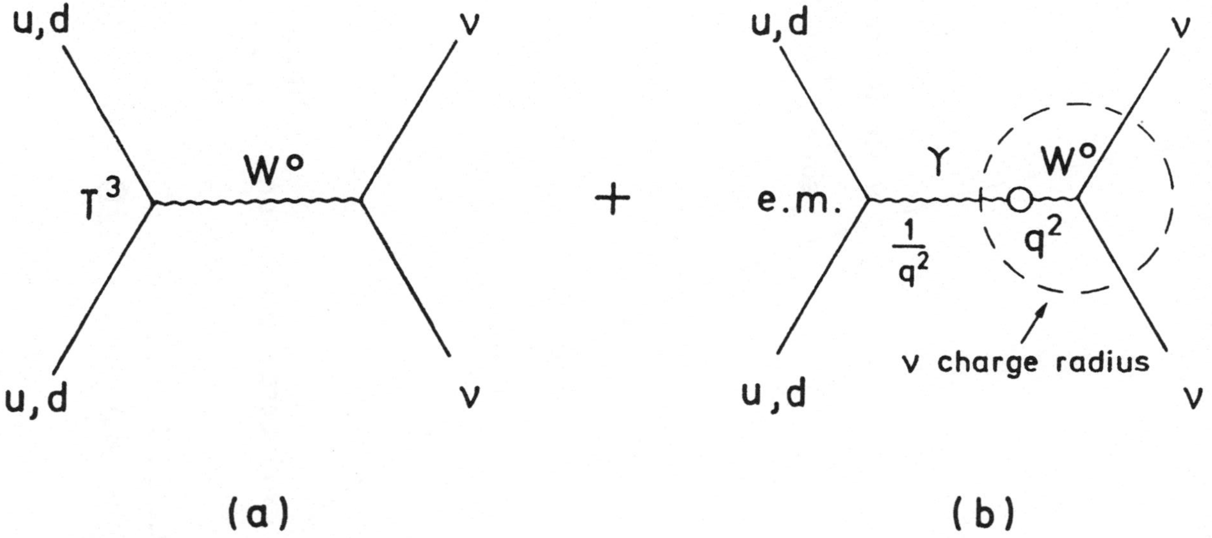

Fig. 1. Simple low-order diagrams for νq scattering

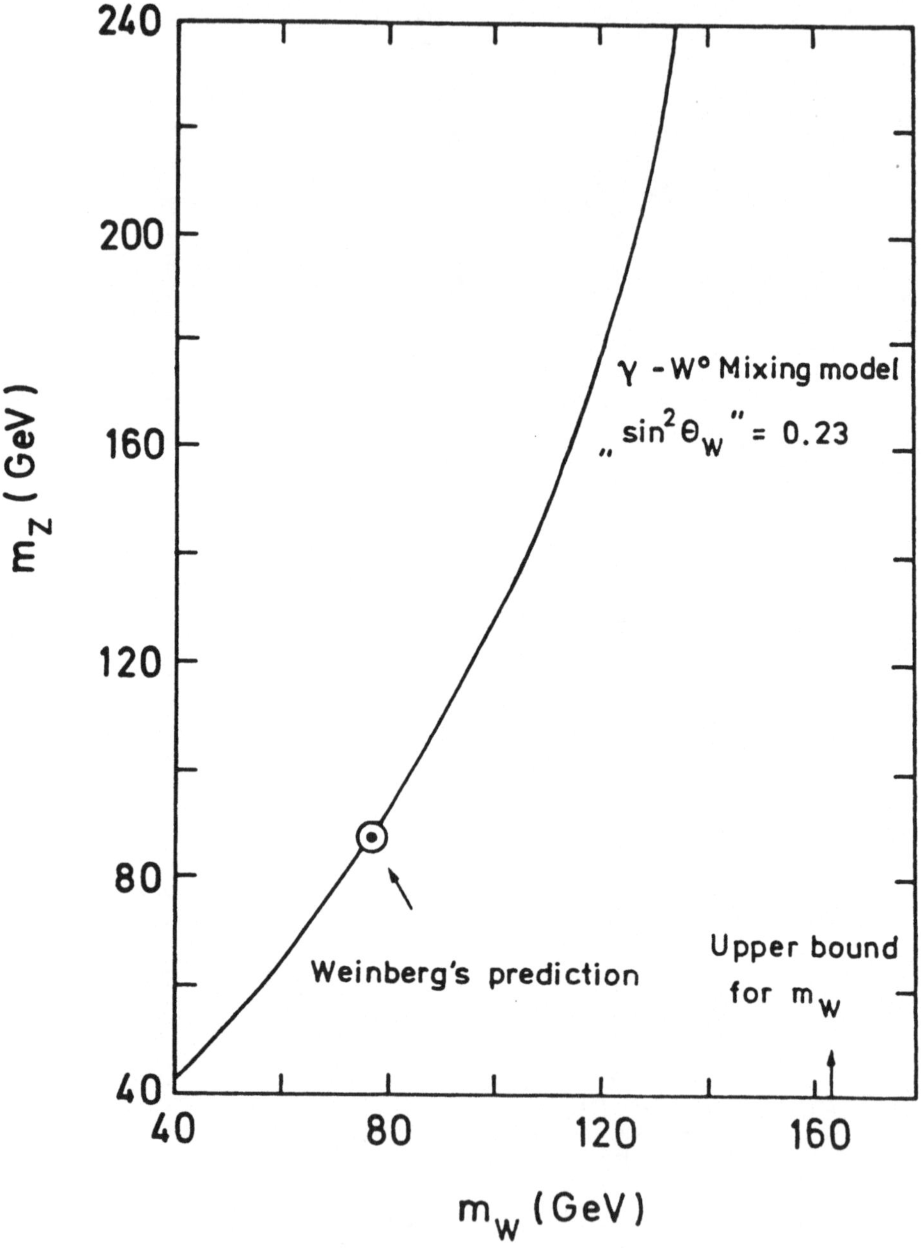

Fig. 2. Weak boson mass relation

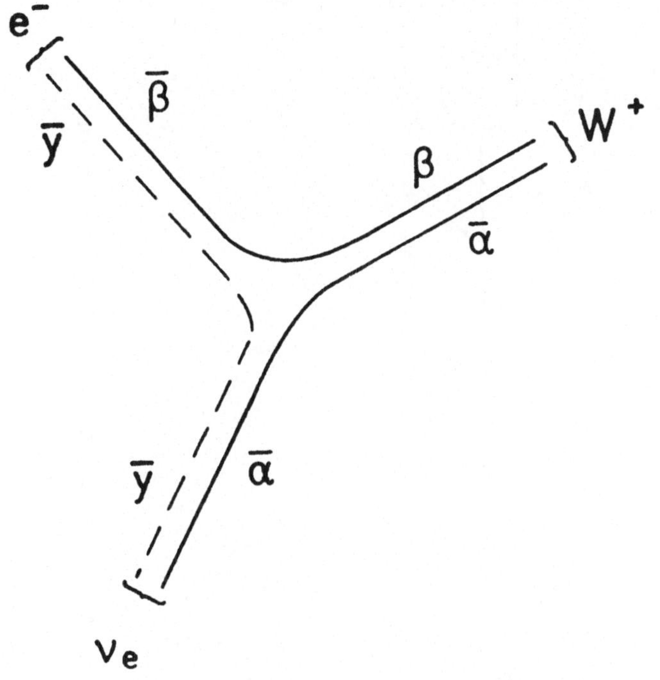

Fig. 3. $\nu_e \leftrightarrow W^+e^-$ vertex

Fig. 4. (a) γ-ρ^{O} junction (b) γ-W^{O} junction

Acta Physica Austriaca, Suppl. XXIV, 307–361 (1982)

NEUTRINO MASSES, NEUTRINO OSCILLATIONS, AND COSMOLOGICAL IMPLICATIONS+

by

F. W. STECKER
NASA/Goddard Space Flight Center
Laboratory for High Energy Astrophysics
Greenbelt, Maryland, U.S.A.

ABSTRACT

Theoretical concepts and motivations for considering neutrinos having finite masses are discussed first. Following this, the experimental situation on searches for masses and oscillations is summarized. This includes a discussion of the solar neutrino problem, reactor, deep mine and accelerator data, tritium decay experiments and double beta-decay data. Finally, the cosmological implications and astrophysical data relating to neutrino masses will be reviewed. Aspects of this topic include the neutrino oscillation solution to the solar neutrino problem, the missing mass problem in galaxy halos and galaxy clusters, galaxy formation and clustering, and radiative neutrino decay and the cosmic ultraviolet background radiation.

+Lectures given at the XXI. Internationale Universitätswochen für Kernphysik, Schladming, Austria, February 25 - March 6, 1982.

"THE QUESTION NOW IS, 'HOW MANY NEUTRINOS CAN DANCE ON THE HEAD OF A PIN?'"

Courtesy of Sidney Harris

CONTENTS

I. INTRODUCTION

Neutrinos, parity violation, Cabibbo mixing, neutral currents, unification and grand unification are some of the physical concepts which come to mind when one thinks of the subject of weak interactions. A surprisingly wide range of phenomena of unexpected subtlety and beauty has unfolded in this century, leading to a deeper, but still incomplete, understanding of the subject. And now the possibility that neutrinos have mass has led to a new connection between particle physics and cosmology. These lectures will be concerned with the various aspects of neutrino mass and their broad and profound implications.

II. WHY SHOULD NEUTRINOS HAVE MASS?

In order to discuss why neutrinos may have mass and how they "get" mass, we must be more specific about defining the character of the neutrino. This character is determined by the field equations which the neutrinos obey and their couplings. In general, a spin one-half fermion obeys the Dirac equation and can be represented by a four-component spinor. The degrees of freedom represent both the particle and the antiparticle, each with two helicity states. For the neutrino, presently known phenomena only relate to two of these components, viz., the left-handed neutrino and the right-handed antineutrino. Thus, presently observed electroweak phenomena are satisfactorily described by the $SU(2)_L \times U(1)$ model of Glashow, Weinberg and Salam (GWS). There are three ways to account for this situation. Either

(1) there are no right-handed neutrinos and no left-handed antineutrinos. This situation can only occur if neutrinos are massless. Otherwise a large enough Lorentz transformation could always transform a left-handed neutrino into a right-handed neutrino. Or

(2) right-handed neutrinos exist but don't participate in $SU(2)_L \times U(1)$ electroweak interactions ("sterile neutrinos"). If these neutrinos exist, we can construct Dirac mass terms of the form

$$\bar{\nu}_R M_D \nu_L + \bar{\nu}_L M_D \nu_R \quad . \tag{2.1}$$

Or

(3) Right-handed neutrinos exist and are really the antimatter (charge conjugate) counterparts of left-handed

neutrinos. As in the case of the π^o boson, this requires that the neutrino be its own antiparticle. The fields which describe it are therefore real and lepton number is not a good (or well defined) quantum number. This is O.K. in modern gauge theory because there is no massless gauge boson associated with conservation of lepton number as the photon is associated with the conservation of charge. Neutrinos of this character are called Majorana neutrinos. Majorana mass terms are of the form

$$\overline{(\nu^c)}_R M_L \nu_L + \overline{(\nu^c)}_L M_R \nu_R$$

$$= \overline{\nu_L^c} M_L \nu_L + \overline{\nu_R^c} M_R \nu_R \tag{2.2}$$

$$= \nu_L M_L \nu_L + \nu_R M_R \nu_R \quad .$$

(This follows from eqs. (2.5) and (2.6) as we shall see. Also, since ν_L is a two-component spinor field ν^α, $\alpha = 1,2$, "$\nu_L\nu_L$" denotes the antisymmetrized combination $1/2\ \varepsilon_{\alpha\beta}\nu_L^\alpha\nu_L^\beta$ where ε is the totally antisymmetric two-dimensional tensor.)

The charge conjugate field in the four-component notation, ν^c, is defined by

$$\nu^c \equiv C\bar{\nu}^T \quad . \tag{2.3}$$

(4) In general, neutrinos can have both Dirac and Majorana masses. The general mass term therefore involves a mass matrix M and is of the form

$$(\overline{\nu_R^c}, \overline{\nu_R}) \begin{pmatrix} M_L & M_D \\ M_D & M_R \end{pmatrix} \begin{pmatrix} \nu_L \\ \nu_L^c \end{pmatrix} \equiv \bar{\psi}_R\, M\, \psi_L \quad . \tag{2.4}$$

It follows from equation (2.2) that M_L and M_R here are mass terms for left-handed and right-handed Majorana neutrinos, whereas M_D is a Dirac mass term. The equivalence of the off-diagonal terms follows from CPT Theorem. The matrix M in equation (2.4), being symmetric, can be diagonalized by an orthogonal transformation so that the two eigenvalues M_1 and M_2 are the masses of Majorana-type neutrinos and the neutrino states participating in the electroweak interactions are mixtures of these Majorana states.

In order to further clarify the interrelationships between the various neutrino fields, one can split up the four-component Dirac field into Weyl fields having chirality (handedness) as follows:

$$\nu_{R,L}^{(W)} = \frac{1}{2}(1 \pm \gamma_5)\nu^{(D)}. \qquad (2.5)$$

Alternatively, we can define a Majorana (self-conjugate) two-component field from the Dirac field:

$$\nu^{(M)} = 2^{-1/2}(\nu^{(D)} + \nu^{(D)c}) . \qquad (2.6)$$

In the limit where the neutrinos are massless, neutrinos from equation (2.4) with left-handed chirality have left-handed helicity, and the antineutrinos have right-handed helicity, i.e.,

$$(\nu_L)^c \equiv (\nu_L^{(W)})^c = (\nu^c)_R . \qquad (2.7)$$

Since we have compared a Majorana neutrino to a π^o boson, using the same analogy we can compare a Dirac neutrino to a K^o boson which has a separate charge conjugate counterpart $\overline{K^o}$. The Dirac neutrino can be constructed from two independent Majorana neutrinos (Hereafter, we will define $\nu^{(D)} \equiv \Psi$ and $\nu^{(M)} \equiv \chi$.) :

$$\Psi = 2^{-1/2}(\chi_1 + i\chi_2)$$
$$\Psi^c = 2^{-1/2}(\chi_1 - i\chi_2) \quad , \qquad (2.8)$$

so that (properly antisymmetrized)

$$\Psi^c\Psi = 1/2(\chi_1\chi_1 + \chi_2\chi_2) \quad . \qquad (2.9)$$

All Majorana fields χ are real and self-conjugate, so that it is only necessary in what follows to denote their chirality, e.g., χ_L, χ_R.

In the past, it has usually been assumed that the mass of the neutrino is identically zero. This assumption was bolstered by the fact that no right-handed neutrinos have been seen. The argument was that if neutrinos had mass, right-handed neutrinos could be produced from left-handed neutrinos by a Lorentz transformation. However, if neutrinos have mass and Majorana character, then such a transformation would be equivalent to changing left-handed neutrinos into right-handed "antineutrinos", which we do know exist. Also, as we have mentioned, right-handed neutrinos could exist and be presently unobserved because they do not participate in standard GWS electroweak interactions.

Given then the possibility that neutrinos have mass, there are now several motivations for considering this possibility very seriously. They are:

A. Some recent experimental indications favoring a non-zero mass for the neutrino.

B. Observational results from astrophysical data and cosmological considerations which could be explained under the hypothesis that neutrinos have mass.

C. Theoretical considerations within the general

framework of grand unified gauge theory leading to the ideas that (1) there is no general gauge principle leading to conservation of lepton number, and (2) grand unified models do not generally conserve lepton number, so that models can be constructed in which neutrinos have Majorana masses. Those of particular interest here contain very heavy right-handed Majorana neutrinos.

Let us now consider some of these grand unified models to see more specifically how neutrinos with non-zero masses[+] arise.

In the standard GWS model, the neutrino is part of a left-handed fermion SU(2) doublet, and the right-handed electron comprises a singlet, i.e.,

$$\binom{\nu_e}{e^-}_L \quad , \quad (\bar{e})_R \quad . \tag{2.10}$$

These arrangements are, of course, duplicated for the other lepton families, μ, ν_μ, τ, ν_τ. At this stage in the unification there is no need for ν_R's or approximately sterile ν's. as there are no right-handed currents. The simplest grand unified model of strong, weak and electromagnetic interactions, viz. the SU(5) model, has the families of left-handed fermions placed in the SU(5) representations of the form

---

[+]We will henceforth refer to them as "massious" neutrinos, i.e., having the property of mass. (The word "massive" in English means heavy or bulky, a term not well suited for neutrinos with $m \simeq 1$ ev.)

$$\bar{5}:\quad \begin{bmatrix} \bar{d}^r \\ \bar{d}^g \\ \bar{d}^b \\ e^- \\ \bar{\nu}_e \end{bmatrix}_L \qquad 10:\quad 2^{-1/2}\begin{bmatrix} 0 & -\bar{u}^b & -\bar{u}^g & u^r & d^r \\ \bar{u}^b & 0 & \bar{u}^r & u^g & d^g \\ \bar{u}^g & -\bar{u}^r & 0 & u^b & d^b \\ -u^r & -u^g & -u^b & 0 & e^+ \\ -d^r & -d^g & -d^b & -e^+ & 0 \end{bmatrix}_L , \tag{2.11}$$

so that there are 15 fundamental fermions (per family) including only one neutrino. Again, this representation admits massless neutrinos. (There is no room for massious neutrinos unless an additional SU(5) singlet is added.)

In the SO(10) model, however, the picture changes. Here, the fermions are grouped into a total of 16 states so that in addition to the 15 fermions of SU(5) there is an additional neutral fermion which is an SU(5) singlet. In order to be consistent with experimental data, this new fermion must be quite heavy.

At the very heavy mass scale corresponding to grand unification, the $SU(2)_L \times U(1)$ electroweak symmetry should hold quite well. Therefore, a Majorana mass term for the neutrino must be constructed from an $SU(2)_L \times U(1)$ invariant. Since the ν_L field is part of an SU(2) doublet,

$$\psi \equiv \binom{\nu}{e^-}_L , \tag{2.12}$$

it cannot by itself (i.e. $\nu_L\nu_L$) be used to construct a gauge invariant Majorana mass. However, by introducing the GWS scalar (Higgs) doublet

$$\Phi \equiv \binom{\phi^+}{\phi^0} \tag{2.13}$$

and by replacing ν_L by the SU(2) gauge invariant form, one can construct a Majorana mass term of the SU(2) singlet form [1]:

$$[\Phi^T \varepsilon \psi]^2 = (\phi^o \nu_L - \phi^+ e_L^-)(\phi^o \nu_L - \phi^+ e_L^-) = \langle\phi^o\rangle^2 \nu_L \nu_L + \dots \tag{2.14}$$

The operator shown in equation (2.14) is of dimension five, so that it is non-renormalizable[2]. It is therefore undesirable to have it as it stands in the fundamental Lagrangian. However, an operator of the form

$$\frac{f\langle\phi^o\rangle^2}{M} \nu_L \nu_L \tag{2.15}$$

can appear in the effective Lagrangian from the exchange of heavy particles of mass M with lepton-number violating couplings. The effective coupling constant is then of the form f/M. The effective theory versus renormalizable theory can be compared with the case of the Glashow-Weinberg-Salam theory (See Figure 1). The interaction terms of the fundamental Lagrangian are renormalizable and the corresponding Majorana neutrino mass is

$$m_\nu = \frac{f\langle\phi^o\rangle^2}{M} \quad . \tag{2.16}$$

In the GWS model, $\langle\phi^o\rangle \sim 300$ GeV. The mass M can have various values depending on the grand unified theory taken. If we take a typical scale $M \sim 10^{14}$ GeV, then the corresponding mass for the neutrino is of the order of 10^{-5} eV.

One scheme for generating a neutrino mass in the SO(10) model was suggested by Gell-Mann, Ramond, and Slansky[3]. For this model, the breakdown of SO(10) and the corresponding symmetry breakdown scenario, the relevant Higgs multiplet, and neutrino mass matrix are indicated in Figure 2.

The final neutrino mass matrix contains a heavy Majorana mass $M \sim 10^{15}$ GeV and a light Dirac mass $m \sim m_q$, the up quark mass induced by the Higgs field ϕ.
The mass matrix can be diagonalized to yield the Majorana mass terms

$$\lambda_1 \chi_1 \chi_1 + \lambda_2 \chi_2 \chi_2 \tag{2.17}$$

where

$$\lambda_{2,1} = \frac{M \pm (M^2 - 4m^2)^{1/2}}{2} \quad , \tag{2.18}$$

so that

$$\lambda_1 \simeq m^2/M$$

$$\lambda_2 \simeq M \quad , \qquad m^2 << M^2 \quad , \tag{2.19}$$

are respectively the masses of a light left-handed and a heavy right-handed Majorana neutrino. Note that in the GWS theory, the quark masses are induced by Yukawa couplings of the form

$$L_Y = h\bar{\Psi}_q \phi \Psi_q \tag{2.20}$$

so that $m_q \sim h <\phi>$ and

$$\lambda_1 = \frac{h^2 <\phi>^2}{M} \tag{2.21}$$

is of the form of eq.(2.15) with $f \sim h^2$.

In the scheme of Gell-Mann, Ramond, and Slansky[3] (GRS), the superheavy right-handed Majorana neutrino obtains its mass from the vacuum expectation value of the 126-plet of

Higgs fields which breaks the $SU(2)_L \times SU(2)_R \times U(1)_{B-L}$ symmetry and is an SU(5) and an SU(2) singlet. There are many other models within the context of grand unified theories which have been explored. One motivation for this has been the size of the neutrino mass obtained. A GRS mass is typically in the range 10^{-4} to 10^{-3} eV. While this range may be significant for the solar neutrino problem, as we shall see, such masses are not large enough to play a significant cosmologcial role, to account for the "missing mass" in galaxies or to account for some experimental results.

Because of the reasons mentioned above, masses in the 1-100 eV range are more "desirable". One possibility suggested by Witten[4] does not involve explicit Higgs fields at the 10^{15} GeV level. In this scheme M is not of the order of 10^{15} GeV because the right-handed Majorana neutrino, not being coupled to a 10^{15} GeV Higgs multiplet, remains massless at the level of SO(10) breakdown. However, the mixing of SO(10)-vector representation Higgs 10-plets and 16-plets in the spinor representation can induce a mass at the two-loop level. The effective mass is lower than $M = 10^{15}$ GeV, the level of SO(10) breaking, and the right-handed Majorana neutrino N is given a mass

$$M_N = (m_q/M_W)\varepsilon(\alpha/\pi)^2 M \quad , \tag{2.22}$$

where M_W is the W-boson mass, ε is the ϕ_{10}-$\phi_{\overline{16}}$ mixing angle $\sim$0.1, and therefore the mass of the N_R would be in the 10^5 - 10^6 GeV range. The corresponding mass of the left-handed ν_e can then be obtained from equation (2.21) using M_N instead of M, since the form of the effective neutrino mass matrix is the same as in the Gell-Mann, Ramond and Slansky model. Within different lepton families, the relevant masses are those of the up-quarks ($I_W = +1/2$). In the GRS case, $m_\nu \propto m_q^2$, from eq. (2.21), whereas in the

Witten Model $m_\nu \propto m_q$, as can be seen from combining eqs. (2.21) and (2.22). Other models[5] break $SU(2)_L \times SU(2)_R \times \times U(1)_{B-L}$ symmetry at a lower energy level, $M_R \simeq M_{B-L}$, which replaces M in equation (2.21). Within the context of SU(5) models, neutrinos cannot be given Dirac masses because there are no right-handed neutrino fields in the basic fermion representations. However, Majorana masses can be included in a non-minimal SU(5) which contains an SU(5) Higgs 15-plet which transforms as an $SU(2)_L$ triplet $(I_W=1)$.[6]The Majorana mass is induced by the vacuum expectation value of an $SU_L(2)$ triplet, which can also be introduced in the SO(10) model as part of a left-right symmetric theory[5].

The GRS mechanism, although it may not be the whole answer, provides a way of explaining, within the context of grand unified theories, why the neutrino mass is much less than other typical Dirac-type fermion masses obtained by Yukawa terms in the GWS Lagrangian involving the ϕ_L fields, i.e.

$$L = h\bar{\psi}_R \phi_L \psi_L \tag{2.23}$$

$$m_f^D \sim h \langle\phi_L\rangle \ .$$

At the same time, the GRS mechanism, through the heaviness of the right-handed Majorana neutrino, $\nu_R \equiv N_L^C$, explains why right-handed neutrinos do not play a significant role in "low energy" physics.

We may generalize our discussion somewhat by noting that the mass matrix of equation (2.4) has Dirac-type off-diagonal terms

$$m = m^D \sim h\langle\phi_L\rangle \tag{2.24}$$

and, with $\overline{\nu_R^c} \equiv \nu_L \equiv \chi_L$ and $\bar{\nu}_R \equiv \nu_L^c \equiv \chi_R$, Majorana-type diagonal terms of the form $M_R\chi_R\chi_R$ and $M_L\chi_L\chi_L$. By diagonalization of the symmetric matrix we obtain the mass eigenstates of the two Majorana-type neutrinos whose wave functions (with m<<M) can be approximated by

$$\nu_L = \chi_L + \frac{m}{M}\chi_R$$

$$N_R = \chi_R - \frac{m}{M}\chi_L \qquad (2.25)$$

with χ_L originally assumed massless. Thus, the resulting left-handed Majorana neutrino gets a very light mass because of the small mixture of heavy right-handed χ_R neutrinos.

We may generalize the formalism of equations (2.4) (2.19) (2.21) and (2.24) to include the mixing of neutrino flavors by writing the mass terms as matrices which mix generations:

$$m = m^D \rightarrow M^D$$

$$M_R \rightarrow M_R \qquad (2.26)$$

$$\frac{m^2}{M_R} = \frac{h^2 <\phi_L>^2}{M_R} \rightarrow M_\nu^D \, M_R^{-1} \, (M_\nu^D)^T$$

so that

$$M_{\nu_L} = M_\nu^{(D)} M_R^{-1} \, M_\nu^{(D)T} \qquad (2.27)$$

(see figure 3). Since the Majorana matrix M_R is real and symmetric and can therefore be diagonalized by an orthogonal transformation O, we can rewrite equation (24) as

$$M_{\nu L} = M^{(D)} \; O \, M_R^{-1} \, O^T M^{(D)T} = M^{(D)} \, \tilde{M}_R^{-1} \, M^{(D)T} \qquad (2.28)$$

where $\tilde{M}_R^{-1}$ is diagonal.

In many unified models where ϕ_L is a color singlet, the matrix

$$M_\nu^{(D)} = M_u \qquad (2.29)$$

where M_u is the mass matrix of up-type quarks.

Thus, at least in some simple versions of grand unified models, the generation mixing in the left-handed neutral lepton sector can be the same as that in the up quark sector. However, this is not a necessary or proven condition. Such generation mixing brings us to the question of neutrino oscillations which we pursue in the next lecture. Some "predicted" neutrino masses are shown in Table I.

Table I. Neutrino Masses (eV)

Flavor	Observed	GRS Model(e)	Witten Model(f)	Left-Right Models(g)
ν_e	< 60(a) or 14-46(b)	$\sim 10^{-5}$	$\sim 5 \times 10^{-3}$	$\lesssim 1.5$
ν_μ	$< 5.2\text{x}10^5$(c)		~ 1	$\lesssim 5.6\text{x}10^4$
ν_τ	$< 2.5\text{x}10^8$(d)		~ 30	$\lesssim 1.8\text{x}10^7$

(a) Ref.20
(b) Ref.21
(c) Lu, D. C., et al., Phys. Rev. Lett 45, 1066 (1980).
(d) DELCO Collaboration
(e) Ref.3
(f) Ref.4
(g) Ref.5

III. NEUTRINO OSCILLATIONS

If neutrinos have mass and the masses of different eigenstates are different, oscillations can result either from (A) generational mixing ("first class") or (B) doublet-singlet mixing ("second class"). Consider, for example, the case where two weak interaction eigenstates, e.g. ν_μ and ν_e, are mixtures of mass eigenstates ν_1 and ν_2 with masses m_1 and m_2. Then this mixing is given by a simple 2-dimensional or orthogonal matrix characterized by a mixing angle θ,

$$\begin{pmatrix} \nu_e \\ \nu_\mu \end{pmatrix} = \begin{pmatrix} \cos\theta & \sin\theta \\ -\sin\theta & \cos\theta \end{pmatrix} \begin{pmatrix} \nu_1 \\ \nu_2 \end{pmatrix} . \quad (3.1)$$

In general, we can have mixing of a larger number of generations. If we define the neutrino wave function $\psi_\nu(t)$ by an N-dimensional column vector in the case of N-generation mixing, and if we label the weak eigenstates by ν_α ($\alpha = e,\mu,\tau,\ldots$) and the mass eigenstates by ν_i ($i = 1,2,\ldots$), the general mixing matrix is an NxN unitary matrix. An NxN complex matrix has $2N^2$ independent parameters. The unitarity condition $U^\dagger = U^{-1}$ eliminates N^2 parameters. Of these $\frac{1}{2}N(N-1)$ can be placed in an orthogonal (rotation) Cabibbo matrix as independent mixing angles. In the case involving Dirac neutrinos (as with quark mixing) 2N-1 relative phases can be absorbed into a redefinition of the fermion fields without any observable effect, leaving $\frac{1}{2}(N-1)(N-2)$ arbitrary phases which can cause CP violation. In the case involving Majorana neutrinos there are N "reality" constraints in place of the (2N-1) relative phases of the Dirac case. (The "real" Majorana fields do not admit any relative phase transformations.) The result is that in the Majorana neutrino case, we are left with more arbitary CP violating phases [7], viz., $\frac{1}{2}N(N-1)$.

Thus

$$|\nu_\alpha> = U_{\alpha i}|\nu_i>$$
$$|\nu_i> = U^{-1}_{i\alpha}|\nu_\alpha> \quad . \tag{3.2}$$

For the $|\nu_i>$, the mass matrix can be diagonalized to a form $M^{(d)} = m_i\delta_{ij}$, so that we obtain N independent Schrödinger equations

$$M^{(d)} = UMU^{-1}$$
$$\Psi_\nu^{(d)} = -i(M^{(d)2} + p^2\ 1)^{1/2}\Psi_\nu \tag{3.3}$$

or

$$\psi_i = -i(p^2 + m_i^2)^{1/2}\psi_i \quad . \tag{3.4}$$

Consider again the two state case given by eq.(30). For a beam of neutrinos of momentum p produced in a weak eigenstate (say ν_e) at time t=0, defining $E_{1,2} = (p^2+m^2_{1,2})^{1/2}$, it follows that the probability to stay in the state ν_e at time t is

$$P(\nu_e \rightarrow \nu_e) = 1 - P(\nu_e \rightarrow \nu_\mu),$$
$$P(\nu_e \rightarrow \nu_\mu) = |<\nu_\mu(t)|\nu_e(0)>|^2$$
$$= |\cos\theta\ <\nu_\mu(t)|\nu_1> + \sin\theta\ <\nu_\mu(t)|\nu_2>|^2 \tag{3.5}$$
$$= |\sin\theta\cos\theta e^{-iE_1t} + \sin\theta\cos\theta e^{-iE_2t}|^2$$
$$= \sin^2\theta\cos^2\theta|e^{-iE_1t} - e^{-iE_2t}|^2$$
$$= \frac{1}{2}\sin^2 2\theta\ [1 - \cos(E_1 - E_2)t] \quad .$$

For

$p \gg m_1, m_2$ then $E_1 \simeq E_2 \simeq p$ and

$$E_1 - E_2 \simeq p[(1 + \frac{m_1^2}{2p^2}) - (1 + \frac{m_2^2}{2p^2})] \simeq \frac{m_1^2 - m_2^2}{2E} . \qquad (3.6)$$

Thus

$$1 - \cos(E_1 - E_2)t \simeq (1 - \cos \frac{m_1^2 - m_2^2}{2E} t)$$

$$= 2\sin^2(\frac{m_1^2 - m_2^2}{4E} t) \qquad (3.7)$$

and

$$P(\nu_e \to \nu_\mu) = \sin^2 2\theta \sin^2 (\frac{m_1^2 - m_2^2}{4E} t)$$

$$P(\nu_e \to \nu_e) = 1 - \sin^2 2\theta \sin^2 (\frac{m_1^2 - m_2^2}{4E} t) . \qquad (3.8)$$

Defining

$$\Delta^2 \equiv m_1^2 - m_2^2 \qquad (3.9)$$

and with Δ^2 in eV^2, E in MeV and ct in meters, eq. (3.8) becomes

$$P(\nu_e \to \nu_e) = 1 - \sin^2 2\theta \sin^2 (1.27 \Delta^2 \frac{L}{E}) . \qquad (3.10)$$

From equation (3.10), it follows that three conditions must exist in order for neutrino oscillations to occur: (A) there must exist at least one non-zero neutrino mass and (B) this mass must be different from the mass of at least one other mass eigenstate so that there exists a $\Delta^2 \neq 0$, and (C) there must be mixing between neutrino flavors so that at least one mixing angle $\theta \neq 0$.

Given these three conditions, there are three distinct

ranges for the oscillation phenomena. In the above units eV^2 m MeV^{-1} they are:

1. $\Delta^2(L/E) \ll 1$. In this case an experiment at distance L with neutrino energy E will not detect oscillations ($\sin^2(1.27\Delta^2 L/E) \ll 1$).

2. $\Delta^2(L/E) \simeq 1$. In this case, there will be significant changes in the detection probability with L provided $\sin^2 2\theta$ is moderately large.

3. $\Delta^2(L/E) \gg 1$. In this case, the oscillations will be on a scale small compared to L and the oscillations will average out to some constant probability < 1.

Oscillation experiments now exist in several ranges of L/E. They may be classified as follows:

A. Solar Neutrino Detection of ν_e

$$L = 1.5 \times 10^{11} \text{m}$$

$$E \simeq 1 - 10 \text{ MeV}$$

$$L/E \simeq 10^{10} - 10^{11} \text{ m/MeV}$$

B. Deep Mine Cosmic Ray ν_μ Detection

$$L \simeq 10^6 - 10^7 \text{ m}$$

$$E \simeq 10^4 - 10^6 \text{ MeV}$$

$$L/E \simeq 1 - 10^3 \text{ m/MeV}$$

C. Reactor Experiments ($\bar{\nu}_e$)

$$L \sim 5 - 10 \text{ m}$$

$$E \sim 5 \text{ MeV}$$

$$L/E \simeq 1\text{-}2 \text{ m/MeV}$$

D. Accelerator Experiments

$$L \simeq 10^3 \text{ m}$$

$$E \simeq 2.5 \times 10^4 \text{ MeV}$$

$$L/E \simeq 4 \times 10^{-2} \text{ m/MeV}$$

The minimum Δ^2 for which an experiment is sensitive is $\Delta^2_{MIN}(eV^2) \sim E(MeV)/L(m)$ in the limit of moderately large mixing, so that

$$(\Delta^2_{MIN})_{SOLAR} < (\Delta^2_{MIN})_{COSMIC\ RAY} < (\Delta^2_{MIN})_{REACTOR} < (\Delta^2_{MIN})_{ACCEL.} \tag{3.11}$$

A. Solar Neutrinos

We first consider the data on solar neutrinos. The solar neutrino experiment[8] uses a large tank of CCl_4 in an underground mine to detect ν_e's via the reaction

$$^{37}Cl + \nu_e \rightarrow {}^{37}A + e^-, \quad E_\nu \geq 0.814 \text{ MeV}. \tag{3.12}$$

The ν capture rate is given in solar neutrino units (SNU's) defined such that 1 SNU $\equiv 10^{-36}$ captures/atom/s. Because of the relatively high threshold reaction for capture by ^{37}Cl, the CCl_4 experiment is most sensitive to ν_e's produced in the sun via the reaction

$$^8B \rightarrow {}^8Be^* + e^+ + \nu_e \tag{3.13}$$

(see Table II). The standard solar model predicts [9] a rate of 8 ± 3.3 (3σ) SNUs with the uncertainties in the calculation being due to the nuclear physics parameters

(2.9 SNU), solar composition (1.3 SNU), solar opacity (0.5 SNU) and the neutrino cross section (0.7 SNU). However, the present data gives a capture rate of 1.9 ± 0.3 (1σ) SNU.

Table II. Solar Neutrino Rates [9]

Source Reaction	Predicted Flux ($10^{10}cm^{-2}s^{-1}$)	E_ν(MeV)	SNU(^{37}Cl)	SNU(^{71}Ga)
$p+p\to d+e^++\nu_e$	6.1	0 - 0.42	0	65.1
$p+e^-\to p+d+\nu_e$	0.015	1.4	0.23	2.4
$^7Be+e^-\to {^7Li}+\nu_e$	0.34	0.86(90%)	1.03	27.6
		0.34(10%)		
$^8B\to {^8Be^*}+e^++\nu_e$	6.0×10^{-4}	0-14	6.48	1.8
$^{13}N\to {^{13}C}+e^++\nu_e$	0.045	0-1.2	0.07	2.4
$^{15}O\to {^{15}N}+e^++\nu_e$	0.035	0-1.7	0.23	3.2
		Total:	8.04	102.4

Thus, the ratio of observed neutrinos to expected solar neutrinos is R = 0.31 ± 0.13 (3σ). It was suggested by Gribov and Pontecorvo [10] that neutrino oscillations could account for this ratio. Such a scenario would require the mixing of at least three neutrino flavors with large mixing angles and $\Delta^2 \gtrsim 10^{-11}$ eV^2.

Because the ^{37}Cl experiment measures neutrinos from a relatively insignificant solar reaction, it has been suggested that other materials such as ^{71}Ga and ^{115}In be used in order to detect the lower energy neutrinos from the basic reaction

$$p + p \to d + e^+ + \nu_e . \tag{3.14}$$

The threshold energy for capture reactions on ^{71}Ga,

i.e.

$$^{71}Ga + \nu_e \rightarrow {}^{71}Ge + e^- , \qquad (3.15)$$

is only 0.236 MeV as compared with 0.814 MeV for ^{37}Cl. The total capture rate expected for ^{71}Ga is 102.4 SNU (see Table II) of which 65.1 SNU is expected from reaction (3.14). Thus a ^{71}Ga experiment will test solar theory and the neutrino oscillation hypothesis at a more sensitive and basic level.

B. Reactor Experiments

Reactor experiments have provided the next possible indication of neutrino oscillations; Reines et al.[11] used a detector with D_2O to look for the charged - and neutral current reactions on deuterium induced by reactor generated $\bar{\nu}_e$'s from ^{235}U, ^{238}U, and ^{239}Pu. The $\bar{\nu}_e$'s have a continuum energy spectrum with typical energies of a few MeV. The relevant reactions were :

$$\bar{\nu}_e + d \rightarrow n + n + e^+ \quad (CC) \qquad (3.16a)$$

$$\bar{\nu}_e + d \rightarrow n + p + \bar{\nu}_e \quad (NC) \qquad (3.16b)$$

$$\bar{\nu}_x + d \rightarrow n + p + \bar{\nu}_x \quad (NC) \qquad (3.16c)$$

Reaction (3.16a) is only induced by $\bar{\nu}_e$'s. However, reactions b and c are equivalent and can be induced by any neutrino flavors. Thus, if $P(\bar{\nu}_e \rightarrow \bar{\nu}_e) < 1$ owing to oscillations, $[R(CC)/R(NC)]_{obs}/[(R(CC)/R(NC)]_{theor} < 1$. Reines et al.[11]reported a depletion of $\bar{\nu}_e$'s to (0.40 ± 0.22) of the expected value. They interpreted this result as indicating $\Delta^2 \sim 1$ eV^2 and $\sin^2 2\theta \sim 1/2$. There has been controversy regarding this result, partly owing to an uncertainty in the $\bar{\nu}_e$ spectrum [12]. A more recent reactor experiment performed by a group at Grenoble[13] found $P(\bar{\nu}_e \rightarrow \bar{\nu}_e) \gtrsim 0.7$ by looking at the reaction $\bar{\nu}p \rightarrow ne^+$ at a distance of 8.7m from the reactor. These results are consistent with no oscillations $P(\bar{\nu}_e \rightarrow \bar{\nu}_e) = 1$ for $\Delta^2 \gtrsim 0.5\ eV^2$ and large mixing angles. However, Silverman and Soni[14] have obtained solutions implying mixing (e.g. $\Delta^2 \sim 0.9\ eV^2$, $\sin^2 2\theta \sim 0.4$) which they argue are a best fit to both reactor results. (see Fig. 3) It should be noted that the solution sets given by these two experiments only overlap on the edges of the 90% confidence limits so that the probability of both results agreeing is $\lesssim 1$ %. Clearly, further work needs to be done to resolve this situation.

C. Deep Mine Experiments

There are two reported results from deep mine experiments looking at cosmic-ray ν_μ's. Here again, the results are mixed. The Kolar Gold field group [15] results give indications of oscillations ($P(\nu_\mu \rightarrow \nu_\mu) = 0.62 \pm 0.17$) for $\Delta^2 \gtrsim 10^{-2}\ eV^2$. However, the Baksan group [15] finds a result

$$P(\nu_\mu \rightarrow \nu_\mu) \gtrsim 0.8.$$

D. Accelerator Experiments

Finally, there have been a large number of accelerator results. One interesting type of experiment is the "beam dump" experiment which detects neutrinos from the decay of short lived ($<10^{-12}$s) charmed mesons (as opposed to π-decay and K-decay neutrinos). These are referred to as "prompt" neutrinos. At the source, the ratio $\nu_e/\nu_\mu = 1$

from the decay of charmed particles. Thus the ratio e/μ produced by prompt neutrinos should be 1 in the case of no oscillation. The measured ratios were reported as shown in Table III.

Table III. Accelerator results [17]

Ratio	Error		Group
0.49	±0.21		CHARM
0.46	(+0.55,-0.22)		BEBC
0.77	±0.18 (STAT.)	±0.24 (SYST.)	CDHS

Here again the results are mixed. Many other results have been obtained by various groups. They have been reviewed by Baltay[17] and others. The remaining results are null results, placing limits on regions of the (Δ^2, sin 2θ) plane allowed to the oscillation parameters. These limits are shown in Figure 4 from Barger[18].

IV. OTHER NEUTRINO MASS EXPERIMENTS

There are two other types of experiments which have given indications of neutrino masses. They are the tritium β-decay endpoint experiment and searches for neutrinoless double β-decay. These experiments both pertain to the mass of the neutrino mass eigenstate connected with ν_e. In the case of the neutrinoless double β-decay, violation of lepton number and therefore the Majorana character of the neutrino also come into the picture.

A. The ^{3}H Decay Experiments

There have been several experiments to study the endpoint of the β-decay spectrum of tritium from the decay

$$^3H \rightarrow {}^3He + e^- + \bar{\nu}_e \ . \qquad (4.1)$$

Until recently, this type of experiment has only placed limits on the mass of $\bar{\nu}_e$. Bergkvist[19]obtained m_ν<55 eV (90%CL), and Simpson et al.[20] found m_ν < 65 eV (95%CL). However, one of the most stimulating results in the field has been the report by Lyubimov et al.[21] that they had measured a neutrino mass

$$14 \text{ eV} \leq m_{\nu_e} \leq 46 \text{ eV} \ . \qquad (4.2)$$

The electron β-decay spectrum is of the form

$$\frac{dN}{dp} \equiv N_\beta(E;Z) = CF_c(E;Z)\ p^2(Q-E)[(Q-E)^2 - m_\nu^2]^{1/2} \qquad (4.3)$$

where C is a constant, $F_c(E;Z)$ is the Coulomb factor

$$F_c(E;Z) \simeq \frac{2\pi Z}{(v/c)}\ [1-\exp(-\frac{2\pi Z\alpha}{(v/c)})]^{-1} \simeq F_c(Z) \ . \qquad (4.4)$$

Q is the total energy released in the decay. For ^3H decay, Q = 18.6 keV.

It follows from equation (4.3) that a convenient way to plot the β-spectrum is in the form of the "Kurie plot" K(E) such that

$$K(E) \equiv [\frac{N_\beta(E)}{F_c p^2}]^{1/2} \propto \{(Q-E)\ [(Q-E)^2 - m_\nu^2]^{1/2}\}^{1/2} \ . \qquad (4.5)$$

Thus, for m_ν = 0, the Kurie plot is a straight line

$$K(E) \propto (Q-E),\ m_\nu = 0 \ , \qquad (4.6)$$

and K(Q) = 0. However for $m_\nu \neq 0$, K(E) takes on a modified form near E = Q as shown in Figure 5.

The shape of the endpoint spectrum is affected by other factors in addition to m_ν. For one thing, the finite energy resolution of the detector spreads out the observed electron energy spectrum and produces an artificial "tail". For another, there is the possibility that the ^3H decays into an excited state of energy ϕ of ^3He rather than the ground state. This will cause an effect similar to that of a finite neutrino mass, since the endpoint energy will be lowered from Q to $(Q-\phi)$. In atomic hydrogen, transitions to the ground state will occur 70% of the time. Another 25% probability is that the transition will be to an n = 2 state with ϕ=41 eV if the ^3H is in atomic form. Note that ϕ is of the order of m_ν. Nobody has solved the molecular transition problem for a complex molecule such as valine, $NH_3CH_3CHCOOH$, so that this is a principal source of uncertainty for the experiment of Lyubimov, et al. [21]which used tritiated valine.

B. Neutrinoless Double β-Decay

The study of double β-decay has long been associated with a test for the Majorana character of the neutrino. The appropriate nuclides for study are those for which the single β-decay process is energetically suppressed. The double β-decay transitions looked for are the second order weak decay

$$(A,Z) \rightarrow (A,Z+2) + 2e^- + 2\bar{\nu}_e \tag{4.7}$$

and the neutrinoless counterpart

$$(A,Z) \rightarrow (A,Z+2) + 2e^- \tag{4.8}$$

which violates lepton number by two units.[22]

Reaction (4.8) can be looked at as the two stage

sequence (in quark language)

$$d_1 \rightarrow u_1 + e^- + \nu^{(M)} \tag{4.9a}$$

followed by

$$\nu^{(M)} + d_2 \rightarrow u_2 + e^- \tag{4.9b}$$

involving two down quarks and a Majorana neutrino $\nu^{(M)}$ (see Fig. 6).

The nuclide for which double β-decay is energetically favorable are even-even nuclides. The relevant transitions are $0^+ \rightarrow 0^+$ to the ground state of the daughter nuclide with also some possibility for $0^+ \rightarrow 2^+$ transitions to the excited state (see Fig. 7). The relevant energy spectra, also shown in Fig. 7, indicate that the electrons carry off the total energy Q in the case of ν-less decay whereas they share the energy with the $\bar{\nu}_e$'s in the standard double β-decay.

It can be seen from Fig. 6 that in order for the neutrino which is emitted in the first stage (4.9a) to be absorbed by the d quark in the second stage (4.9b) of the neutrinoless decay, a spin flip must occur. This can either be accomplished by the neutrino mass and/or by the existence of right-handed weak currents with a strength $j_R \equiv \eta j_L$. Transitions of the form $0^+ \rightarrow 2^+$ are produced solely by the right-handed current mechanism[23]. Thus, the study of ν-less double β decay provides not only a test for lepton number violation and neutrino masses, but also one for right-handed weak currents. The theory for this process has been given in great detail recently [22,23] and will not be detailed here.

There are two categories of double β decay measurements which have been carried out, <u>viz.</u>, geochemical and laboratory. The geochemical measurements consist of the analysis of ores of known age ($\sim 10^9$ yr) which are rich in the parent nuclide where one looks for traces of the daughter nuclide. The daughter nuclides most amenable to analysis of this type are the noble gases. Thus, good measurements are available for the lifetimes of the decays $^{130}Te \rightarrow {}^{130}Xe$, $^{128}Te \rightarrow {}^{128}Xe$ and $^{82}Se \rightarrow {}^{82}Kr$. Of course, in this type of experiment only the lifetimes are measured, not the electron energy distribution, so that one cannot tell directly whether or not neutrinoless decay has occured. However, different lifetimes are calculated for the $2\bar{\nu}_e$ and $0\bar{\nu}_e$ decays owing to the fact that the lifetime depends on a phase space factor involving a function $f(\hat{m},\eta)$ where $\hat{m} \equiv m_\nu/m_e$.

Several groups have measured $T_{1/2}(^{130}Te \rightarrow {}^{130}Xe)$ and obtained values in the range $\sim(2 \pm 1) \times 10^{21}$ yr both from geochemical data and laboratory data. For the decay of ^{82}Se, there appears to be an unfortunate conflict between the geochemically obtained lifetime ($\sim 2 \times 10^{20}$ yr) and that found experimentally ($\sim 10^{19}$ yr).

One method for determining the neutrino mass $m_\nu = \hat{m}\, m_e$ has been to study the ratio of lifetimes of ^{128}Te and ^{130}Te. Letting ρ be the ratio of $2\bar{\nu}_e$ to $0\bar{\nu}_e$ decay matrix elements, Rosen obtained the following condition on $\hat{m}$ and η [22]:

$$\hat{m}^2 + 0.093\, \hat{m}\eta + 0.15\, \eta^2 = 1.5 \times 10^{-9}\, \rho^2 . \qquad (4.10)$$

Thus,

$$\hat{m} = \frac{m_\nu}{m_e} = 3.9 \times 10^{-5}\rho, \qquad \eta = 0$$

$$0 \leq \hat{m} \leq 4 \times 10^{-5}\rho, \qquad \eta \leq 10^{-4}\rho \qquad (4.11)$$

with estimates for ρ of 0.5 and 1.2. Such estimates give m_ν in the range of 10 to 40 eV with $\eta \lesssim 10^{-4}$. The limits on

η could be greatly strengthed by nonobservations of $0^+ \rightarrow 2^+$ transitions.

Various other calculations of m_ν from double β-decay have been reviewed by Rosen [22]. Here gain, as in the case of neutrino oscillations, one finds conflicting results:

$m_\nu = 34$ eV [Ref.23]

$m_\nu \lesssim 15$ eV [Ref.24]

The experimental situation needs to be clarified.

C. Internal Bremsstrahlung in Electron Capture

De Rujula [25] suggested a new method for obtaining m_{ν_e}. He has pointed out that radiative orbital electron capture reactions involving neutrino emission from neutron deficient nuclides could be used to determine m_ν. Here, the spectrum of the emitted photons would take the place of the electron spectrum in the ^3H decay experiment. No experiments of this sort have yet been attempted.

V. CONCLUSIONS REGARDING THE EXPERIMENTAL SITUATION

As we have seen, there are conflicting data within the various categories of neutrino mass experiments. Phillips [26] and Barger [18] have pointed out additional problems in reconciling the data among these categories.

Within the spirit of these discussions, one example of the type of puzzling relationships obtained is outlined below:

Suppose (A) $m_{\nu_e} > 14$ eV (Lyubimov, et al.[21];

(B) ν-mass eigenstates are highly non-degenerate (as in grand unified models - see Section II).

Then (C) $\Delta^2 >> 1\ eV^2$.

But (D) from the Grenoble reactor experiment[13] Δ^2 (probably) $< 1\ eV^2$,

unless (E) Θ is small.

But (F) if Θ is small, oscillations don't solve the solar ν problem.

However (G) there are still loopholes in these arguments;

so (H) ?!?

VI. NEUTRINOS AND COSMOLOGY

In this last lecture, I will discuss the possible role of neutrinos in cosmology. This is another quite active field of investigation at present, having many facets. I will stress here primarily the gravitational effects of a "neutrino dominated universe" within the context of the hot big-bang cosmology.

The hot big-bang model is now quite familar and the basic relations describing it may be found in many places[27] We will consider here that it rests on two main pieces of evidence: (1) the Hubble relation showing that the distant galaxies are receeding from us at velocities proportional to their distance,

$$v_{rec}(km/s) = H_0(km/s/Mpc)\,r(Mpc), \tag{6.1}$$

where $50 \leq H_0 \leq 100$ in these units and 1 megaparsec (Mpc) $\simeq 3 \times 10^{24}$ cm; and (2) the universe is filled with thermal blackbody radiation at a temperature $T = 2.8 \pm 0.1$ K.

From these two relations come the conclusions that (A) the universe is expanding (as implied also by the Einstein gravitational equations sans cosmological term) and (B) it was in a much hotter as well as denser state in the past. Most workers would also add (3) the data on the ^{4}He and ^{2}H abundances (implying primordial nucleosynthesis) as additional evidence of the hot big-bang model. This argument most likely has an "element" of truth. However, I do not consider this evidence to be on the same footing with (1) and (2) because it involves additional assumptions and may be inherently self-contradictory in its simplest form [28]. (Many things have been "deduced" from the ^{4}He and ^{2}H data, e.g., the number of neutrino flavors, and the student should approach these arguments with academic scepticism. I will, therefore, not repeat them here.)

The Einstein equations are second order differential equations. With a homogeneous isotropic metric (called the Robertson-Walker metric) they can be solved to give a scale size, R, as a function of cosmic time, t, in terms of two parameters, the "deceleration parameter"

$$q_O = - \frac{\ddot{R}R}{(\dot{R})^2} \tag{6.2}$$

and the expansion rate

$$H_O = \frac{\dot{R}}{R} \tag{6.3}$$

where the subscript O refers to the present time (redshift z = O). (Throughout this discussion, we will assume that the cosmological term, or equivalently Einstein's cosmological constant Λ=O.) The gravitational deceleration parameter can be replaced by a mass parameter

$$\Omega = 2q_O, \ \Lambda = O , \tag{6.4}$$

where we denote by

$$\Omega = \frac{\rho}{\rho_c} \tag{6.5}$$

the fraction of the critical mass density needed to close the universe gravitationally. The critical density can be determined in the Newtonian limit by equating the potential and kinetic energies of a test particle:

$$G\left(\frac{4\pi}{3} R^3\right) \frac{\rho_c}{R} = 1/2 \cdot (H_0 R)^2 \tag{6.6a}$$

or

$$\rho_c = \frac{3H_0^2}{8\pi G} \tag{6.6b}$$

so that

$$\Omega = \frac{8\pi G\rho}{3H_0^2} \tag{6.7}$$

Observationally, from studies of q_0, it is found that $\Omega < 2$. From studies of total matter density in galaxy clusters it is found that $\Omega > 0.02$.

The neutrino contribution to Ω, which we will call Ω_ν, can be calculated in the context of the hot big-bang model. We assume that at some time $t < t_\nu$, corresponding to a temperature $T > T_\nu$, photons, electrons, positrons and neutrinos were in thermal equilibrium. The temperature T_ν when this situation last occurred was when the ν-e interaction rate was equal to the expansion rate of the universe, $T_\nu \sim 1$ MeV. Shortly thereafter at $T \sim m_e \simeq 1/2$ MeV, the electrons and positrons went out of thermal equilibrium and annihilated

$$e^+ e^- \rightarrow 2\gamma \tag{6.8}$$

with all of the energy release going into the photons, the neutrinos having decoupled. At $T > T_\nu$, the ratio of neutrinos to photons was

$$\frac{n_\nu}{n_\gamma} = \frac{\omega_\nu}{\omega_\gamma} \frac{\int_0^\infty FdE}{\int_0^\infty BdE} = 3/4 \cdot f \qquad (6.9)$$

where $\omega_\gamma = 2$ is the number of photon degrees of freedom, ω_ν is the number of neutrino degrees of freedom (taken to be 2 per flavor $\times$ f = the number of flavors) which were in thermal equilibrium with the photons at $T_\nu \simeq 1$ MeV (only ν_L and $\bar{\nu}_R$ meet this criterion), and the factor of 3/4 comes from the ratio of the integrals over the Fermi-Dirac function F(E;T) and the Planck function B(E;T) in equation (6.9). For $T < m_e$, additional photons are added from the e^+e^- annihilation. The new factor multiplying the photon number is determined by the additional entropy per unit volume added to the photon component and is 11/3. Thus, for $T_\gamma < m_e$

$$\frac{n_\nu}{n_\gamma} = \frac{3/4 \cdot f}{11/4} = \frac{3}{11} f \; . \qquad (6.10)$$

At the present time

$$n_\gamma = 400 \left(\frac{T}{2.7}\right)^3 \text{cm}^{-3} \; , \qquad (6.11)$$

a number which is obtained from the fact that the effect of redshift $z \equiv \Delta\lambda/\lambda$ on the Planck function is to shift the temperature

$$B[(1+z)T] \rightarrow B(T) \; . \qquad (6.12)$$

Thus, from equations (6.10) and (6.11)

$$n_\nu = 110 \; f \left(\frac{T}{2.7K}\right)^3 \text{cm}^{-3} \; , \qquad (6.13)$$

and the total mass density divided by the closure density is (from equations (6.7) and (6.13))

$$\Omega_\nu = 0.01\ h_0^{-2} \sum_f m_f\ (\mathrm{eV}) \tag{6.14}$$

where $h_0 \equiv H_0/100$ km/s/Mpc so that $1/2 \lesssim h_0 \lesssim 1$.

We may compare Ω_ν with the various values of Ω deduced from the gravitational dynamics of galaxies and groups of galaxies at various scales. From these measurements, it has been found that the ratio of gravitational mass (i.e. all mass) to luminosity M/L scales roughly linearly with scale size r over a wide range of r up to ∿1Mpc. Figure 8 shows some results together with an analytic fit to M/L of the form

$$\frac{M}{L} = \mu_0\ [1 - \exp\ (-r/\Lambda)] \tag{6.15}$$

which serves roughly to define a scale size ∿ 3 Mpc which appears to be characteristic of the non-luminous mass in the universe[29]. This size is interestingly close to the gravitational clustering size[30]. As we shall see, it is characteristic of the Jeans mass one obtains from neutrinos with m_ν ∿10-30 eV.

We note that it follows from equation (6.14) that $\Omega_\nu = 1$ for $25 \lesssim \sum m_\nu \lesssim 100$ eV and, from the lower limit on Ω in baryons, it is possible for ν's to gravitationally dominate the universe[31] if $1/2\ \mathrm{eV} \lesssim \sum m_\nu \lesssim 2$ eV. Hereafter we will assume that this is the case. And we will further assume for simplicity (and also because grand unified models favor a neutrino mass hierarchy similar to that in other fermion families) that one neutrino mass eigenstate dominates, i.e.

$$\sum_f m_f = \sum_f m_i \simeq \text{"}m_\nu\text{"}\,. \tag{6.16}$$

Thus, neutrino masses similar to those which we discussed in previous lectures could gravitationally dominate or even close the universe. It has also been pointed out by various workers that massious neutrinos could play an important role in producing the largest scale structure in the universe[32].This is basically because perturbations of neutrinos on a large enough scale (see below) can survive and grow, whereas in a hot dense universe plasma baryon perturbations are damped by the high viscosity of the thermal blackbody radiation.

For a collisionless gas of neutrinos, the gravitational trapping scale is determined by the virial theorem with a thermal velocity dispersion. Gravitational trapping occurs for scales greater than the Jeans length λ_J such that

$$\frac{G\rho_\nu \lambda^3}{\lambda} > \langle v^2 \rangle \simeq 3.6 \; T_\nu/m_\nu \tag{6.17a}$$

or

$$\lambda > \lambda_J = \left(\frac{\langle v^2 \rangle}{G\rho_\nu}\right)^{1/2} \tag{6.17b}$$

with the corresponding Jeans mass

$$M_J \equiv \frac{4\pi}{3} \rho_\nu \left(\frac{\lambda_J}{2}\right)^3 . \tag{6.18}$$

For relativistic neutrinos, perturbations can exist on the scale of the horizon size

$$\lambda_{\nu J} \simeq \lambda_H \simeq ct \tag{6.19}$$

below which they decay by collisionless (Landau) damping owing to the fact that the thermal motion of the neutrinos smears out irregularities. This process is effective until the ν's become non-relativistic at $T_{NR} \simeq m_\nu/3$, below which pressure effects become unimportant.

Since, in the hot big-bang model for $\Omega z >> 1$, $\lambda_H = ct \propto T^{-2/3}$, it follows that $t_{NR} \propto m_\nu^{-2/3}$ and the maximum neutrino Jeans mass

$$M_\nu^{max} \propto \lambda_H^3 \propto t_{NR}^3 \propto m_\nu^{-2} \; . \tag{6.20}$$

Plugging in the numbers, one finds

$$M_\nu^{max} = 4 \times 10^{18} \; m_\nu^{-2} \; (\mathrm{eV}) \; M_\Theta \tag{6.21}$$

in solar mass units.

If this mass scale is the size of galaxy clusters, 10^{15}-10^{16} M_Θ, the corresponding neutrino mass required is in the range 20 eV $\lesssim m_\nu \lesssim$ 65 eV.

Tremaine and Gunn[33] have related the observational parameters of nonluminous mass in galaxy "halos" and rich clusters of galaxies to derive another astrophysically related requirement on m_ν. For simplicity, let us assume non-degeneracy and consider only the heaviest mass eigenstate. Then, from Fermi-Dirac statistics

$$n_\nu \leq 2 \int^{m_\nu v_{esc}} \frac{p^2 dp}{2\pi^2} = \frac{m_\nu^3 \, v_{esc}^3}{3\pi^2} \tag{6.22}$$

where v_{esc} is the gravitational escape velocity. Thus

$$\rho_\nu \leq \frac{m_\nu^4 \, v_{esc}^3}{3\pi^2} \tag{6.23}$$

and the maximum neutrino mass density is proportional to m_ν^4. This sets a lower limit requirement on m_ν in order to account for non-luminous ("missing") mass. Tremaine and Gunn have modified this argument by considering the neutrinos to be distributed in isothermal gas spheres with Maxwellian velocities and central density ρ_0. The numbers are basically the same (within a factor of $2^{1/4}$) but the descriptive parameters are now the core radius r_c and maximum velocity, where r_c is given by

$$r_c = \frac{9\sigma^2}{8\pi G\rho_0} \ , \tag{6.24}$$

σ being the 1-dimensional velocity dispersion. Numerically, one obtains

$$m_\nu \gtrsim 30 \text{ eV} \left(\frac{\sigma}{300 \text{ km/s}}\right)^{-1/4} \left(\frac{r_c}{10 \text{ kpc}}\right)^{-1/2} . \tag{6.25}$$

To explain the rotation curve of (velocity versus galactocentric distance) of our own galaxy [34] and others [35] with a massious neutrino halo would then require $m_\nu \gtrsim 15\text{-}30$ eV, and a typical galaxy cluster mass distribution could be explained by neutrinos with $m_\nu \gtrsim 4\text{-}8$ eV.

Finally, we note one other possible piece of astrophysical evidence regarding neutrino mass from observations of the cosmic ultraviolet background spectrum at high galactic latitudes[29,36].De Rujula and Glashow[37] pointed out that the decay of a massious neutrino from a heavier mass eigenstate ν' to a lighter one ν, i.e.,

$$\nu' \rightarrow \nu + \gamma \tag{6.26}$$

could be detectable through the decay of cosmic neutrinos producing photons in the ultraviolet range. The photon energy

$$E_0 = \frac{m'^2 - m^2}{2m} \tag{6.27}$$

or, in the hierarchy approximation $m' \gg m$,

$$E_0 \simeq \frac{m'}{2} . \tag{6.28}$$

The diffuse line intensity of ν-decay photons from

the galactic halo neutrinos is given by the integral along the line of sight[38] of the telescope

$$I_\lambda = \frac{1}{4\pi\tau\Delta\lambda} \int n' d\ell \qquad cm^{-2}s^{-1}sr^{-1}\mathring{A}^{-1} \tag{6.29}$$

where τ and n' are the lifetime and density of ν' neutrinos. The line with $\Delta\lambda$ is

$$\Delta\lambda = \frac{\langle v^2\rangle^{1/2}}{c} \lambda_o , \quad \lambda_o = E_o^{-1} \tag{6.30}$$

so that $\Delta\lambda/\lambda_o \sim 10^{-3}$ for galactic halo neutrinos.

If the mass of a galaxy cluster is assumed to be mainly from ν' neutrinos, then the number of neutrinos in the source is given by

$$N' = \frac{2 \times 10^{66}(M_S/M_\odot)}{m'(eV)} \tag{6.31}$$

and the flux from the source is

$$F_\lambda = \frac{N'}{4\pi R_S^2 \tau\Delta\lambda} \quad cm^{-3}s^{-1}\mathring{A}^{-1} \tag{6.32}$$

where R_S is the distance of the source.

There should also be a cosmic isotropic background component of radiation from the decay of ν's at all redshifts. The spectrum is a smeared out continuum which is roughly a power-law in wavelength for $\lambda \gtrsim \lambda_O$ and which vanishes for $\lambda<\lambda_O$. More precisely

$$I_\lambda = 7.8 \times 10^{28}\, h_O^{-1}\, \tau^{-1} \frac{\lambda_O^{3/2}}{\lambda^{5/2}} [1-(\Omega-1)(1-\frac{\lambda_O}{\lambda})]^{-1/2} cm^{-2}s^{-1}sr^{-1}\mathring{A}^{-1}, \quad \lambda \geq \lambda_O , \tag{6.33}$$

as obtained by taking various cosmological factors into account[36,38,39]. Lower limits on $\tau(m_\nu)$ obtained from astrophysical data [29] using equation (88) are shown in Fig. 9.

It turns out that there is an enhancement in the cosmic ultraviolet spectrum at high galactic latitudes which has been observed at $\lambda_o \sim 1700$ Å. This would correspond to $E_O \simeq 7$eV and $m \simeq 14$ eV from equation (6.28). The implied neutrino lifetime of 2×10^{17} yr is higher than that predicted for the standard GWS model, however, such lifetimes are possible within the context of composite models of quarks and leptons. A detailed discussion is given elsewhere[29]. Further measurements of this ∿1700 Å feature with much higher wavelength resolution will be required in order to determine if this feature is indeed from neutrino radiative decay.

To sum up this section, we see that the astrophysical data all hint at (but do not prove) cosmological neutrino masses in the 10-100 eV range. Note the similar numbers given below:

A) From the "missing mass" in galaxy clusters ($\Omega \simeq \Omega_\nu \gtrsim 0.4$)
$m_\nu \gtrsim$ 10-40 eV.

B) From the Jeans mass for galaxy cluster formation
$m_\nu \simeq$ 20-65 eV.

C) To explain the galaxy cluster mass distribution
$m_\nu \gtrsim$ 4-8 eV.

D) To explain the galaxy halo mass distribution
$m_\nu \gtrsim$ 15-30 eV.

E) To explain the 1700Å ultraviolet background feature
$m_\nu \simeq$ 14-15 eV.

Of course, all of these indications are consistent with the mass results obtained by Lyubimov, et al.[21], 14 eV $\lesssim m_{\bar{\nu}_e} \lesssim$ 46 eV. But again we have a puzzle because the simplest grand unified models would predict that the mass eigenstate associated with $\bar{\nu}_e$ would have the <u>lightest</u> mass whereas the cosmological interpretations would pertain to the eigenstate with the <u>heaviest</u> mass.

VII. CONCLUSIONS

Many avenues of investigation have opened up for addressing the problem of neutrino masses with a whole host of future investigations planned and perhaps new surprises to come. This is as it should be considering the great importance of this topic for many basic questions ranging from unified field theories to cosmology. The phenomena involved indeed range from structure on the smallest scales - composite models of quarks and leptons - to those on the largest scales - clustering and "super-clustering" of galaxies.

As we have seen, despite all of the many areas of investigation, we only possess hints rather than answers to our questions. While ideas such as the Gell-Mann, Ramond, and Slansky model may be pointing us in the right direction theoretically, it is far from a complete picture. In addition, the generation problem is at least as puzzling here as it is for the other fermions. Questions have been raised regarding the standard solar neutrino model, and detection of the dominant pp neutrinos must await a new generation of experiments. Reactor, deep mine and accelerator experiments have defined limits to the oscillation parameters, but mixed, possibly conflicting, results in these areas leave us with more unanswered questions. The double β-decay experiments also give conflicting results among themselves. The ^3H decay results are indeed exciting. But here there is uncertainty in the molecular physics and the results themselves raise questions about the theoretical framework and the neutrino mass hierarchy. Long-lived neutrinos with masses above 100 eV would create conflicts with the astrophysical observation that $q_o < 1$. If ν_e has an associated mass $\sim 30 \pm 16$ eV, what of ν_μ? Or ν_τ? Finally, the astrophysical data provide only hints. The existence of 10-100 eV neutrinos could help provide many answers to cosmological questions - but do such neutrinos exist?

ACKNOWLEDGMENT

I would like to thank Robert W.Brown, Richard Holman, and S. Peter Rosen for helpful discussions. Special thanks to Sidney Harris for providing me with the cartoon which appears at the front of these lectures and for permission to reproduce it.

SUPPLEMENTAL READING

It has been my purpose in these lectures to try to present a large number of the basic arguments pertaining to various aspects of the neutrino mass problem. For this reason, I have tried to limit the number of specific references rather than compile an extensive review of the literature. Thus, many significant papers have not been referenced explicitly. (I hope that my colleagues will bear my purpose in mind so that little offense will be taken.)

However, more specialized recent papers and reviews cover specific parts of the literature more intensively. Further details of the topics discussed here may also be found in these works. I list below a few of these by topic (again not a complete listing) as recommended supplemental reading.

A) <u>Theory of Neutrino Mass</u>:

P. Langacker, "Grand Unified Theories and Proton Decay" [6].

C. Wetterich, "Neutrino Masses and the Scale of B-L Violation",

Nucl. Phys. B187, 343 (1981).

L. Wolfenstein, "Lepton and Baryon Number Nonconservation Neutrino Mass", Proc. 1981 Intl. Conf. on Neutrino Physics and Astrophysics ("ν81"), Maui, Hawaii, Ed. R.J. Cence, E. Ma and A. Roberts, Univ. Hawaii Press 2, 329 (1981).

B) Neutrino Oscillations and Solar Neutrinos

J.N. Bahcall, "Solar Neutrinos: Rapporteurs Talk", Proc. "ν81" ibid. 2, 253 (1981).

C. Baltay, "Experimental Results on Neutrino Oscillations and Lepton Nonconservation" [16].

V. Barger, "Neutrino Oscillation Phenomena"[17].

D. Silverman, and A. Soni, "Reactor Experiments and Neutrino Oscillations" [13].

C) Double Beta Decay

S.P. Rosen, "Lepton Non-Conservation and Double Beta Decay: Constraints on the Masses and Couplings of Majorana Neutrinos" [21].

D) Cosmology and Background Radiation

S. Weinberg, Gravitation and Cosmology [26].

F.W. Stecker, Cosmic Gamma Rays [38].

F.W. Stecker and R.W. Brown, "Astrophysical Tests for Radiative Decay of Neutrinos and Fundamental Physics Implications" [28].

E) Neutrino Dominated Universe

Dorochkevich, et al., "Cosmological Impact of the Neutrino Rest Mass" [31].

H. Sato, "The Early Universe and Clustering of Relic Neutrinos" [31].

P.J.E. Peebles, "The Mass of the Universe"[29].

REFERENCES

1. S. Weinberg, Phys. Rev. Lett. 43, 1566; E. Witten, Talk presented at First Workshop on Grand Unification, UTP 80/A031 (1980).
2. C. Itzykson and J.B. Zuber, Quantum Field Theory, McGraw-Hill Pub. Co., New York, 1980.
3. M. Gell-Mann, P. Ramond, and R. Slansky, in "Supergravity", Proc. of the Supergravity Workshop, Stonybrook, New York, ed. P. Van Nieuwenhausen and D.Z. Freedman, North-Holland Pub. Co., Amsterdam 1979.
4. E. Witten, Phys. Lett. 91B, 81 (1980).
5. R.N. Mohapatra and G. Senjanovic, Phys. Rev. Lett 44, 912 (1980); Phys. Rev. D23, 165 (1981); A. Masiero and G. Senjanovic, Phys. Lett. 108B, 191 (1982).
6. See, e.g., P. Langacker, Phys. Repts. 72, 185 (1981).
7. L.F. Li and F. Wilczek, Preprint NSF-ITP-81-56.
8. R. Davis, Jr., D.S. Harmer, and K.C. Hoffman, Phys. Rev. Letters 20, 1205 (1968).
9. J.N. Bahcall, Proc. Neutrino 81 Intl. Conf. on Neutrino Physics and Astrophysics (ν81) Maui, Hawaii, ed. R.J.Cence, E. Ma, and A. Roberts, Univ. Hawaii Press, 1, 1 (1981).
10. B. Pontecorvo, Sov. Phys. JETP 53, 1717 (1967).
11. F. Reines, H.W. Sobel, and E. Pasierb, Phys. Rev. Lett. 45, 1307 (1980).

12. R. Feynman and P. Vogel, unpublished.
13. H. Kwon et al., Phys. Rev. D24, 1097 (1981).
14. D. Silverman and A. Soni, Proc. ν81, ibid-, 2, 17 (1981).
15. M.F. Crouch et al., Phys. Rev. D18, 2239 (1978).
16. Boliev et al., Proc. ν81, ibid. 1, 283 (1981).
17. C. Baltay, Proc. ν81, ibid., 2 295 (1981).
18. V. Barger, Proc. ν81, ibid., 2, 1 (1981).
19. K.E. Bergkvist, Nucl. Phys. B39, 317 (1972).
20. J.J. Simpson, Phys. Rev. D23, 659 (1981).
21. V.A. Lyubimov, E.G. Novikov, V.Z. Nozik, E.F. Tretyakov, V.S. Kosik, Phys. Lett 94B, 266 (1980).
22. S.P. Rosen, Proc. ν81, 2, 76 (1981).
23. M. Doi, T. Kotani, H. Nishiura, K. Okuda, and E. Takasugi, Prog. Theor. Phys. 66, 1765 (1981).
24. W.C. Haxton, G.J. Stephenson, Jr., and D. Strottman, Phys. Rev. Lett. 47, 153 (1981).
25. A. De Rujula, CERN Preprint TH-3045 (1981).
26. R.J.N. Phillips, Nucl. Phys. B188, 459 (1981).
27. E.G. Weinberg, Gravitation and Cosmology, John Wiley and Sons, New York, 1972.
28. F.W. Stecker, Phys. Rev. Lett. 44, 1237 (1980); 46, 517 (1981); N.C. Rana, Phys. Rev. Lett. 48, 209 (1982); J.F. Rayo, M. Peimbert, and S. Torres-Peimbert, Astrophys. J., in press (1982).
29. F.W. Stecker, Proc. ν81, ibid., 1, 124 (1981); F.W. Stecker and R.W. Brown, Astrophys. J., in press (1982).
30. R.J.E. Peebles, Proc. Tenth Texas Symp.n Relativistic Astrophysics, ed. R. Ramaty and F.C. Jones, Ann., N.Y. Acad. Sci. 375, 157 (1981).
31. S.S. Gershtein and Ya.B. Zel'dovich, JETP Lett. 4, 174, (1966); G. Marx and A.S. Szalay, Proc. ν72, 1, 123; R. Cowsik and J. McClelland, Phys. Rev. Lett. 29, 699 (1972); D.N. Schramm and G. Steigman, Gen. Rel., Grav. 13, 101 (1981).
32. A.S. Szalay and G. Marx, Astron. Astrophys. 49, 437 (1976); J.R. Bond, G. Efstathiov, and J. Silk, Phys.

Rev. Lett. 45, 1980 (1980); A.G. Doroshkevich et al., Proc. Tenth Texas Symp., ibid, 32; H. Sato, Proc. Tenth Texas Symp., ibid, 43; M. Davis, M. Lecar, C. Pryor, and E. Witten, Astrophys. J. 250, 423 (1981).

33. A. Tremaine and J.E. Gunn, Phys. Rev. Lett. 42, 407 (1979).
34. L. Blitz, Astrophys. J. 231, L115 (1979).
35. D. Burstein, V.C. Rubin, N. Thonnard, and W.K. Ford, Jr., Astrophys. J. 253, 70 (1982).
36. F.W. Stecker, Phys. Rev. Lett. 45, 1460 (1980).
37. A. De Rujula and S.L. Glashow, Phys. Rev. Lett. 45, 942 (1980).
38. F.W. Stecker, Cosmic Gamma Rays, Mono Book Co., Baltimore, 1971.
39. R. Kimble, S. Bowyer, and P. Jacobsen, Phys. Rev. Lett. 46, 80 (1981).

FIGURE CAPTIONS

Fig. 1. Effective weak interaction and neutrino mass terms and renormalizable theories.

Fig. 2. Scheme for breakdown of SO(10) and Gell-Mann, Ramond, and Slansky model.

Fig. 3. Solution regions for the Reines, et al. (UCI) and Kwon et al. (ILL) data, and the best fit solutions (black areas) obtained by Silverman and Soni [13].

Fig. 4. Limits on $\sin^2 2\theta$ and Δ^2 as summarized by Barger [17].

Fig. 5. Kurie plots for $m_\nu=0$ and $m_\nu \neq 0$ shown with and without tails (T) owing to the energy resolution of the detector.

Fig. 6. Feynman diagram: neutrinoless double β decay.

Fig. 7. (a) Transition level diagram and (b) electron energy spectrum for double β decay. For neutrinoless double β decay the spectrum of $E_1 + E_2$ is a spike at

$E_1 + E_2 = Q$ as shown. With accompanying neutrinos sharing the energy, the spectrum is spread out owing to the phase space factor.

Fig. 8. Plot of M/L as a function of astronomical distance scale showing data on a fit to the analytic form of equation (6.15).

Fig. 9. Theoretical model predictions for $\tau(m_\nu)$ and astrophysical lower limits on $h_O\tau(E_O)$[28].(It is assumed that $m_\nu = 2E_O$.) The limits marked SB_F (Stecker and Brown) were obtained directly from cosmic photon fluxes. The limits MS_I (Melott and Sciama) and SB_I (Stecker and Brown) are from ionizing flux limits. The point S is obtained from the $\sim$1700 Å feature. The limits marked SCC and SCV were obtained by Shipman and Cowsik from observations of the Coma cluster and Virgo cluster. Limits obtained from other observations of Coma and Virgo by Henry and Feldman are labled HC and HV, respectively. (See [28] for a complete reference list.)

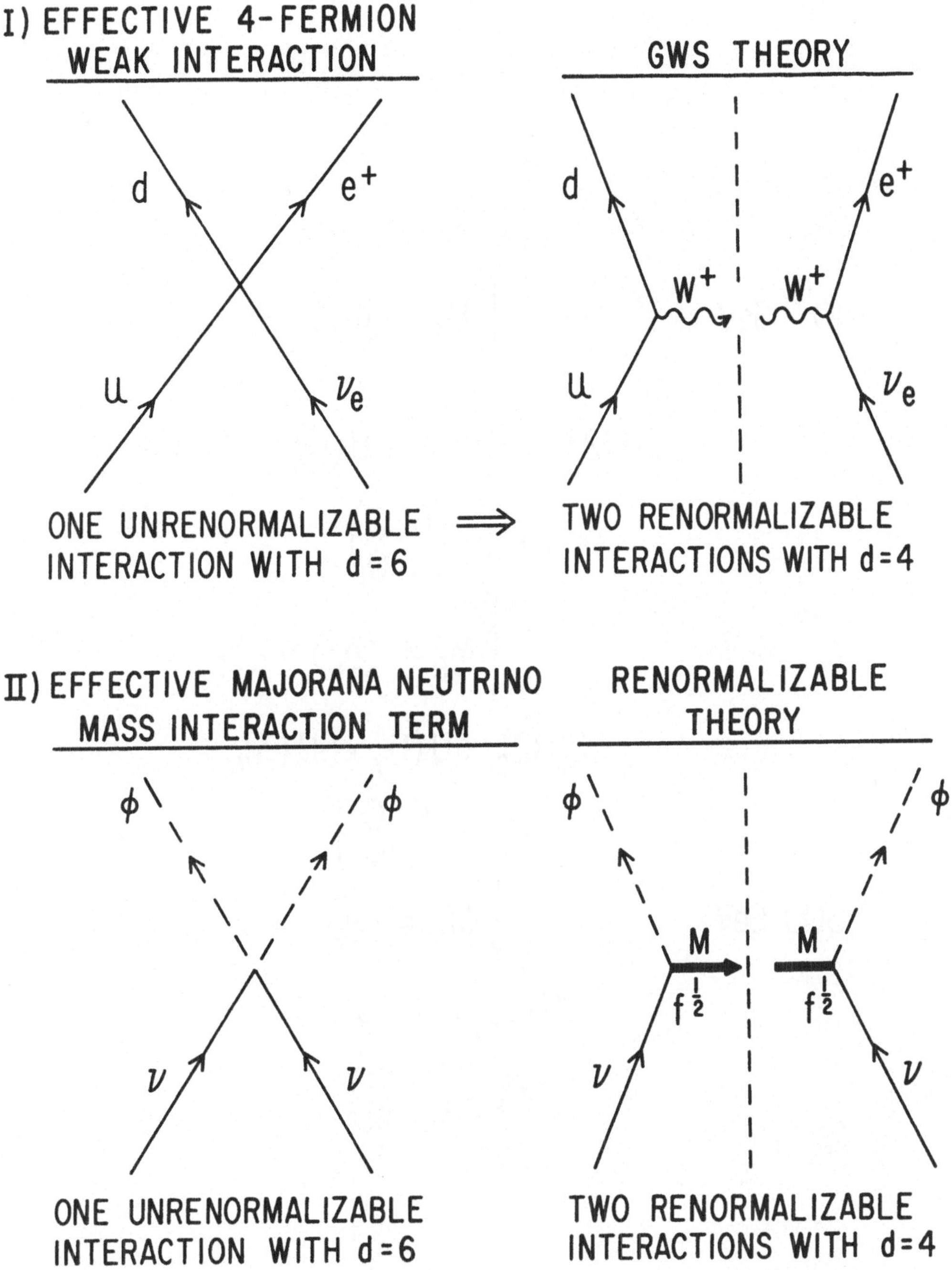
I) EFFECTIVE 4-FERMION WEAK INTERACTION
GWS THEORY
d
e+
u
νe
W+
W+
ONE UNRENORMALIZABLE INTERACTION WITH d=6
⟹
TWO RENORMALIZABLE INTERACTIONS WITH d=4
II) EFFECTIVE MAJORANA NEUTRINO MASS INTERACTION TERM
RENORMALIZABLE THEORY
φ
φ
ν
ν
M
M
f½
f½
ONE UNRENORMALIZABLE INTERACTION WITH d=6
TWO RENORMALIZABLE INTERACTIONS WITH d=4

Fig. 1

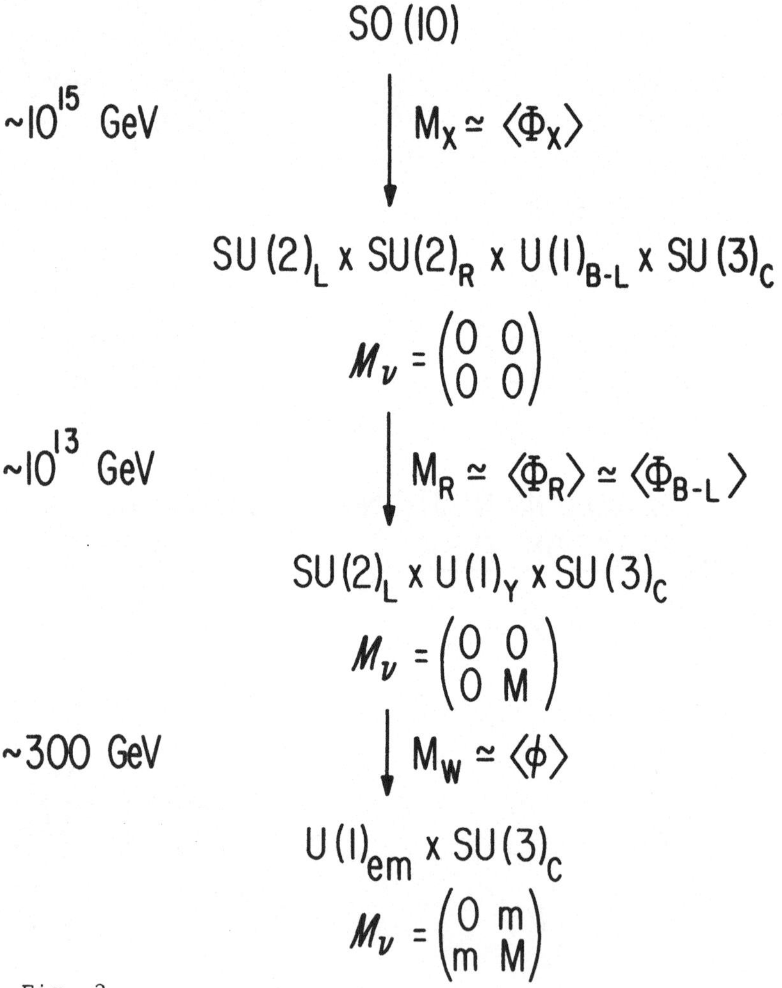
SO (10)
~10^{15} GeV
$M_X \simeq \langle \Phi_X \rangle$
$SU(2)_L \times SU(2)_R \times U(1)_{B-L} \times SU(3)_C$
$\mathcal{M}_\nu = \begin{pmatrix} 0 & 0 \\ 0 & 0 \end{pmatrix}$
~10^{13} GeV
$M_R \simeq \langle \Phi_R \rangle \simeq \langle \Phi_{B-L} \rangle$
$SU(2)_L \times U(1)_Y \times SU(3)_C$
$\mathcal{M}_\nu = \begin{pmatrix} 0 & 0 \\ 0 & M \end{pmatrix}$
~300 GeV
$M_W \simeq \langle \phi \rangle$
$U(1)_{em} \times SU(3)_C$
$\mathcal{M}_\nu = \begin{pmatrix} 0 & m \\ m & M \end{pmatrix}$

Fig. 2

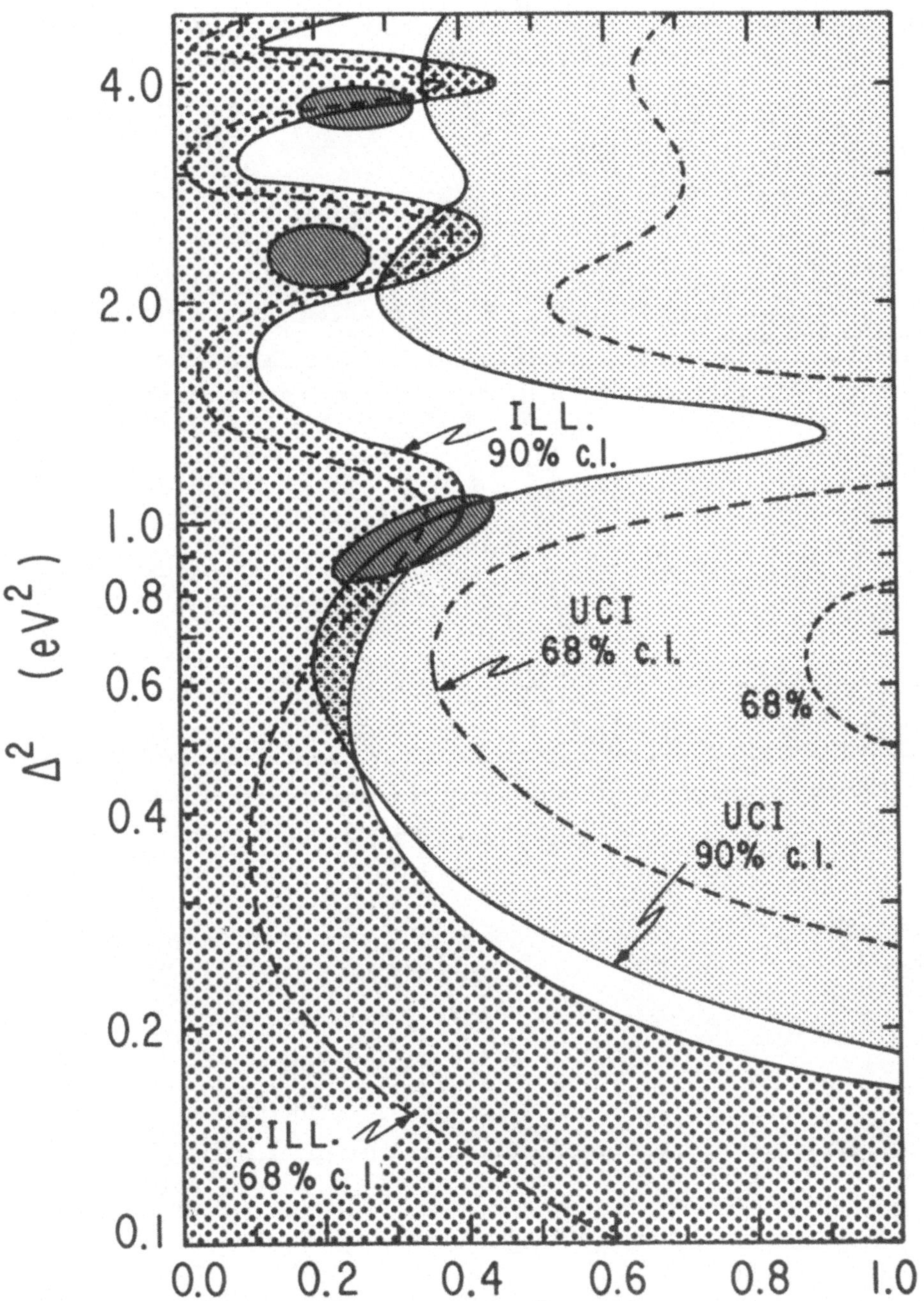
4.0
2.0
1.0
0.8
0.6
0.4
0.2
0.1
Δ^2 (eV^2)
0.0
0.2
0.4
0.6
0.8
1.0
Sin^2 2θ
ILL.
90% c.l.
UCI
68% c.l.
68%
UCI
90% c.l.
ILL.
68% c.l.

Fig. 3

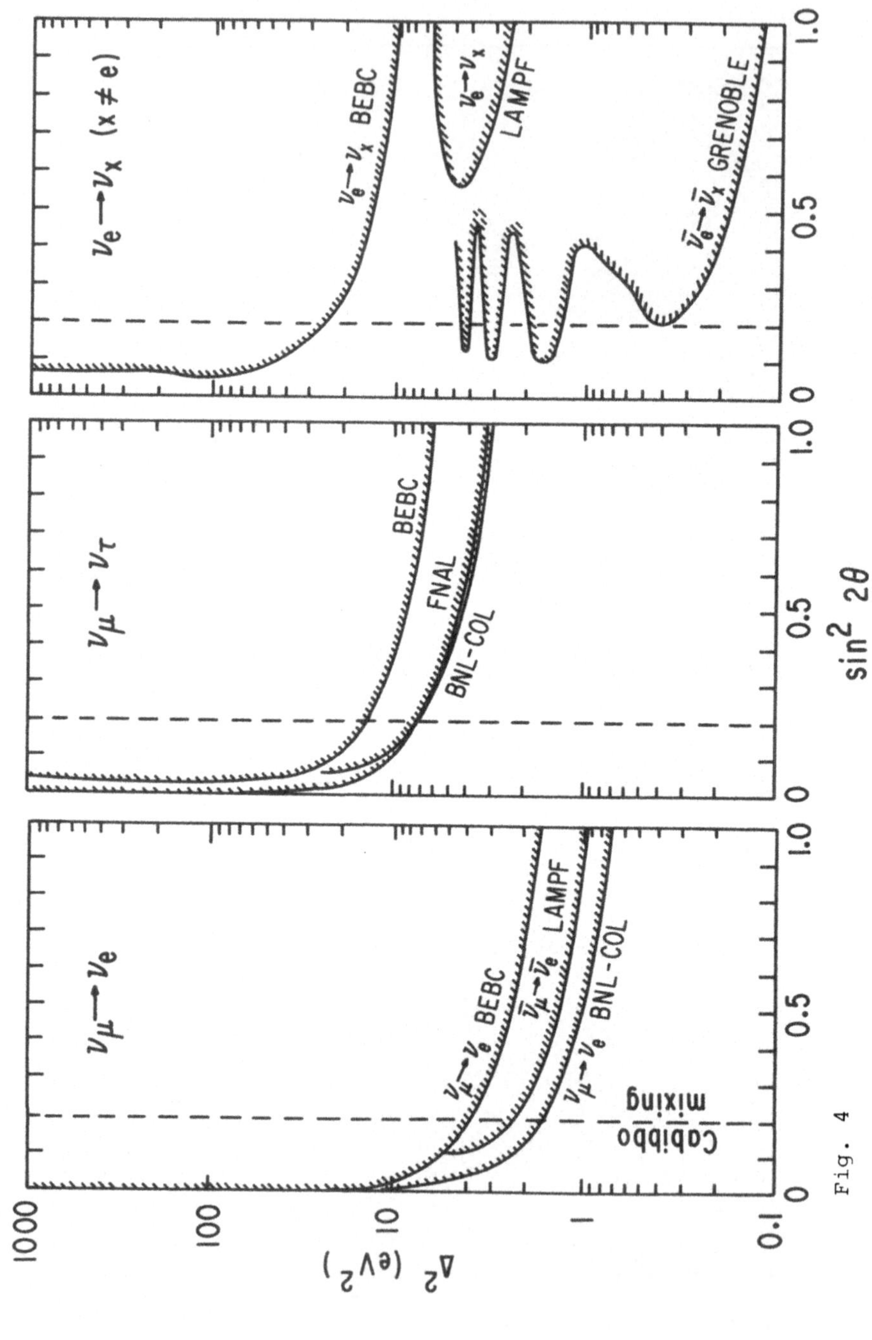

$\nu_e \rightarrow \nu_x$ $(x \neq e)$
$\nu_e \rightarrow \nu_x$ BEBC
$\nu_e \rightarrow \nu_x$
LAMPF
$\bar{\nu}_e \rightarrow \bar{\nu}_x$ GRENOBLE
$\nu_\mu \rightarrow \nu_\tau$
BEBC
FNAL
BNL-COL
$\nu_\mu \rightarrow \nu_e$
$\nu_\mu \rightarrow \nu_e$ BEBC
$\bar{\nu}_\mu \rightarrow \bar{\nu}_e$ LAMPF
$\nu_\mu \rightarrow \nu_e$ BNL-COL
Cabibbo mixing
1000
100
10
1
0.1
Δ^2 (eV^2)
0
0.5
1.0
$\sin^2 2\theta$

Fig. 4

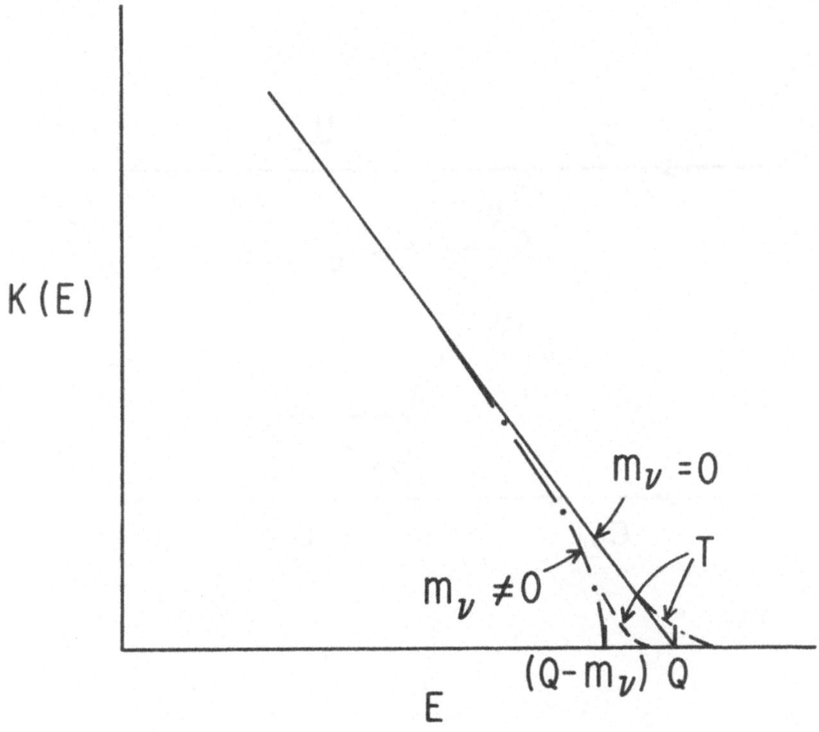

Fig. 5

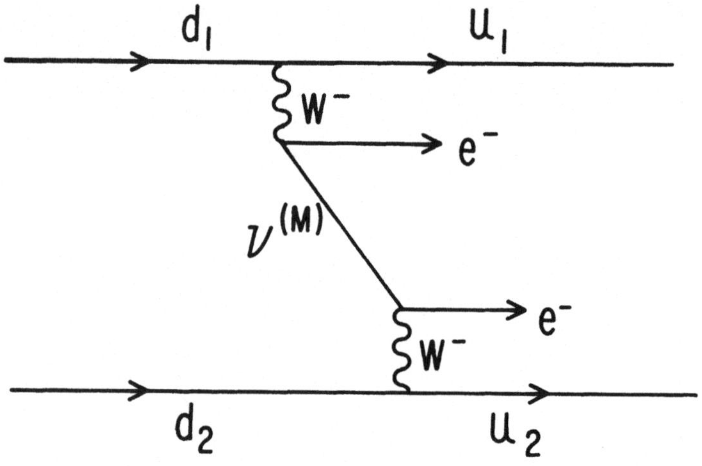

Fig. 6

a)

b)

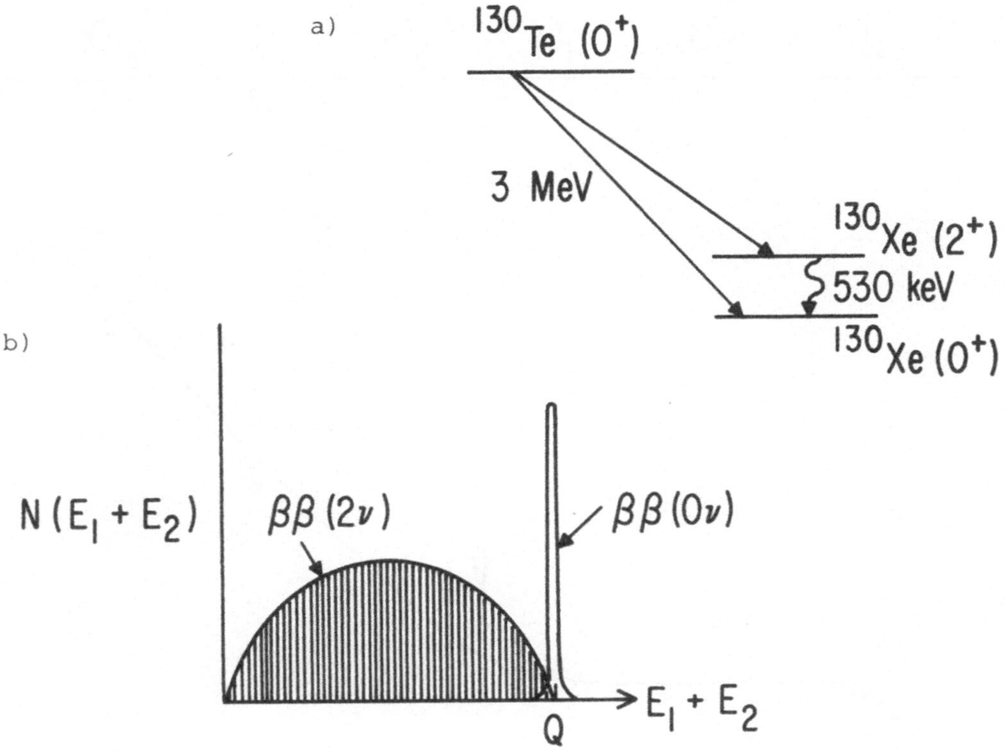

Fig. 7

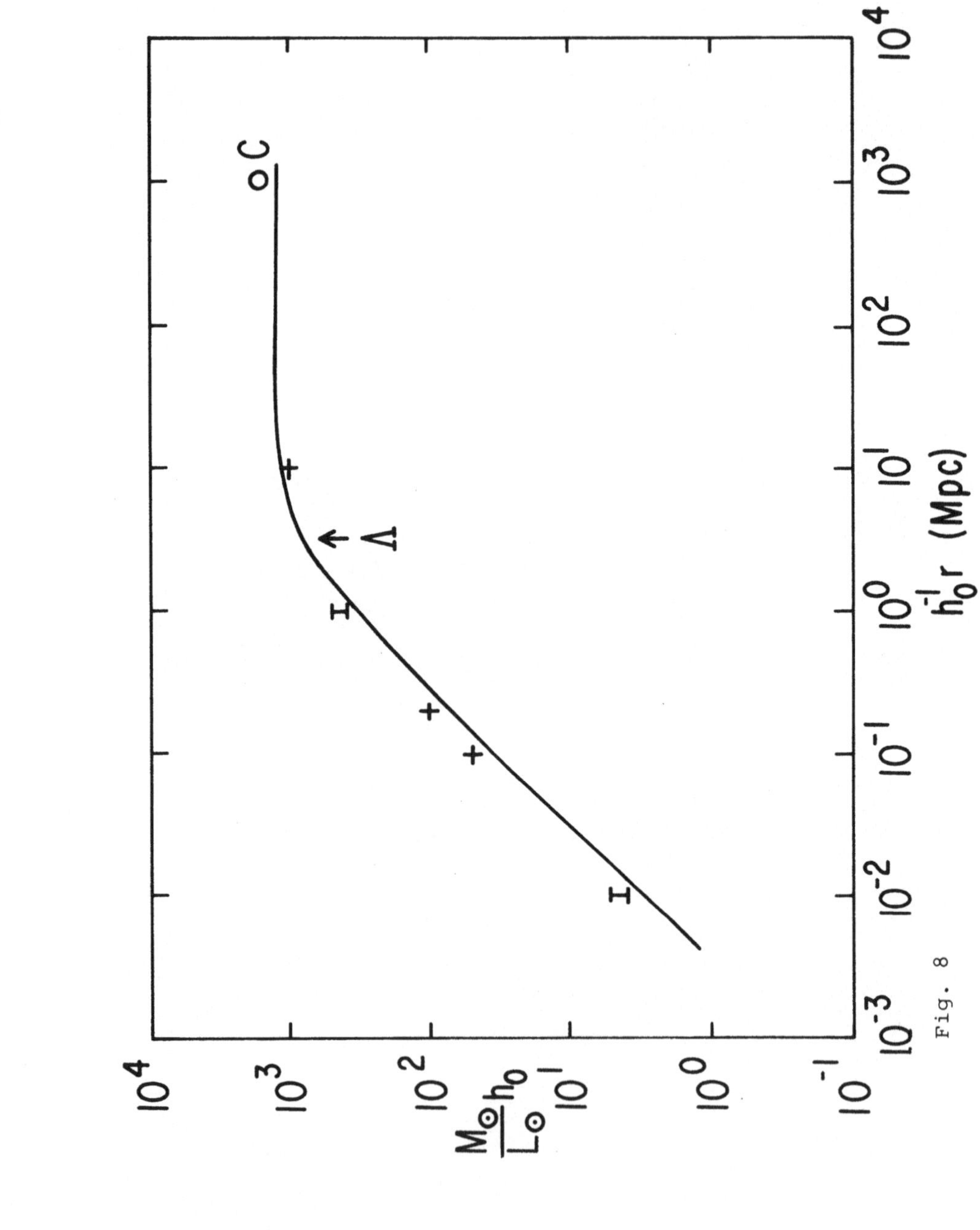
10^4
10^3
10^2
$\frac{M_\odot h_0}{L_\odot}$ 10^1
10^0
10^{-1}
10^{-3}
10^{-2}
10^{-1}
10^0
10^1
10^2
10^3
10^4
$h_0^{-1} r$ (Mpc)
C
Λ

Fig. 8

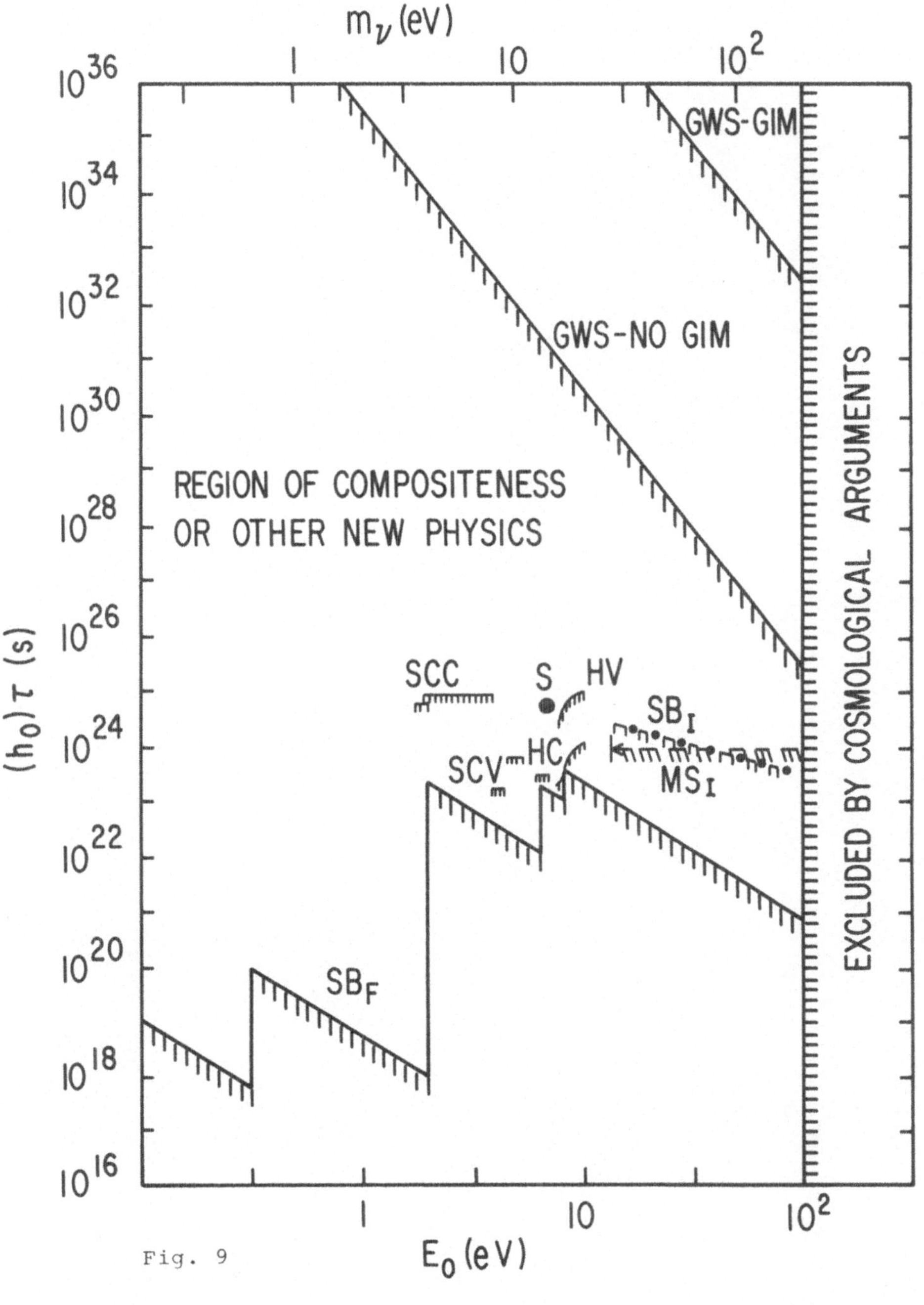
m_ν (eV)
1
10
10^2
10^{36}
10^{34}
10^{32}
10^{30}
10^{28}
10^{26}
10^{24}
10^{22}
10^{20}
10^{18}
10^{16}
$(h_0)\tau$ (s)
GWS-GIM
GWS-NO GIM
REGION OF COMPOSITENESS
OR OTHER NEW PHYSICS
EXCLUDED BY COSMOLOGICAL ARGUMENTS
SCC
S
HV
SB_I
SCV
HC
MS_I
SB_F
1
10
10^2
E_0 (eV)

Fig. 9

Acta Physica Austriaca, Suppl. XXIV, 363–392 (1982)

SUPERSYMMETRIC GAUGE MODELS OF THE FUNDAMENTAL INTERACTIONS+

by

R. BARBIERI
Scuola Normale Superiore
and
INFN, Sezione di Pisa, Italy

INTRODUCTION

These lectures consist of two distinct parts.

The first part has an introductory character[1]. I mainly settle the formalism for the N =1 globally supersymmetric gauge Lagrangians. In particular I choose to illustrate the convergence properties of these Lagrangians by giving an explicit calculation of the order $\hbar$ corrections to the effective potential in the general situation. The content of the "non-renormalization" theorems clearly emerges.

The second part - devoted to the realistic model building - is addressed primarily to the reader with the necessary background and interested in the further development of the theory. I describe a supersymmetric SU(3) × SU(2) × U(1) model[2] where supersymmetry is

+Lectures given at the XXI. Internationale Universitätswochen für Kernphysik, Schladming, Austria, February 26 - March 6, 1982.

spontaneously broken à la Fayet-Iliopoulos[3]. The necessary ξ-term is related to a $\tilde{U}(1)$ factor of the gauge group broken at superhigh energy M (of the order of the cut-off scale) and decoupled, as the goldstone field, from the "light sector" of the model. The difficulty[4] of the light scalar quark partners is avoided. Supersymmetry is broken at the geometrical mean of the physical scales M_W and M.

1. GENERALITIES ON SUPERSYMMETRY

1.1 Motivations: a look at the one-loop effective potential

The construction of a finite realistic field theory is the dream of particle physics. By the qualification "finite" one means a theory where all infinities are absent, even those absorbable by the renormalization procedure. It is a fact that the successes of quantum field theory are associated with Lagrangians whose symmetries play a crucial role in the cancellation of divergences naively expected on dimensional grounds. The increase in predictive power related to the suppression of infinities is a most important and well known achievement.

From this point of view, the softening of quantum divergences which occurs in supersymmetric field theory models both with and without gravity is a very remarkable property, which motivates most of the efforts to construct realistic supersymmetric models of particle interactions. The focus on the "hierarchy" problem[5] as well as - to some extent - on the problem of the cosmological constant [6] are aspects of this general attitude.

The main feature of supersymmetry is its fermionic character, namely its property of relating bosons to fermions. The potential importance of this property in softening quantum divergences is readily appreciated by considering, for example, the classical calculation of the order $\hbar$ quantum corrections to the effective potential V of a general gauge field theory.

Let us consider a general Lagrangian function of a set of real gauge vector boson fields A_μ^α, of complex (left handed) Weil spinors ψ_{Li} and of real scalars ϕ_a ,

$$L\,(A_\mu^\alpha,\ \psi_{Li},\ \phi_a) =$$

$$= -\frac{1}{4} F_{\mu\nu}^\alpha F_{\mu\nu}^\alpha + i\,\bar{\psi}_{Li}\not{\partial}\psi_{Li} + \frac{1}{2}\partial_\mu\phi_a\partial_\mu\phi_a$$

$$+ \frac{1}{2} A_\mu^\alpha A_\mu^\beta M_{V\alpha\beta}^2(\phi) - V(\phi)$$

$$- \frac{1}{2} M_{Fij}(\phi)\ \psi_{Li}{}^T C^{-1}\psi_{Lj} + \text{h.c.}\ . \tag{1}$$

Eq. (1) defines the zero-loop vector boson and fermion mass matrices $M_{V\alpha\beta}^2(\phi)$, $M_{Fij}(\phi)$ as functions of the scalar fields, together with the mass matrix of the scalar fields themselves,

$$M_{Sab}^2(\phi) = \frac{\partial^2 V(\phi)}{\partial\phi_a\partial\phi_b}\quad . \tag{2}$$

In terms of these matrices the order $\hbar$ corrections to the classical potential $V(\phi)$, including the zero point energy, are given by

$$V^{(1)}(\phi) = \frac{1}{2}\ \sum_B N_B \int \frac{d^3p}{(2\pi)^3}\sqrt{\vec{p}^{\,2} + M_B^2(\phi)}$$

$$- \frac{1}{2}\ \sum_F N_F \int \frac{d^3p}{(2\pi)^3}\sqrt{\vec{p}^{\,2} + M_F^+ M_F(\phi)}\ . \tag{3}$$

In the boson contribution the weight factor N_B properly counts the various degrees of freedom (N_B = 2 for a massless vevtor, N_B = 3 for a massive vevtor, and N_B = 1 for a real scalar), whereas in the fermionic contribution N_F = 2 for any two-component spinor. Cutting the integration momentum $\vec{p}$ in a sphere of radius Λ one gets

$$V^{(1)}(\phi) = \frac{1}{64\pi^2}\,[4\Lambda^4(\sum_B N_B - \sum_F N_F) +$$

$$+ 4\Lambda^2(3\,\mathrm{Tr}\,M_V^2 - 2\mathrm{Tr}\,M_F^+M_F + \mathrm{Tr}\,M_S^2)$$

$$- \lg\Lambda^2(3\mathrm{Tr}\,M_V^4 - 2\mathrm{Tr}(M_F^+M_F)^2 + \mathrm{Tr}\,M_S^4)$$

$$+ 3\mathrm{Tr}\,M_V^4(\lg M_V^2 - 2\lg 2 - \tfrac{1}{2}) - 2\mathrm{Tr}(M_F^+M_F)^2(\lg M_F^+M_F - 2\lg 2 - \tfrac{1}{2})$$

$$+ \mathrm{Tr}\,M_S^4(\lg M_S^2 - 2\lg 2 - \tfrac{1}{2})]\ . \tag{4}$$

The order $\hbar$ corrections have introduced a quartic divergence in the constant part of $V(\phi)$ - ultimately contributing to the cosmological constant - as well as quadratic divergences in the curvature of $V(\phi)$ or in the squared masses of some scalars. Needless to say, if the original Lagrangian is renormalizable, all these divergences would be absorbable in a redefinition of the parameters appearing in L. On the other hand, the request of automatic cancellation of the various divergences would require the following relations to hold identically in ϕ :

$$\sum_B N_B - \sum_F N_F = 0\ , \tag{5a}$$

$$3\,\mathrm{Tr}\,M_V^2(\phi) - 2\,\mathrm{Tr}\,M_F^+M_F(\phi) + \mathrm{Tr}\,M_S^2(\phi) = 0\ , \tag{5b}$$

$$3\,\mathrm{Tr}\,M_V^4(\phi) - 2\,\mathrm{Tr}(M_F^+M_F(\phi))^2 + \mathrm{Tr}\,M_S^4(\phi) = 0\ . \tag{5c}$$

Eq. (5a) says that the overall number of bosonic degrees of freedom must be equal to the number of fermionic degrees of freedom. Taken at the minimum of the potential, $\phi_a = \langle\phi_a\rangle$, Eqs. (5b, 5c) are relations among the zero-loop masses of the bosons, vectors and scalars, and the fermions of the theory. It is clear that only a theory with a symmetry connecting bosons to fermions has a chance of having Eqs. (5) automatically satisfied. We shall see how the sum rules involved in Eqs. (5) look like in some explicit supersymmetric Lagrangians.

1.2 Supersymmetry charges

The simplest Lagrangian exhibiting a supersymmetry is the free Lagrangian for a complex massless scalar ϕ and a 2-component Weil spinor ψ_L ,

$$h = |\partial_\mu \phi|^2 + i\, \bar{\psi}_L \not{\partial}\, \psi_L \quad , \tag{6}$$

as shown by the fact that the vector spinor current

$$J_\mu = \partial_\alpha \phi^* \gamma^\alpha \gamma_\mu \psi_L \tag{7}$$

is conserved by virtue of the equations of motion. Of course the non-trivial fact that makes this observation interesting is the possibility of introducing interactions among the fields and still keep conserved a suitable current. Examples will be given in the following.

Already from (7) one sees that the spinorial supersymmetry charge

$$Q = \int d^3\kappa\, J_o \tag{8}$$

has mass dimension 1/2, as it should be since the supersymmetry transformation[+] maps a scalar into a spinor field

---

[+]The supersymmetry transformation is not explicitly written down to avoid the introduction of the auxiliary fields, which is strictly needed for the purposes of these lectures. The reader should however be aware of the fact that supersymmetry can in general be defined as an off-mass-shell linear symmetry transformation of a Lagrangian only after the introduction of auxiliary non-propagating fields.

and viceversa. For the same reason the transformation involves derivatives of the fields, which makes plausible the basic anticommutation relation

$$\{Q_\alpha, \bar{Q}_\beta\} = \frac{1}{2}(1+\gamma_5)\gamma^\mu_{\alpha\beta}P_\mu \tag{9}$$

together with the standard commutation relations

$$[P_\mu, Q_\alpha] = 0 \quad , \qquad [M^{\mu\nu}, Q_\alpha] = -(\sigma^{\mu\nu}Q)_\alpha \ . \tag{10}$$

Eq. (9) has a well known and very important consequence, holding for any globally supersymmetric theory,

$$H = \frac{1}{2}\sum_\alpha (Q_\alpha Q^*_\alpha + Q^*_\alpha Q_\alpha) \ . \tag{11}$$

Eq. (11) tells us that

$$\langle\psi|H|\psi\rangle \geq 0 \tag{12}$$

for any state $|\psi\rangle$. Even more, it says that supersymmetry is unbroken if and only if a state exists, to be identified with a ground state of the theory, annihilated by the supersymmetry charges,

$$Q|0\rangle = \langle 0|Q^+ = 0 \ . \tag{13}$$

This property is the basis of the difficulties and the subtleties encountered when trying to obtain a spontaneous breakdown of supersymmetry.

1.3 Spontaneous breakdown of supersymmetry and the Goldstone fermion

If supersymmetry is spontaneously broken,

$$\langle 0|H|0\rangle > 0 \ ,$$

or more precisely,

$$\langle 0|\{Q_\alpha, \bar{J}_{\mu\beta}\}|0\rangle = (\tfrac{1}{2}(1+\gamma_5)\gamma_\mu)_{\alpha\beta}F^2 , \tag{14}$$

where F is a non-vanishing constant of mass dimension 2. In this case, as in the spontaneous breakdown of a bosonic symmetry, a Goldstone particle exists which is a spinor, since it is connected to the vacuum by the supersymmetry current

$$\langle 0|J_{\mu\alpha}|\psi_\beta\rangle = (\tfrac{1}{2}(1+\gamma_5)\gamma_\mu)_{\alpha\beta}F . \tag{15}$$

In the symmetric or in the spontaneously broken case, Ward identities associated to the conservation of J_μ hold. The general matrix element of J_μ between a fermion of momentum p_1 and a scalar of momentum p_2 is given by

$$\langle p_2, B|J_\mu|p_1,F\rangle =$$

$$= (A_1\gamma_\mu + A_2\, p_{1\mu} + A_3\, p_{2\mu} + A_4\, \not{p}_2\gamma_\mu + A_5\, \not{p}_2 p_{2\mu} + A_6\, \not{p}_2 p_{1\mu})u \tag{16}$$

where $\not{p}_1 u = m_F u$. The current conservation in the symmetric limit implies

$$A_1 = m_F A_4 , \qquad A_2 = A_3 , \qquad A_5 = A_6 . \tag{17}$$

On the other hand, in the spontaneously broken case, the matrix element of the current receives a singular contribution from the goldstone exchange

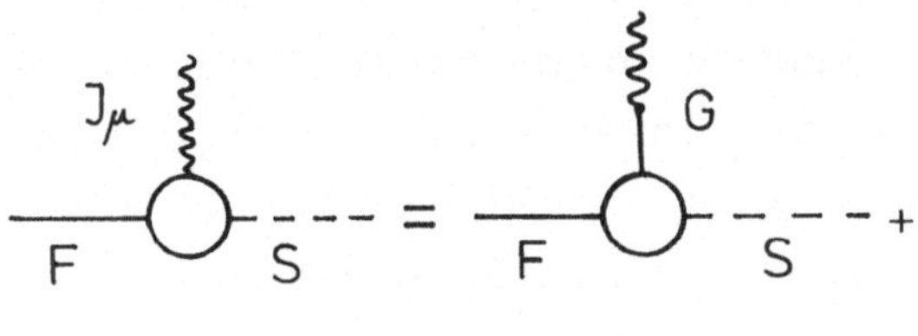

+ (non-singular terms) (18)

given in general by

$$<p_2,B|J_\mu|p_1,F>_{Sing} = F\gamma_\mu \frac{i}{\not{q}} (a + b\,\not{p}_2)u\,. \tag{19}$$

Current conservation now gives

$$-A_1 m_F + A_4 m_S^2 + \frac{1}{2}(A_2 + A_3)\ (m_S^2 - m_F^2) = iFa,$$

$$A_1 - m_F A_4 + \frac{1}{2}(A_5 + A_6)\ (m_S^2 - m_F^2) = iFb\ . \tag{20}$$

To first order in the breaking, inserting (17) in (20), one obtains

$$(A_2 + A_4)\ (m_S^2 - m_F^2) = iFa\ ,$$

$$A_5(m_S^2 - m_F^2) = iFb\ . \tag{21}$$

Furthermore, in the case of a perturbative point-like theory, from the free field current (7) one has $A_2 = i$, $A_4 = A_5 = 0$, so that

$$a = \frac{m_S^2 - m_F^2}{F}\ , \qquad b = 0\ . \tag{22}$$

This is the relation between the supermultiplet mass splitting $m_S^2 - m_F^2$, the goldstone "decay constant" F, and the goldstone coupling to the supermultiplet a.

1.4 N = 1 globally supersymmetric gauge theories

The gauge Lagrangians with a global supersymmetry generated by only one fermionic charge Q_α are the simplest supersymmetric Lagrangians, out of which one may attempt to construct realistic models of particle interactions. In doing so one leaves out the consideration of extended supersymmetric models ($N \geq 2$) as well as the inclusion of gravity, obtained by going to locally supersymmetric Lagrangians. This in turn might be a drawback, in as much as the convergence properties of the supersymmetric Lagrangians increase with N, and as the coupling to gravity might ultimately have effects also on the "low energy" Lagrangian.

In any event, following a criterium of simplicity, let us consider the general structure of the N = 1 supersymmetric Lagrangian based on a gauge group G. It involves minimally the vector multiplet

$$v^{\alpha} \equiv \binom{A^{\alpha}_{\mu}}{\lambda^{\alpha}}, \tag{23}$$

consisting of the gauge vector bosons A^{α}_{μ} and of their fermionic partners λ^{α}, and possibly chiral multiplets

$$S_a \equiv \binom{\psi^a}{s_a} \tag{24}$$

which contain the fermions ψ_a (2-component Weil spinors as the "gauginos" λ^{α}), together with the complex scalar fields s_a, transforming under G as the representation $\underline{r}$. a is the index running on $\underline{r}$, as α runs on the adjoint representation. The general Lagrangian of the fields v^{α}, S_a consists of two parts:

i) the minimal supersymmetric gauge coupling

$$L^d = -\frac{1}{4}F^{\alpha}_{\mu\nu}F^{\alpha}_{\mu\nu} + i\,\bar{\lambda}\not{D}\lambda + i\,\bar{\psi}\not{D}\psi + (D_{\mu}s)^{+}(D_{\mu}s)$$
$$-\frac{1}{2}D_{\alpha}D_{\alpha} - \sqrt{2}\,g^{\alpha}\lambda^{\alpha T}C^{-1}s^{+}T^{\alpha}\psi + \text{h.c.}, \tag{25}$$

where T^{α} are the generators of G as acting on $\underline{r}$ with coupling constants g^{α}, and

$$D_{\alpha} = g^{\alpha}s^{+}T^{\alpha}s; \tag{26}$$

ii) a possible additonal "f-term" which is constructed

from the "superpotential" $f(s_a)$ - an arbitrary gauge invariant polynomial in the fields s_a of a most degree - as

$$L^f = -\frac{1}{2} f_{ab}\psi^{a^T} C^{-1}\psi^b + h.c. - f_a f_a^* \tag{27}$$

where

$$f_a = \frac{\partial f}{\partial s_a}, \quad f_{ab} = \frac{\partial^2 f}{\partial s_a \partial s_b}, \quad f_{abc} = \frac{\partial^3 f}{\partial s_a \partial s_b \partial s_c}. \tag{28}$$

In (26), restricted to the generators T^α of isolated U(1)-factors of the group G, a constant term of mass dimension 2 can be added so that

$$D_\alpha \to g^\alpha (s^+ T^\alpha s + \xi^\alpha). \tag{29}$$

As in the case of a normal gauge Lagrangian, also the supersymmetric Lagrangians have global bosonic continuous symmetries other than the gauge symmetry itself. For a reducible representation

$$\underline{r} = n_1\, \underline{r}_1 \oplus n_2\, \underline{r}_2 \oplus \ldots \oplus n_p \underline{r}_p \tag{30}$$

of the chiral supermultiplet S_a, the minimal coupling Lagrangian L^d has a global symmetry group

$$G_g = \prod_{i=1,..p} U(n_i) \otimes U(1). \tag{31}$$

The $U(n_i)$ factors act in an obvious way on the chiral supermultiplets, whereas the U(1) acts only on the fermionic fields as follows,

$$\lambda^\alpha \to e^{i\theta}\lambda^\alpha, \qquad \psi^a \to e^{-i\theta}\psi^a. \tag{32}$$

By introducing a generic f-term one can normally break explicitly all these symmetries. Out of the $U(n_i)$ factors only the symmetries of the superpotential $f(s_a)$

remain unbroken. On the other hand an extra U(1) factor is also present in the full Lagrangian

$$L = L^d + L^b \tag{33}$$

if, among the U(1)'s in $\prod_i U(n_i)$, there is one which, acting on the chiral supermultiplets

$$s_a \rightarrow e^{iq_a\alpha} s_a \,, \tag{34}$$

transforms the superpotential $f(s_a)$ by an overall phase

$$f(e^{iq_a\alpha} s_a) = e^{i\alpha} f(s_a) \;. \tag{35}$$

In that case it is readily seen that the full Lagrangian (33) is invariant under the product of the two transformations (32) and (34) for $\alpha = 2\theta$. This U(1) global symmetry is called in the literature R-symmetry.

Given the full Lagrangian (33), it is easy to characterize the general ways of obtaining a supersymmetry breaking by the minimization of the tree level scalar potential

$$V = \frac{1}{2} D_\alpha D_\alpha + f_a f_a^* \;. \tag{36}$$

From what was said in Sec. 1.2, supersymmetry is unbroken if and only if V = 0 at the minimum, or

$$D_\alpha = 0 \,, \qquad f_a = 0 \;. \tag{37}$$

In the absence of ξ^α-terms the D_α are quadratic in the fields s_a and therefore $D_\alpha = 0$ is always trivially solved by $s_a = 0$. So is the second set of equations $f_a = 0$ in the case where none of the components s_a is a singlet under G, again because the gauge invariant superpotential f is at least quadratic in s_a.

The possibility of breaking supersymmetry by having a non-vanishing D_α related to a U(1)-factor of G via the introduction of a ξ-term has been realized by Fayet and Iliopoulos [3]. Alternatively the possiblity of forcing some f_a to be always different from zero by introducing a singlet of the gauge group G corresponds to the O'Raifeartaigh models [7]. There is no other way of breaking supersymmetry at the tree level or, in fact, in any order of perturbation theory as we shall see.

1.5 Mass sum rules

Before starting to consider the construction of possible realistic models of the fundamental particle interactions, I want to come back to the question of the infinities that arise in the Lagrangians of the type described in the previous Section and show how supersymmetry acts in reducing them. General results [8] valid to all orders of perturbation theory are already known since some time and they are normally referred to as "non-renormalization theorems". Most of them are formulated in terms of the very powerful concept of "superfield". Here I choose to illustrate them by simply considering the explicit calculation of the one-loop corrections to the general effective potential.

First of all notice that the request of automatic cancellation of the quartic divergence in the constant part of V, expressed in (5a), is automatically met in any supersymmetric theory [9]. As for the possible quadratic and/or logarithmic divergence, referring to Sec. 1.1, one first needs to know the general tree level mass matrices as functions of the scalar fields. From the definitions, Eq.(1) and Eqs. (25-29), one obtains

$$M^2_{V\alpha\beta} = g_\alpha g_\beta s^+\{T^\alpha, T^\beta\}s \quad , \tag{38}$$

$$M_{Fij} = \begin{bmatrix} -f_{ab} & -\sqrt{2}\, g_\beta (s^+T^\beta)_a \\ -\sqrt{2}\, g_\alpha (s^+T^\alpha)_b & 0 \end{bmatrix} \quad , \quad \begin{matrix} i = a,\alpha \\ j = b,\beta \end{matrix} \quad , \tag{39}$$

to be inserted in

$$(\psi^{a^T} \ \lambda^{\alpha^T})C^{-1} \, M_F \begin{pmatrix} \psi^b \\ \lambda^\beta \end{pmatrix} \quad , \tag{40}$$

and finally

$$M^2_S = \begin{pmatrix} M & N \\ N^+ & M^+ \end{pmatrix} , \tag{41}$$

$$M_{ab} = f^*_{ac} f_{cb} + g_\alpha T^\alpha_{ab} D^\alpha + g^2_\alpha (T^\alpha s)_a (s^+ T^\alpha)_b \quad , \tag{42}$$

$$N_{ad} = f^*_{abc} f_c + g^2_\alpha (T^\alpha s)_a (T^\alpha s)_d \quad , \tag{43}$$

to be inserted in

$$(s^*_a \ s_c) \, M^2_S \begin{pmatrix} s_b \\ s^*_d \end{pmatrix} \quad . \tag{44}$$

It is now straightforward to obtain the quadratic mass sum rule [10]

$$3 \, \mathrm{Tr}\, M^2_V - 2\, \mathrm{Tr}\, M^+_F M_F + \mathrm{Tr}\, M^2_S = 2\, g_\alpha D^\alpha \mathrm{Tr}\, T^\alpha \tag{45}$$

and the quartic mass sum rule [11]$^{+}$

$$3\ \mathrm{Tr}\ M_V^4 - 2\ \mathrm{Tr}\ (M_F^+ M_f)^2 + \mathrm{Tr}\ M_S^4 =$$

$$= 2g^2(\sum_A T(A) - 3C_2(G))D^\alpha D^\alpha + 4g^3 \sum_A C_2(A) s^*_{Ai} T^{\alpha Ai}_{Aj} s_{Aj} D^\alpha + 2\ f_c X_{cd} f^*_d$$

$$- g\ s^+\{X, T^\alpha\} s D^\alpha + 2g^2 \sum_A C_2(A)[-4f^*_{Ai} f_{Ai} - 2f_{AiCj} f^*_{Ai} s_{cj} - 2f^*_{AiCj} f_{Ai} s^*_{cj}]\ , \tag{46}$$

where the following convention have been used,

$s_a = s_{Ai}$ (A: irr. repres. of G, i: element of A),

$$T(A)\delta^{\alpha\beta} = \sum_{ij} T^{\alpha Ai}_{Aj}\ T^{\beta Aj}_{Ai}\ ,$$

$$C_2(A)\delta^i_k = \sum_{\alpha j} T^{\alpha Ai}_{Aj} T^{\alpha Aj}_{Ak}\ ,$$

$$X_{cd} = \sum_{ab} f^*_{abc} f_{abd}\ . \tag{47}$$

On the right hand side of Eq. (46), the first line comes from the pure gauge coupling, the second line from the pure "chiral" coupling, and finally the third one is the interference term. The quartic mass sum rule takes the final form of Eq. (47) - particularly apt to discuss the "non-renormalization" theorems - by making use of the identity

$$f_a T^{\alpha a}_{\ \ b}\ s_b = 0 \tag{48}$$

which expresses the gauge invariance of the superpotential $f(s_a)$ and holds identically for any s_a.

$^{+}$For ease of notation the gauge group is taken to be semi-simple so that there is only one gauge coupling constant $g^\alpha \equiv g$.

1.6 The "non-renormalization" theorems illustrated

To illustrate the "non-renormalization" theorems let us stick, for simplicity, to the divergent part $V_\infty^{(1)}$ (s) of the overall one-loop potential $V^{(1)}$. By inserting Eqs. (45, 46) into the general expression (4), it is now easy to see that [11]

$$V(s) + V_\infty^{(1)}(s) = \hat{V}(\hat{s}) = \frac{1}{2}\hat{D}_\alpha\hat{D}_\alpha + \hat{f}_a\hat{f}_a^* \tag{49}$$

in terms of the following renormalized quantities⁺:

$$f = \hat{\eta}_a\hat{s}_a + \hat{m}_{ab}\hat{s}_a\hat{s}_b + \hat{g}_{abc}\hat{s}_a\hat{s}_b\hat{s}_c$$

$$\hat{\xi}^\alpha = \xi^\alpha + \delta\xi^\alpha$$

$$\hat{g} = (1+\rho)g$$

$$\hat{s}_a = (z^{1/2})_a^{a'} s_{a'}$$

$$\hat{\eta}_a = (z^{1/2})_a^{a'} \eta_{a'}$$

$$\hat{m}_{ab} = (z^{1/2})_a^{a'} (z^{1/2})_b^{b'} m_{a'b'}$$

$$\hat{g}_{abc} = (z^{1/2})_a^{a'}(z^{1/2})_b^{b'} (z^{1/2})_c^{c'} g_{a'b'c'} \quad , \tag{50}$$

where

$$\delta\xi^\alpha = \frac{\Lambda^2}{(4\pi)^2} \cdot (\mathrm{tr}\, T^\alpha)$$

$$\rho = 2k\, g^2(3C_2(G) - \sum_A T(A))$$

$$(z^{1/2})_a^{a'} = \delta_a^{a'} + k(X_a^{a'} - 2g^2 C_2(A)\delta_a^{a'})$$

⁺In the renormalization of ξ^α a covariant cut-off Λ has been introduced rather than the one of Eq. (4).

$$(Z^{1/2})_a^{a'} = \delta_a^{a'} - k(X_a^{a'} - 4g^2C_2(A)\delta_a^{a'})$$

$$k = \frac{1}{64\pi^2} \lg \Lambda^2. \qquad (51)$$

Several interesting observations can be made at this point.

The only parameters that are not multiplicatively renormalized and receive a quadratic renormalization are the ξ^α-terms associated with isolated U(1)-factors in G. In fact, even these parameters are not renormalized at all if the relative generators T^α are traceless. Notice in particular that the dimensional parameters η_a and m_{ab} introduced in V through the superpotential f receive only a logarithmic multiplicative renormalization.

Actually, if one switches off the gauge coupling all the parameters introduced in V through the superpotential term receive only a renormalization consistent with the pure wave function renormalization of the scalar fields s_a through the universal $z = Z^{-1}$ factors. In the presence of the gauge couplings, even though $z \neq Z^{-1}$, the same relation is maintained among the renormalization of the mass parameters η_a, m_{ab} and of the dimensionless couplings g_{abc}. The fact that one loses the connection with the wave function renormalization is due to the non-gauge-invariant nature of this concept.

All these relationships clarify the remarkable non-renormalization properties of the general N=1 globally supersymmetric gauge theories. While I have concentrated on discussing the infinite renormalizations of the various quantities, the same properties hold, with the appropriate qualifications, for the finite renormalizations as can still be seen from the general form of the one-loop corrections to the effective potential in Eq. (4) and the explicit form of the mass matrices (Eqs. (38-44)). In particular it is easily realized that no Coleman-Weinberg phenomenon is able to break supersymmetry. This is because at the same point $\langle s_a \rangle$ where the lowest-order potential vanishes - corresponding to no tree level supersymmetry breaking - also the one-loop corrected potential does.

I want to recall finally the further improvement of the convergence properties of the Lagrangians with an extended ($N \geq 2$) supersymmetry. To the extent that these Lagrangians can always be viewed as N=1 Lagrangians with constrained superpotential terms, one can use all the previous explicit results to study or control the various statements that can be made. As an example, a remarkable fact is the complete non-renormalization of the mass terms introduced in any N=2 supersymmetric Lagrangian [11], analogous to the non-renormalization of the gauge coupling constant in the N=4 supersymmetric Yang-Mills theory [12].

2. REALISTIC MODEL BUILDING

2.1 Introduction

In the process of building realistic models of particle interactions one is confronted with a general difficulty first pointed out by Fayet [4]. For a gauge group $G = SU(3)_c \times \tilde{G}$, by specializing the general mass matrices, Eqs. (38-44) in the quark and in the scalar-quark sectors, one readily finds that the mass matrix

$$\sum_{\alpha} g_{\alpha} T^{\alpha}_{ab} \langle D^{\alpha} \rangle \qquad (52)$$

needs to be positive definite over all quarks of a given charge and helicity [2,13]. Otherwise one scalar-quark partner will be degenerate with or lighter than the down quark itself, at least at the level of the tree masses. Unfortunately, this is bound to happen in the standard $SU(3)_c \times SU(2) \times U(1)$ model since its possible generators with non-zero $\langle D^{\alpha}\rangle$, T_{3L} and the hypercharge Y, have vanishing traces over quarks, and so has (52).

To remedy this problem Fayet has proposed [14] the enlargement of the standard group to include an extra generator $\tilde{Y}$ with a definite sign over all matter multiplets and a nonvanishing $\langle D\rangle$-term contributing to (52) and actually dominating over all other possible contributions from T_{3L} and Y. The actual group is $SU(3) \times SU(2) \times U(1) \times \tilde{U}(1)$ with the ξ-term associated with the new $\tilde{U}(1)$-factor used to trigger the breaking of supersymmetry.

The main point [2,15] that I want to make here is that in order to remedy the above mentioned difficulty of the tree level mass matrix, the $\tilde{U}(1)$-factor need not be a low energy phenomenon, but can rather be broken at superlarge energies (M) relative to the breaking scale ($\sim M_W$) of the electroweak group. In spite of the fact that the auxiliary D-field accociated with it becomes superheavy ($\sim$M) and decouples from low energy ($\sim M_W$), still its nonvanishing vacuum expectation value $\langle D\rangle \sim M_W^2$ can contribute a positive squared mass for the light scalars, which is what is needed[+].

The important feature of the model [2] where this situation is realized is that supersymmetry is broken at the intermediate scale $\sqrt{M_W M}$. This in turn introduces splittings among scalar and fermion masses of order $\sqrt{M_W M}$ in the superheavy

[+] Fayet has mainly focused [4] on the possibility that the mass of the $\tilde{U}(1)$ gauge boson be light, indeed even subtantially lighter than the W-mass itself. This in turn requires the $\tilde{U}(1)$ gauge coupling constant to be much smaller than the weak or the electromagnetic ones. I am not in the position of judging the internal consistency of this point of view, since I have not seen an explicit model discussed in detail. The attitude that is taken here is essentially opposite to Fayet's.

A peculiar way of using the $\tilde{U}(1)$-factor to break supersymmetry has been suggested by Slavnow [16]. It corresponds essentially to taking the limit $g\to 0$ for the $\tilde{U}(1)$ coupling constant. In the resulting theory however supersymmetry is explicitly broken, although softly.

($\sim M$) sector of the model, but not in the light ($\sim M_W$) sector because of a decoupling of the goldstone from the light sector itself. In the notation of Eq. (22)

$$(m_S^2 - m_F^2)_{light} = F\, a_{light} \tag{53}$$

with $F \simeq M_W M$ and

$$a_{light} \simeq \frac{M_W}{M} \quad . \tag{54}$$

This is the mechanism which keeps the appropriate Higgs fields light ($\sim M_W$). By virtue of the non-renormalization theorems the decoupling occurs to all orders of perturbation theory.

The breaking of the Fayet $\tilde{U}(1)$-gauge generator at a heavy scale rather than at relatively low energy appears to be an unavoidable feature of any consistent model with physically acceptable tree level masses. The first reason for that has to do with the fact that it seems difficult to break supersymmetry at all à la Fayet-Iliopoulos, in the absence of Yukawa couplings or rather with Yukawa couplings not satisfying the lower bound[+] $g_Y \gtrsim M_W/M$ where M is the breaking scale of the $\tilde{U}(1)$ generator [15]. Furthermore, a "low energy" $\tilde{U}(1)$ factor requires a cancellation of the ABJ anomalies associated with this generator among the "low energy" fermions and in turn an extension of the chiral multiplets, again making it difficult to break supersymmetry [2]. Both these points will be illustrated in the following Sections.

[+]The actual lower bound, in a specific model, will contain also some gauge coupling which is neglected here.

2.2 An explicit $SU(3)\times SU(2)\times U(1)\times \tilde{U}(1)$ model

The chiral super-multiplet structure of the model [2] is given in Table 1. The $\tilde{U}(1)$-charge of all normal matter multiplets is taken to be equal to 1. The two Higgs doublets H, H^c have $\tilde{U}(1)$-charge -2 in order to have gauge-invariant Yukawa-like couplings to normal matter. Finally the four extra fields R, R^c, S, P transform only under the $\tilde{U}(1)$ factor and are mainly introduced to trigger the supersymmetry breaking à la Fayet-Iliopoulos. To that purpose a ξ-term is also introduced in the $\tilde{U}(1)$ D-term.

The superpotential f is given by+

$$f = \alpha Q u^c H^c + \beta Q d^c H + \gamma L e^c H + \delta P L H^c + \lambda H H^c S + \mu R R^c \tag{55}$$

with obious contractions of the group indices. Here only the parameter μ has mass dimension 1, whereas all the other terms correspond to dimensionless Yukawa-like couplings.

To understand the symmetry breaking pattern of the model, it is useful to consider it having first set $\mu=0$ in Eq. (55). In this case the model, and in particular the scalar potential, has only one dimensional parameter, ξ, entering into the $\tilde{U}(1)$ D-factor,

$$D = g(-2|H|^2-2|H^c|^2-4|R^c|^2 +4|R|^2+4|S|^2+|P|^2+|Q|^2+|u^c|^2+ + |d^c|^2+|L|^2+|e^c|^2+\xi). \tag{56}$$

ξ, which is taken to be positive, receives a quadratically divergent renormalization

---

+Henceforth we shall use the same notation to denote both a chiral supermultiplet and its scalar component.

$$\hat{\xi} = \xi + g \frac{\Lambda^2}{(4\pi)^2} \text{Tr } \tilde{y} = \xi + 12g \frac{\Lambda^2}{(4\pi)^2} . \tag{57}$$

Its "natural" value is therefore of order Λ^2 itself, where Λ is the energy cut-off of the model.

For $\mu = 0$ it is easy to see that supersymmetry is unbroken. The supersymmetric minima, among other equations, have to satisfy

$$\frac{\partial f}{\partial u^c} = \alpha \; Q \; H^c = 0 \qquad \frac{\partial f}{\partial P} = \delta \; L \; H^c = 0$$

$$\frac{\partial f}{\partial d^c} = \beta \; Q \; H = 0 \qquad \frac{\partial f}{\partial S} = \lambda \; H \; H^c = 0 \; .$$

$$\frac{\partial f}{\partial e^c} = \gamma \; L \; H = 0 \tag{58}$$

These equations require either that all the nonvanishing SU(2)-doublets of the model be parallel to each other or $H = H^c = 0$. The first situation is compatible with the condition of vanishing of the D-term associated with SU(2) if and only if

$$H = H^c = L = Q = 0 \; . \tag{59}$$

At this point the vanishing of the $SU(3)_c$ and U(1) D-terms requires also

$$u^c = d^c = e^c = 0 \tag{60}$$

and one is left with only the SU(3)× SU(2)× U(1) singlets R, R^c, S, P as possibly nonzero fields. The same conclusion is reached, with some more work, in the case $H = H^c = 0$. It is now easy to see that demanding the vanishing of the f-terms in the potential does not impose conditions on R, R^c, S, P, whereas D = 0 in Eq. (56) requires

$$-4|R^c|^2 + 4|R|^2 + 4|S|^2 + |P|^2 + \xi = 0 \ . \qquad (61)$$

Eqs. (59-61) define all the supersymmetric minima of the model for $\mu = 0$. Any one of these minima gives rise to the breaking of $\tilde{U}(1)$ only, occurring at a mass $M \sim g\sqrt{\xi}$, with the formation of a massive vector supermultiplet. All the other states are massless.

It is important to notice at this point that equalling to zero any one of the dimensionless couplings in f, Eq.(55), would introduce other supersymmetric minima than those defined in Eqs. (59-61), corresponding to the disappearance of some of the constraints in Eq. (58). This in turn would be disastrous for the model under consideration, since these same supersymmetric minima would not be removed by switching on the other mass parameter μ.

2.3 The minimum of the potential and the decoupling of the light states

Let us introduce now the term μRR^c in f, Eq. (55), which in turn corresponds in the scalar potential V to a mass μ for the fields R and R^c. Since now the field R^c can no longer take a nonzero vacuum expectation value consistently with unbroken supersymmetry because of

$$\frac{\partial f}{\partial R} = \mu R^c = 0 \ , \qquad (62)$$

Eq. (61) is no more soluble for $\xi > 0$ and supersymmetry is broken. The minimization of the potential now forces all the fields having a negative $\tilde{U}(1)$-charge, R^c, H, H^c, to acquire a nonzero v.e.v., whereas the positively charged ones all remain at zero. In the variables H, H^c, R^c, the potential is

$$V(H, H^c, R^c) = \frac{1}{2} g_2^2 (H^+ \frac{\sigma_i}{2} H + H^{c+} \frac{\sigma_i}{2} H^c)^2 +$$
$$+ \frac{1}{2} g_y^2 (- \frac{1}{2}|H|^2 + \frac{1}{2} |H^c|^2)^2 +$$
$$+ \frac{1}{2} g^2 (-2|H|^2 - 2|H^c|^2 - 4|R^c|^2 + \xi)^2 +$$
$$+ \lambda^2 |H\, H^c|^2 + \mu^2 |R^c|^2 \quad . \tag{63}$$

For the range of values of the parameters

$$g_2^2 > 2\lambda^2 \quad ,$$
$$4g^2\xi > \mu^2 (1 + \frac{8g^2}{\lambda^2}) \quad , \tag{64}$$

it reaches its minimum at

$$H = \begin{pmatrix} \frac{\mu}{\sqrt{2}\lambda} \\ 0 \end{pmatrix} \qquad H^c = \begin{pmatrix} 0 \\ \frac{\mu}{\sqrt{2}\lambda} \end{pmatrix} \tag{65}$$

$$R^{c2} = \frac{\xi}{4} - \mu^2 \frac{8g^2 + \lambda^2}{16 g^2 \lambda^2} \quad . \tag{66}$$

At this point

$$D_{SU(2)} = D_y = 0 \quad , \qquad D = \frac{\mu^2}{4g} \quad , \tag{67}$$

and

$$f_R = \mu (\frac{\xi}{4} - \mu^2 \frac{8g^2 + \lambda^2}{16 g^2 \lambda^2}) \tag{68}$$

is the only nonvanishing "f-term".

By working out the quadratic part of the overall V around this point one finds that all the other fields are stabilized at zero. The positive nonvanishing D-term for the additional $\tilde{U}(1)$ factor in G is the crucial ingredient.

In particular all the normal matter superpartners receive a positive squared mass of the order of the W-mass itself.

The expectation values of the fields H, H^c break the electroweak group in the correct way at a mass of order μ. On the other hand the supersymmetry breaking, although also controlled by μ, takes place at a mass scale of the order of the geometrical mean of the two scales μ and $\sqrt{\xi}$. In fact, at the minimum,

$$V \equiv F^2 = \frac{1}{2} D^2 + f_R^2 \simeq f_R^2 \simeq \frac{\mu^2 \xi}{4} \quad , \tag{69}$$

where F is the goldstone "decay constant" defined in Sec. 1.3.

Several comments are in order. The first has to do with the would-be supersymmetric minima when switching off some of the Yukawa-like couplings in (55). Around those points, to first order in these couplings - generically denoted by α - the potential has the value

$$V \sim \alpha^2 \xi^2 \, , \tag{70}$$

to be compared with the value at the wanted minimum (69). Since at those points the symmetry breaking pattern is not the correct one - color and/or electromagnetism get broken - one ends up with the lower bound [15]

$$\alpha \gtrsim \frac{\mu}{\sqrt{\xi}} \simeq \frac{\lambda g}{g_2} \frac{M_W}{M} \, . \tag{71}$$

In turn this is a lower limit on all the fermion masses,

$$m_F \gtrsim \frac{\lambda g}{g_2^2} \frac{M_W^2}{M} \tag{72}$$

including the neutrino. The neutrino receives a mass, specifically through the coupling δ LPH^c, of the Dirac type, with no breaking of lepton number. The fermion component of the P-supermultiplet can be considered as a right-handed neutrino.

Another comment concerns the problem of the decoupling of the fields with a tree level mass of order μ from the superheavy sector of the theory^{+}. This is particularly relevant in view of the fact that supersymmetry is broken at intermediate masses $\sim\sqrt{M_W M}$. Although the light fermion fields are kept at M_W by suitable chiral symmetries, their scalar superpartners - among which is the standard Higgs field - might be split by an amount of the order of the supersymmetry breaking itself. Fortunately this is not the case because the auxiliary f_R-field, whose v.e.v. is responsible for the relatively large supersymmetry breaking, is only coupled to the superheavy R_c-multiplet, as seen by rewriting the f-term contribution to the scalar Lagrangian as

$$V^f = - |f_R|^2 + f_R \mu R^{c*} + f_R^* \mu R^c - |f_{R_c}|^2 + f_{R_c} \mu R^* + f_{R_c}^* \mu R \,. \tag{73}$$

Radiative corrections will induce couplings of f_R to light multiplets, but only through the suppression of an intermediate superheavy R^c scalar field. In fact, after the

$^{+}$The possibility of proving the decoupling of the scales M_W and M was unclear to me when I gave the lectures. This and other points will be the subject of a paper in preparation [15]. On this and related matters I have profited of stimulating conversations with Stuart Raby who has built a model, together with S. Dimopoulos, where an analogous decoupling phenomenon takes place. I understand that also L. Susskind has formulated a "decoupling theorem" (private communication by S. Raby).

integration of the superheavy degrees of freedom the effective Lagrangian of the light fields will still be a supersymmetric Lagrangian, apart from mass terms of order μ and interactions weighted by atleast a power of μ^2/ξ. It is this decoupling of the intermediate supersymmetry breaking from the light sector of the model that keeps the Higgs doublet from receiving a mass of the order of the supersymmetry breaking itself. Splittings of order $(\mu\sqrt{\xi})^{1/2} \sim (M_W M)^{1/2}$ occur only among the members of the heavy supermultiplets, to which the goldstone field

$$\psi_g \simeq \psi_R \qquad (74)$$

is mainly coupled.

A final point [2] concerns the problem of the ABJ anomalies. The triangle diagrams with at least one superheavy U(1)-gauge boson will introduce anomalies into the model. Formally one could dispose of those anomalies by taking $\xi \to \infty$. In this way one would remain with a supersymmetric Lagrangian involving the light field and an explicit soft supersymmetry breaking from the mass terms. More reasonably, ξ has to be kept at the cut-off scale of the model, where the anomaly problem is not, a priori, a severe one. For example, the introduction of gravity might change it qualitatively [16]. On the contrary, the anomaly constraint would be compelling for the consistency of a theory with the $\tilde{U}(1)$ factor of G broken at low energy. This in turn may lead to problems in obtaining the breaking of supersymmetry and/or of the gauge group in the physical direction [2].

2.4 Conclusions and perspectives

The "naturality" problem of the Fermi scale suggests an extension of the standard strong and electroweak gauge theory to an N=1 globally supersymmetric model, where the weak interaction scale does not receive quadratic renormalization. The difficulty of the light scalar quark partners - met in the simplest SU(3)× SU(2)× U(1) version - is solved by extending the gauge group to include an extra $\tilde{U}(1)$-factor "naturally" broken at superheavy energies (the cut-off scale). Supersymmetry is broken at the geometric mean of the two basic scales of the model, the Fermi mass and the cut-off. A decoupling occurs of the light particles, sitting at $\sim M_W$, from the heavy sector and the goldstone field.

The Lagrangian described in the previous Sections, taken as a realistic model of the strong and electroweak interactions, predicts a detailed spectrum for essentially all the new particles introduced to implement supersymmetry. Among them there are massless gluinos λ^a (the fermionic partners of the gluons) and an almost massless Peccei-Quinn axion, which are probably in conflict with experiments. The reason for their existence has to do with the two continuous global symmetries that the overall Lagrangian possesses other than the gauge symmetry and supersymmetry itself: an R-invariance which remains unbroken and - by acting nontrivially on all gaugino fields - forbids a gluino mass term $\lambda^{a^T} C^{-1} \lambda^a$; a Peccei-Quinn symmetry which gets broken together with the electroweak group. Both of these problems could be taken care of [2] by introducing a chiral superfield N with no gauge interaction at all and a "chiral" interaction from the extra superpotential term

$$f_N = \eta R^c S\, N + f(N) \, . \qquad (75)$$

A suitably chosen polynomial in the singlet field, f(N), gives rise to an explicit breaking of both the above symmetries.

A more interesting possibility is that these

phenomenological problems could be resolved by a suitable embedding of the model described into a grand unified scheme and/or by the inclusion of gravity through a locally supersymmetric action. Insisting on a purely perturbative discussion, the SU(3)× SU(2)× U(1) × $\tilde{U}(1)$ Lagrangian could not arise from a complete unification into a semi-simple group. There is in fact no way of completing the multiplets of Table 1 to form an overall representation of a semi-simple group because: i) $Tr\tilde{Y} \neq 0$ over the fermions of Table 1; ii) the additional massive fermions should be vector-like under $\tilde{Y}$. Adding some "light" fermions to those of Table 1 to make $Tr\tilde{Y} = 0$ would not help, because in that case a ξ-term could not be generated. On the other hand, a unification is conceivable of the standard interactions in a semi-simple group (SU(5),...) with $\tilde{U}(1)$ as an external factor broken at superlarge energies (M_{Pl}).

ACKNOWLEDGEMENTS

I have profited of very stimulating conversations with Sergio Ferrara, Luciano Maiani, Dimitri Nanopoulos, and Stuart Raby.

Table 1

$G = SU(3) \times SU(2) \times U(1) \times \tilde{U}(1)$

	SU(3)	SU(2)	U(1)	$\tilde{U}(1)$
Q_i^α	3	2	1/6	1
u_i^c	$\bar{3}$	1	-2/3	1
d_i^c	$\bar{3}$	1	1/3	1
L^α	1	2	-1/2	1
e^c	1	1	1	1
H^α	1	2	-1/2	-2
$H^{c\,\alpha}$	1	2	1/2	-2
S	1	1	O	4
R	1	1	O	4
R^c	1	1	O	-4
P	1	1	O	1

The chiral supermultiplet content of the model of Ref.[2] with the relative transformation properties under G.

REFERENCES

1. For a review on supersymmetry see, for instance, P. Fayet and S. Ferrara, Phys. Rep. 32C (1977) No.5, 249.
2. R. Barbieri, S. Ferrara and D.V. Nanopoulos, to appear in Z. Phys. C - Particles and Fields.
3. P. Fayet and J. Iliopoulos, Phys. Lett, 51B (1974) 61.
4. P. Fayet in: "Unification of the Fundamental Particle Interactions", eds. S. Ferrara, S. Ellis and P. van Nieuwenhuizen (Plenum Press, N.Y., 1980), p.587; and

references therein.

5. E. Gildener and S. Weinberg, Phys. Rev. D13 (1976) 3333; S. Weinberg, Phys. Lett. 82B (1979) 387.
6. A. Linde, JETP Lett. 19 (1974) 183; M. Veltman, Phys. Rev. Lett. 34 (1975) 777.
7. L. O'Raifertaigh, Nucl. Phys. B96 (1975) 331.
8. A probably partial list of references includes: J. Wess and B. Zumino, Phys. Lett. 49B (1974) 52; J. Iliopoulos and B. Zumino, Nucl. Phys. B76 (1974) 310; S. Ferrara, J. Iliopoulos and B. Zumino, Nucl. Phys. B77 (1974) 413; S. Ferrara and O. Piguet, Nucl. Phys. B93 (1975) 261; M. Grisaru, W. Siegel and M. Rocek, Nucl. Phys. B159 (1979) 420; W. Fischer, H. Nilles, J. Polchinski, S. Raby and R. Susskind, Phys. Rev Lett. 47 (1981) 757.
9. B. Zumino, Nucl. Phys. B89 (1975) 535.
10. S. Ferrara, R. Girardello and F. Palumbo, Phys. Rev. D20 (1979) 403.
11. R. Barbieri, S. Ferrara, L. Maiani, F. Palumbo and C. Savoy, CERN preprint Ref. TH. 3282 (April 1982).
12. K. Stelle, Ecole Normale Sup. preprint LPTENS 81/24 (1981); and references therein.
13. S. Dimopoulos and H. Georgi, Nucl. Phys. B193 (1981) 150.
14. P. Fayet, Phys. Lett. 69B (1977) 482.
15. R. Barbieri, S. Ferrara and D.V. Nanopoulos, paper in preparation.
16. A. Slavnov, Sov. Phys. Usp. 21(1978) 240; and references therein.
17. R. Barbieri, S. Ferrara and D.V. Nanopoulos, and K. Stelle, CERN preprint Ref. TH. 3243 (February 1982).

Acta Physica Austriaca, Suppl. XXIV, 393–474 (1982)

CP-NONCONSERVATION[+]

by

R.H. DALITZ
Dept. of Theoretical Physics
Oxford University, Oxford, Great Britain

1. INTRODUCTION

The standard model of the Electroweak Interactions, a gauge theory based on an SU(2) × U(1) symmetry spontaneously broken, has been remarkably successful in accounting for a very wide range of data on these interactions. Consequently, in these lectures concerning CP-nonconservation in the weak interactions, we shall work entirely within the framework of this model, since this is the most economical approach until such time as this framework may be shown to be too restrictive for the interpretation of the observed facts. Unfortunately, the positive facts which we have concerning CP-nonconservation effects have been few in number for quite some time, although there appears hope of obtaining further facts from observations on the New Mesons, as well as from more precise experiments on the neutral K mesons, in the next few years, as we shall discuss below.

The plan of the lectures is as follows. In Sec. 2, we discuss in a general way the various mechanisms by which CP-nonconservation effects can arise within the SU(2) × U(1) framework, given the quarks and leptons now known tu us. Sec. 3

[+] Lectures presented at the XXI. Internationale Universitäts-wochen für Kernphysik, Schladming, Austria, Feb.25-March 6, 1982.

then gives a phenomenological discussion of the CP-nonconservation phenomena known for the K^o-$\bar{K}^o$ complex, leading on to a discussion in Sec. 4 of its interpretation, within the standard model, and of the aims of further experiments under way. In Sec. 5, we discuss the situation concerning the New Mesons D^o, B^o and T^o and the possibility for the observation of CP-nonconservation effects in their decay processes. In Sec. 6, we consider recent estimates for the magnitude of the electric dipole moment for the neutron, in relation to the experimental accuracy expected in the current round of experiments.

2. CP-NONCONSERVATION WITHIN THE SU(2) × U(1) STANDARD MODEL

We shall not review the SU(2) × U(1) gauge theories of the electroweak interactions in detail here, but refer the reader to the parallel lectures given by Dr. G. Ecker at this School [1], and to the lectures given recently by Dr. J. Ellis elsewhere [2]. There are three places in these theories where the question of CP-nonconservation can arise.

2.1 The Quark Mixing Matrix

Here we confine attention to the case where the theory is based on one Higgs doublet (ϕ^+, ϕ^o), and some number n_g (generations) of quark doublets (Q_u, Q_d) where Q_u and Q_d denote the up and down states within each of these doublets. When the theory is spontaneously broken, the Higgs bosons $(\phi^+, (\phi^o-\bar{\phi}^o)/\sqrt{2}, \phi^-)$ generate masses for the $W^\pm$ and Z^o bosons, in a well-known way [3], and are absorbed into the longitudinal components of these heavy weak bosons. Only one physical Higgs boson H survives, given by

$$H = (\phi^o + \bar{\phi}^o)/\sqrt{2} - v \,, \qquad (2.1)$$

where v denotes the vacuum expectation value

$$v = \langle 0|\phi^0 + \bar{\phi}^0|0\rangle/\sqrt{2} \quad . \tag{2.2}$$

The quarks gain masses through the quark-Higgs Yukawa couplings, initially of the general form

$$L_{\underset{\sim}{\bar{q}q}} = \Sigma_{qq'}\{F_{qq'}(\bar{q}_L\cdot\phi q'_R) + \text{h.c.}\} \; , \tag{2.3}$$

where the wavy underline refers to weak isospin, as a result of the non-zero expectation value (2.2). After the spontaneous symmetry breaking, the terms of (2.3) which survive may be written in the form

$$-\frac{H+v}{v}\{\bar{Q}_{uL}(F_u)Q_{uR} + \bar{Q}_{dL}(F_d)Q_{uR} + \text{h.c.}\} \; , \tag{2.4}$$

which includes both the quark mass matrix and the coupling of the physical Higgs-particle H with the quarks Q_u and Q_d, which we may hereafter consider as column matrices in generation space. The mass matrices F_u and F_d may now be diagonalized in generation space by suitable unitary transformations L_u on Q_{uL}, R_u on Q_{uR}, etc., giving expression (2.4) the form

$$-\frac{H+v}{v}\{\bar{q}_{uL}(L_u^\dagger F_u R_u)q_{uR} + \bar{q}_{dL}(L_d^\dagger F_d R_d)q_{dR}\} \; , \tag{2.5}$$

where the quark mass eigenstates have been denoted by (q_u, q_d), q_u and q_d being column vectors in generation space. For the quarks known at present, we have $n_g = 3$, and

$$\text{(a)}\; q_u = \begin{bmatrix} u \\ c \\ t \end{bmatrix} \qquad \text{(b)}\; q_d = \begin{bmatrix} d \\ s \\ b \end{bmatrix} \quad . \tag{2.6}$$

Of course, the top quark t is not yet known, but it is a necessary partner to the b quark in this scheme. The diagonalized mass matrices are then

$$\text{(a)} \qquad M_u = (L_u F_u R_u) = \begin{bmatrix} m_u & \cdot & \cdot \\ \cdot & m_c & \cdot \\ \cdot & \cdot & m_t \end{bmatrix}$$

$$\text{(b)} \qquad M_d = (L_d F_d R_d) = \begin{bmatrix} m_d & \cdot & \cdot \\ \cdot & m_s & \cdot \\ \cdot & \cdot & m_b \end{bmatrix} \quad . \qquad (2.7)$$

From (2.5), it is apparent that these transformations diagonalize the $\bar{q}qH$ coupling at the same time as the mass matrix, and that the coupling coefficient is (m_q/v) for quark type q, thus proportional to the mass of the quark.

These transformations do not modify the neutral weak currents and their couplings to the Z^o weak boson and to the photon γ. In the initial basis Q, these currents were diagonal in flavour, and they remain so. The charged current interactions are modified in form as follows:

$$W^{\dagger}_{\mu}\bar{Q}_{dL}\gamma^{\mu}Q_{dL} + \text{h.c.} \rightarrow W^{\dagger}_{\mu}\bar{q}_{uL}\gamma^{\mu}(L^{\dagger}_{u}L_{d})q_{dL} + \text{h.c.}. \qquad (2.8)$$

The matrix in brackets is a generalized Cabibbo mixing matrix, which we shall denote by U^C, thus

$$U^C(q) = L^{\dagger}_{u}L_{d} \quad . \qquad (2.9)$$

We are already familiar with this matrix for the case $n_g = 2$. After suitable choices for the phases of the quark states, $U^C_2(q)$ takes the real form

$$U^C_2(q) = \begin{bmatrix} \cos\theta & -\sin\theta \\ \sin\theta & \cos\theta \end{bmatrix} \qquad (2.10)$$

where θ denotes the Cabibbo angle.

It was first pointed out by Kobayashi and Maskawa [4] that the matrix $U_3^C(q)$ for $n_g = 3$ cannot generally be transformed to a real form. The quark phases are readily chosen to make the first row and the first column of $U_3^C(q)$ real, and the general form of the matrix may then be written

$$U_3^C(q) = \begin{bmatrix} c_1 & -s_1c_3 & -s_1s_3 \\ s_1c_2 & c_1c_2c_3 - s_2s_3e^{i\delta} & c_1c_2s_3 + s_2c_3e^{i\delta} \\ s_1s_2 & c_1s_2c_3 + c_2s_3e^{i\delta} & c_1s_2s_3 - c_2c_3e^{i\delta} \end{bmatrix} \quad (2.11)$$

where we have used the abbreviated notation $c_i = \cos\theta_i$, $s_i = \sin\theta_i$, for $i = 1,2,3$. There are now three Cabibbo-like angles $(\theta_1,\theta_2,\theta_3)$ and one phase angle δ. We shall discuss our empirical knowledge of these angles in Appendix A.

In all field theories today, the interaction Lagrangian is necessarily invariant under the product CPT of the operations of charge conjugation (C), space inversion (P), and Wigner time-reversal (T). Under the operation CPT, the element of the reaction matrix K (cf. Appendix I) for the transition $a \to b$ is transformed into that for the corresponding antiparticle transition $\bar{a} \to \bar{b}$, and CPT-invariance leads us to the equality

$$K(\bar{a} \to \bar{b};\ \bar{\underline{p}},\bar{\underline{\sigma}}) = K(a \to b;\ \underline{p},\underline{\sigma}) \ , \quad (2.12)$$

where $\underline{p}$ and $\underline{\sigma}$ denote the momenta and spins of all the particles in the initial state a and the final state b, the bar notation being used to specify the corresponding quantities for the antiparticles in the charge-conjugated states $\bar{a}$ and $\bar{b}$. The S-matrix-elements for the two processes (2.12) are not ncessarily the same, owing to the effects of the initial and final state interactions; this point is illustrated by explicit calculation in Appendix B. With CPT-invariance, the consequences of CP-violation are conjugate to those of time-reversal (T) non-invariance. The interaction (2.8) for $n_g = 3$, with $U_3^C(q)$ given by (2.11), is Hermitean, but it is not T-invariant, in general, since the time-reversal operator T replaces $U_3^C(q)$ by U_3^{C*} and $U_3^{C*} \neq U_3$ when $\delta \neq 0$. The interaction (2.8) is not T-invariant and hence not CP-invariant, when $\delta \neq 0$, and this CP-noninvariance will then show itself in the matrix-elements for physical processes in a sufficiently high order of approximation, as we shall discuss briefly just below.

We note first that, if any one of the angles θ_i is zero, there exists a transformation in generation space which will reduce this space into the sum of two subspaces, with $n_g = 1$ and $n_g = 2$ respectively, and so give $U_3^C(q)$ the form

$$U_3^C(q) = \begin{bmatrix} I & 0 \\ 0 & U_2^C \end{bmatrix} \quad . \tag{2.13}$$

In other words, setting one mixing angle to zero decouples one generation from the other two. However, we know already that U_2^C can always be made real and that there are then no T-violation effects predicted. From this remark and the remarks made in the last paragraph, we can conclude that any T-violating effect we calculate must have as a factor the product

$$s_1 s_2 s_3 \sin\delta \quad . \tag{2.14}$$

In the same way, we may remark that any physical effect giving T-violation in consequence of the finite imaginary part of U_3^C must involve no less than four U_3^C-matrix elements, no three of which come from the same row or column. If this condition is not satisfied, then it would be possible, by transforming the quark field phases, to make real all of the four elements involved, in which case the effect calculated can only be zero. At energies appropriate to strange particle decays, the mechanism giving rise to T-violation effects therefore necessitates virtual transitions to and from massive intermediate states, involving at least a c or $\bar{c}$ quark, and the effect is correspondingly suppressed in magnitude, by a further mass factor beyond the factor (2.14) already mentioned. This is the origin of the hope that substantially larger T-violation effects may be found in experimental situations involving higher energies and heavier quarks. We should also mention explicitly the fact that the relative magnitude of a T-violating effect can be larger than the factor (2.14) would suggest, if it can be compared with a T-invariant effect whose rate already has one or more of these s_i factors. We return to this point in Sec. 5.

It is also useful to note that, if any two of the Q_u quarks, or of the Q_d quarks, were to have the same mass, then there could be no T-violation effects arising from U_3^C. The strong interactions of the two degenerate quarks then have an additional SU(2) symmetry, which allows sufficient additional freedom in defining the quark field phases to remove the phase δ from the matrix U_3^C. We also note the symmetry of the interaction (2.8) under the simultaneous transformations $Q_u \leftrightarrow Q_d$ and $\theta_2 \leftrightarrow \theta_3$. These symmetries are not appropriate to the physical situation, of course, but they provide useful checks on detailed calculations.

2.2 The Neutrino Mixing Matrix

A similar discussion can be given for the leptons and their electroweak interactions. However, if the neutrinos are all assumed to be massless, then, after the charged lepton masses have been diagonalized, a unitary transformation may be made on the neutrino fields which makes the leptonic couplings of the $W^{\pm}$ bosons diagonal, one definite neutrino $(\nu_e, \nu_\mu, \nu_\tau)$ being associated with each of the leptons (e, μ, τ). Only if the three neutrino masses are all unequal, can the neutrino Cabibbo mixing matrix $U_3^C(\nu)$ have a non-zero phase and so lead to T-violations for the leptonic interactions. At present, there is much discussion concerning the possibility of small non-zero masses for the neutrinos, both on the basis of direct measurement [6-8] and from astrophysical considerations [9,10], and it appears that this possibility is not yet excluded. If this is the case, no reason is known why the neutrinos $(\nu_e, \nu_\mu, \nu_\tau)$ should not have different masses. With a non-diagonal $U_3^C(\nu)$, oscillations of neutrino type would occur as function of time, so that a neutrino beam initially ν_μ would be found to have components ν_e and ν_τ downstream. The possibility of such neutrino oscillations was first mentioned by Pontecorvo [11] in 1967.

The effects of a non-diagonal neutrino mixing matrix $U_3^C(\nu)$ have been examined by Fritzsch [12], with the conclusion that no observable effects would result in any leptonic processes, except for those related with neutrino oscillations. Cabibbo [13] then pointed out that non-diagonal neutrino oscillation effects, i.e. the observation of processes such as $\nu_\mu \to \nu_e$ as opposed to observation of the ν_μ content of a beam initially ν_μ as function of distance from source, could be strongly affected by T-violation arising from a non-zero phase δ in the matrix $U_3^C(\nu)$. The most detailed treatment of these effects has been given recently by Barger et al.[14]. Since these T-violation effects affect only the leptonic sector, and even then only off-diagonal neutrino oscillation effects, we relegate further discussion of them to Appendix C.

2.3 CP Nonconservation in the Higgs Sector

The electroweak interactions may be carried, in part, by the exchange of Higgs bosons. The standard model is based on just one Higgs doublet, and, in its spontaneous symmetry-breaking, the charged Higgs component ϕ^+ is absorbed into the longitudinal and scalar components of the W^+ weak boson, so that no physical charged Higgs particles remain in the final form of the model. However, if the standard model is extended to include n_H Higgs doublets, only one combination of the charged Higgs fields $\{\phi_i^+\}$ is absorbed into W^+, and (n_H-1) physical charged Higgs particles remain finally.

Quite early, Lee [15] pointed out that T-violating effects could arise through spontaneous symmetry breaking for models with $n_H = 2$. His detailed remarks were based on a modified Georgi-Glashow model which was subsequently ruled out by the data on neutral weak currents. However, the main point was that, in an $n_H = 2$ model, there are two vacuum expectation values, v_1 and v_2, defined by

$$v_i = \langle 0|(\phi_i^o + \bar{\phi}_i^o)|0\rangle/\sqrt{2} \quad . \tag{2.15}$$

If v_1 and v_2 have different phases, the spontaneously-broken theory includes a relative phase which cannot be transformed away and which will give rise to T-violation effects in hadronic weak processes. The most serious physics objection to this proposal was that the model then generally predicted the existence of flavour-changing neutral weak currents, whereas we know now that the empirical evidence [16] gives a rather small upper limit for the strength of the $\Delta s = +1$ weak current $s \to d$, for example. This objection could be met, in principle, by making the Higgs particles sufficiently massive, since the neutral Fermi coupling generated by Higgs boson exchange and off-diagonal in generation space has strength $\simeq G_F m_q m_{q'}/m_H^2$, whereas the dominant weak interactions are those due to $W^\pm$ and Z^o exchange, which have strength G_F and lead naturally to flavour conservation for the neutral weak currents. Weinberg [17] subsequently discussed an $n_H = 2$ theory based on the standard $SU(2) \times U(1)$ model, but these $n_H = 2$ models all link together the smallness of the T-violating effects which are observed and the smallness of the flavour-changing neutral weak currents, which have not yet been observed. This was not an attractive relationship to reach, and no other motivation existed for increasing the number of Higgs doublets beyond $n_H = 1$, so that this did not appear to be a promising line of development. At least experiment might be able to exclude a large class of such $n_H = 2$ theories in due course, when the upper limits on neutral weak currents are lowered significantly further.

As Weinberg [17] emphasized, the attractiveness of the Higgs-sector origin for T-violation is that the strength of this violation is measured by $G_F(m_q m_{q'}/m_H^2)$. The "milliweak" character of T-violation then has a natural explanation; it reflects the fact that Higgs particles are much heavier than the light quarks. For this explanation to hold, it is necessary that the T-violating interactions arise only through Higgs-boson exchange; if they received contributions from other sources, these would generally be much larger and so outweigh the Higgs-sector contribution. With this problem in mind, Branco [18] then investigated the question whether finite T-violating interactions and exact flavour conservation for the neutral weak currents, the latter situation being referred to as <u>natural flavour conservation</u> (NFC), could both hold true in any standard model extended to have $n_H \geq 2$. Weinberg [17] had already commented on the properties of theories with $n_H = 3$, based on the standard model, but it had not been clear whether or not these theories necessarily predicted neutral flavour-changing weak currents. Branco [19] first demonstrated that T-invariance necessarily holds in any $n_H = 2$, $SU(2) \times U(1)$ theory which is spontaneously broken and obeys NFC. He then developed in full detail a specific $SU(2) \times U(1)$ model with n_g arbitrary and with $n_H = 3$, which had these two properties. Its Higgs potential $V(\phi_i)$ was taken to be even in each of the Higgs fields separately, whilst the $\bar{Q}Q\phi_i$ coupling was taken to be zero for $i = 3$, and to be symmetrc under the transformation $\phi_1 \to -\phi_1$ with $Q_{dR} \to -Q_{dR}$. These symmetries are sufficient to restrict the model to NFC, whilst leaving $V(\phi_i)$ with a considerable number of free parameters. After spontaneous symmetry-breaking, the vacuum expectation values for the neutral Higgs fields are denoted by

$$v_i e^{i\theta_i} = \langle 0|(\phi_i^o + \bar{\phi}_i^o)|0\rangle/\sqrt{2} \quad . \tag{2.16}$$

Only the phase differences have physical significance, so there are five parameters involved in (2.16). The quark mass matrix is then diagonalized in the usual way, leading to a charged weak current of the form (2.8), but with a generalized Cabibbo mixing matrix U^C necessarily real. The reality of U^C follows from the fact that the quarks are coupled with only two Higgs doublets. Branco had already proved that there could be no T-violating interaction arising from a standard model with $n_H = 2$ and NFC, and, although the model now under consideration has $n_H = 3$, its KM matrix U^C must obey the same general constraints as those for the $n_H = 2$ case with NFC, since the third Higgs doublet gives no contribution. Hence, the requirement of NFC (which motivated the absence of $\bar{Q}Q\phi_3$ coupling), together with this spontaneous CP symmetry-breaking, leads to the conclusion that the $W^{\pm}$, Z^o and γ coupling with quarks are necessarily T-invariant, whatever the value of n_g.

The Higgs potential of this model must be constrained to have its minimum at the potential values given by the equations (2.15). This leads to a determination of the phase angles $(\theta_1-\theta_2)$ and $(\theta_3-\theta_2)$, and there is generally a T-violating solution, as well as T-invariant solutions. The charged Higgs-boson mass matrix is then written down explicitly and diagonalized One Higgs particle remains massless, having the form

$$G^+ = (\sum_i v_i \phi_i^+)/v \quad , \tag{2.17}$$

where $v = \sqrt{(\sum_i v_i^2)}$, but is absorbed into the longitudinal and scalar components of $W^{\pm}$. The other two Higgs particles, H_1^+ and H_2^+, acquire mass and appear as physical scalar particles. The Higgs mass diagonalization matrix U_3^H may be written in the KM form, by suitable choice of the Higgs field phases. The exchange of charged Higgs bosons then leads to an effective Fermi interaction whose T-violating part has the form

$$H_{\bar{T}} = (\sqrt{2}\, G_F \frac{vv_3}{v_1 v_2} (\frac{1}{m_1^2} - \frac{1}{m_2^2}) \cos\,\xi) \cdot$$

$$\cdot\ (\bar{Q}_{uL} U_3^C(q) M_d Q_{dR})(\bar{Q}_{dL} U_3^C(q) M_u Q_{uR}) \qquad (2.18)$$

where (m_1, m_2) are the masses of the Higgs particles (H_1^+, H_2^+), (M_u, M_d) are the (diagonal) mass matrices for the quarks (Q_u, Q_d), and $\cos\xi$ is an explicit function of the constant coefficients in $V(\phi_i)$ and of the amplitudes (v_1, v_2, v_3). We note that the dominant factors in (2.18) are $G_F\ m_q^2/m_1^2$, where H_1^+ denotes the lightest Higgs particle, as expected. That the expression (2.18) vanishes for $m_1 = m_2$ is a useful check; in this situation there is an additional SU(2) symmetry for the strong interactions in the (H_1^+, H_2^+) space, and this additional freedom allows the Higgs field phases to be chosen such that $\delta = 0$ in the Higgs mass mixing matrix U_3^H.

This model illustrates the possibility of understanding the "milliweak" character of the known CP and T violating effects observed to date, in terms of the magnitude of the Higgs particle masses relative to the quark masses. All of the T-violation in this class of models arises through the Higgs boson interactions with quarks, the quark mass mixing matrix itself being real for all n_g.

3. THE PHENOMENOLOGY OF THE K^o-$\bar{K}^o$ COMPLEX

3.1 The Parameters of the K_S and K_L States

We first outline the physical situation, giving the empirical values known for the parameters [16]. The K^o-$\bar{K}^o$ complex is characterized by two lifetimes. The shorter lifetime is $(0.892\pm0.002) \times 10^{-10}$ sec. and defines the K_S state. The longer lifetime, $(5.18\pm0.04) \times 10^{-8}$ sec., defines the K_L

state. The mass difference between these two states is well known; the K_L state is the heavier (cf. Fig. 1), their mass difference being

$$-\delta m = m_L - m_S = (0.535 \pm 0.002) \times 10^{10} \text{ sec.}^{-1} \quad . \tag{3.1}$$

The outstanding decay modes for K_S are to the $\pi\pi$ final states; (68.5±0.25)% to $\pi^+\pi^-$, and (31.3±0.25)% to $\pi^0\pi^0$. Indeed, no other decay modes are known, apart from the decay $K_S^0 \to \pi^+\pi^-\gamma$, whose branching ratio is about 0.2%. The K_L state has many decay modes, dominantly to 3π, $\pi^{\pm}e^{\mp}\nu_e$, and $\pi^{\pm}\mu^{\mp}\nu_{\mu}$ final states, but also to both $\pi\pi$ states, and it is of course these last transitions which have made it so apparent that there is a CP-violating interaction influencing the K_S and K_L states and their decay processes. Much effort has been devoted to the measure of the amplitudes for these $\pi\pi$ decay processes, which data can be summarized briefly as follows [16]:

$$\eta_{+-} = \frac{A(K_L \to \pi^+\pi^-)}{A(K_S \to \pi^+\pi^-)} = (2.274 \pm 0.022) \times 10^{-3} \cdot \mathrm{Exp}(i(44.6 \pm 1.2)^{\circ}) \tag{3.2a}$$

$$\eta_{00} = \frac{A(K_L \to \pi^0\pi^0)}{A(K_S \to \pi^0\pi^0)} = (2.33 \pm 0.08) \times 10^{-3} \cdot \mathrm{Exp}(i(54 \pm 5)^{\circ}) \quad . \tag{3.2b}$$

To discuss these phenomena, we consider the mass matrix in the space spanned by the K^0 and $\bar{K}^0$ states, for which states we use the labels 1 and 2, respectively. The elements of this matrix are given by

$$M_{ij} - \frac{i}{2}\Gamma_{ij} = \langle i|H(\Delta s=\pm 2)|j\rangle + \sum_n \frac{\langle i|H_{wk}(\Delta s=\pm 1)|n\rangle\langle n|H_{wk}(\Delta s=\pm 1)|j\rangle}{m_K - (E_n - i\varepsilon)} \quad . \tag{3.3}$$

The interaction $H(\Delta s = \pm 2)$ is included in (3.3) for completeness, since we wish to refer to it later in connection with the Superweak Model, but it is of course zero for all $SU(2) \times U(1)$

electroweak gauge theories. The terms M_{ij} and Γ_{ij} are both complex, in general, and they are defined as follows:

$$M_{ij} = \langle i|H(\Delta s=\pm 2)|j\rangle + \text{P.V.} \sum_n \frac{\langle i|H_{wk}(\Delta s=\pm 1)|n\rangle\langle n|H_{wk}(\Delta s=\pm 1)|j\rangle}{m_K - E_n}, \tag{3.4}$$

where the sum over intermediate states is an integral, and P.V. denotes that the principal value integral is to be taken. The matrix M_{ij} is clearly Hermitian.

$$\Gamma_{ij} = 2\pi \sum_n \delta(E_n - m_K)\langle i|H_{wk}(\Delta s=\pm 1)|n\rangle\langle n|H_{wk}(\Delta s=\pm 1)|j\rangle \ . \tag{3.5}$$

Here the sum integrates over all possible intermediate states on the energy shell. This is a finite integral, and Γ_{ij} is a Hermitian matrix. The mass matrix thus has the form

$$\begin{bmatrix} M_{11} - \frac{i}{2}\Gamma_{11} & M_{12} - \frac{i}{2}\Gamma_{12} \\ M_{21} - \frac{i}{2}\Gamma_{21} & M_{22} - \frac{i}{2}\Gamma_{22} \end{bmatrix} \ . \tag{3.6}$$

This matrix has two eigenvalues, $m_L - \frac{i}{2}\Gamma_L$ and $m_S - \frac{i}{2}\Gamma_S$, with corresponding eigenstates which we denote for convenience as K_L and K_S, respectively. We shall not give the full algebraic expressions, at this point, but use the following expressions:

$$|K_S\rangle = p_S|K^o\rangle + q_S|\bar{K}^o\rangle \ , \tag{3.7a}$$

$$|K_L\rangle = p_L|K^o\rangle - q_L|\bar{K}^o\rangle \ , \tag{3.7b}$$

involving four complex parameters, i.e. eight real parameters. The normalization conditions for the K_L and K_S states,

$$|p_L|^2 + |q_L|^2 = |p_S|^2 + |q_S|^2 = 1 \ , \tag{3.8}$$

reduce these to six real parameters. The absolute phases of K_L and K_S states, and the phase of $\bar{K}^o$ relative to K^o, are not measurable in principle and may therefore be chosen by convention. This reduces the number of real parameters to three, and these may be chosen to give the coefficients in (3.8) the following convenient forms [21]:

$$p_S = \cos\alpha_S \cdot e^{i\theta/2}, \qquad q_S = \sin\alpha_S \cdot e^{-i\theta/2}, \tag{3.9a}$$

$$p_L = \cos\alpha_L \cdot e^{-i\theta/2}, \qquad q_L = \sin\alpha_L \cdot e^{i\theta/2}, \tag{3.9b}$$

involving only three real parameters, where the angles α_L and α_S can be chosen to lie in the first quadrant, while θ can be chosen to lie in the first or fourth quadrant . The states K_L and K_S are not orthogonal; in fact,

$$\langle K_S|K_L\rangle = p_S^* p_L - q_S^* q_L$$

$$= \cos(\alpha_S+\alpha_L)\cos\theta - i\cos(\alpha_S-\alpha_L)\sin\theta \ . \tag{3.10}$$

This quantity $\langle K_L|K_S\rangle$ is also given by the following unitarity relation, due to Bell and Steinberger [22]:

$$i\{(m_S-\tfrac{i}{2}\Gamma_S)-(m_L-\tfrac{i}{2}\Gamma_L)^*\}\langle K_L|K_S\rangle =$$

$$= 2\pi\sum_n \delta(E_n-m_K)\langle K_L|H_{wk}|n\rangle\langle n|H_{wk}|K_S\rangle \ , \tag{3.11}$$

which they derived directly from a statement that the rate of decrease of the net normalization of any state $(a_S K_S^o + a_L K_L^o)$ must equal the net transition rate from this state to all other states n with the same energy. If we denote by $\Gamma_L(n)$ and $\Gamma_S(n)$ the partial widths for K_L and K_S decay to the final state n, so that

(a) $\quad \Gamma_S(n) = 2\pi\sum_n \delta(E_n-m_K)\,|\langle n|H_{wk}|K_S\rangle|^2 \ ,$

(b) $$\Gamma_L(n) = 2\pi\sum_n \delta(E_n - m_K)\,|\langle n|H_{wk}|K_L\rangle|^2 \quad , \qquad (3.12)$$

then the right hand side of (3.11) may be written

$$\sum_n \exp(i\phi_n)\sqrt{(\Gamma_S(n)\Gamma_L(n))} \quad , \qquad (3.13)$$

where ϕ_n denotes the phase difference between $\langle n|H_{wk}|K_L\rangle$ and $\langle n|H_{wk}|K_S\rangle$. This expression is clearly bounded in magnitude by $\sum_n \sqrt{(\Gamma_S(n)\Gamma_L(n))}$. Taking this bound together with (3.11) leads to the following bound for $\langle K_S|K_L\rangle$:

$$|\langle K_L|K_S\rangle|^2 \le 4\Big(\sum_n \sqrt{(\Gamma_S(n)\Gamma_L(n))}\Big)^2/\{4(m_S-m_L)^2 + (\Gamma_S+\Gamma_L)^2\} \quad . \qquad (3.14)$$

One class of experiments of particular interest may be illustrated by the following reactions:

$$\bar{p} + p \rightarrow \begin{cases} K^o + \text{(particles with net strangeness -1)}, & (3.15a) \\ \bar{K}^o + \text{(particles with net strangeness +1)}. & (3.15b) \end{cases}$$

We know that C-invariance holds to a very good approximation for the strong interactions, hence the initial K^o amplitude in (3.15a) must equal in magnitude that for $\bar{K}^o$ in the charge conjugate reaction (3.15b). The observation of $K_S \rightarrow \pi^+\pi^-$ decays close to the source following these reactions then measures the intensity of the K_S component in K^o (for reaction (3.15a)) or in $\bar{K}^o$ (in reaction (3.15b)), since $K_L \rightarrow \pi^+\pi^-$ is a rare process, especially near the target. Inverting the relations (3.7), using the forms (3.9), leads to the conclusion that, to an excellent approximation, the ratio

$$R = \frac{(K_S^o \rightarrow \pi^+\pi^- \text{ in reaction (3.15a)})}{(K_S^o \rightarrow \pi^+\pi^- \text{ in reaction (3.15b)})} = \tan^2\alpha_L \qquad (3.16)$$

gives a measure of the parameter $\alpha_L = (1+\sigma_L)\pi/4$, $\pi/4$ being the value expected if there were no CP violation in $(K^0, \bar{K}^0)$ decay. Empirically [23], σ_L has the value 0.0±0.01.

From (3.10) and the empirical limit for the right hand side of (3.14), we have

$$|\langle K_L|K_S\rangle|^2 = \cos^2(\alpha_S+\alpha_L) + \sin^2\theta\,\cos^2(\alpha_S-\alpha_L) < 4\times 10^{-5} \quad . \tag{3.17}$$

With our above knowledge of α_L, and of the domains accessible to α_L and α_S, the inequality (3.17) gives quite strong limits on $\alpha_S = (1+\sigma_S)\pi/4$ and θ. Collecting these limits together, we have

$$\sigma_S = 0.0\pm 0.02, \qquad \sigma_L = 0.0\pm 0.01, \qquad |\theta| \leq 0.006. \tag{3.18}$$

In view of the form of (3.17), there are of course quite strong correlations between the values permitted for these three parameters, but it is remarkable that these strong limits on the parameters have been obtained without invoking any of the symmetries CTP, CP or T for the weak interactions. The parameter values determined are very close to the values they would have ($\sigma_S = \sigma_L = \theta = 0$) in a CP-invariant theory.

The time development of a beam known to be K^0 at source, as in (3.15a), or $\bar{K}^0$ at source, as in (3.15b), is also of central importance. After time t, the initially K^0 beam will have a $\bar{K}^0$ component with amplitude

$$A(K^0 \to \bar{K}^0;t) = \sin\alpha_S\ \sin\alpha_L\cdot(\phi_S-\phi_L)/\Delta \ , \tag{3.19a}$$

where $\phi_i = \mathrm{Exp}(-i(m_i-\frac{i}{2}\Gamma_i)t)$, and $\Delta = (\sin(\alpha_S+\alpha_L)\cos\theta -$ $- i\,\sin(\alpha_S-\alpha_L)\sin\theta)$, while the initially $\bar{K}^0$ beam will have a K^0 component with amplitude

$$A(\bar{K}^0 \to K^0;t) = \cos\alpha_S\ \cos\alpha_L\cdot(\phi_S-\phi_L)/\Delta \quad . \tag{3.19b}$$

Hence the probability, after time t, that a K^o state has changed into a $\bar{K}^o$ is $(\tan^2\alpha_S \tan^2\alpha_L)$ times the probability that a $\bar{K}^o$ state has changed into a K^o, independent of t. Measurement of this ratio, taken together with (3.16), would allow a direct determination of α_L. If it is other than unity, this ratio would demonstrate directly both CP-noninvariance and T-violation at the same time, independent of the validity of CPT.

The assumption of CPT invariance gives the equalities

$$\text{(a)}\ M_{11} = M_{22}, \qquad \text{(b)}\ \Gamma_{11} = \Gamma_{22} \tag{3.20}$$

for the diagonal elements of the mass matrix (3.6). In fact, Kabir [24] has pointed out that the above analysis provides a stringent test of these equalities (3.20), using the identity

$$(M_{11}-\tfrac{i}{2}\Gamma_{11}-M_{22}+\tfrac{i}{2}\Gamma_{22}) = (m_S-\tfrac{i}{2}\Gamma_S-m_L+\tfrac{i}{2}\Gamma_L)\cdot \left\{\frac{\sin(\alpha_S-\alpha_L)\cos\theta - i\sin(\alpha_S+\alpha_L)\sin\theta}{\sin(\alpha_S+\alpha_L)\cos\theta + i\sin(\alpha_S-\alpha_L)\sin\theta}\right\} . \tag{3.21}$$

Using the values of α_S, α_L and θ obtained above, the right hand side of (3.21) is well approximated by

$$(m_S - m_L - \tfrac{i}{2}(\Gamma_S-\Gamma_L))(\alpha_S - \alpha_L - i\theta) , \tag{3.22}$$

and we conclude that

$$|(M_{11} - \tfrac{i}{2}\Gamma_{11} - M_{22} + \tfrac{i}{2}\Gamma_{22})| \leq 0.016\ \Gamma_S = 2.4 \times 10^{-16}\ m_K . \tag{3.23}$$

This provides a very strong test of CPT for the hadronic interactions, and a significant test of CPT invariance for the weak interactions.

With (3.20), the mass matrix (3.6) may now be written

$$M - \frac{i}{2}\Gamma \quad + \begin{bmatrix} 0 & M_{12} - \frac{i}{2}\Gamma_{12} \\ M_{12}^{*} - \frac{i}{2}\Gamma_{12}^{*} & 0 \end{bmatrix} \tag{3.24}$$

where we have

$$\text{(a)} \quad M = (m_S + m_L) \quad , \qquad \text{(b)} \quad \Gamma = (\Gamma_S + \Gamma_L) \tag{3.25}$$

and we introduce the notations

$$\text{(a)} \quad \delta m = m_S - m_L \; , \qquad \text{(b)} \quad \delta\Gamma = \Gamma_S - \Gamma_L \quad . \tag{3.26}$$

These quantities are given by the equation

$$\delta m - \frac{i}{2}\delta\Gamma = 2\sqrt{\{(M_{12} - \frac{i}{2}\Gamma_{12})(M_{12}^{*} - \frac{i}{2}\Gamma_{12}^{*})\}} \quad . \tag{3.27}$$

The eigenstates are then, explicitly,

$$|K_{S(L)}\rangle = \{(1+\varepsilon)|K^{o}\rangle \underset{(-)}{+} (1-\varepsilon)|\bar{K}^{o}\rangle\}/\sqrt{\{2(1+|\varepsilon|^{2})\}} \quad , \tag{3.28}$$

where ε is defined by the equation

$$\frac{1+\varepsilon}{1-\varepsilon} = \{\frac{M_{12} - \frac{i}{2}\Gamma_{12}}{M_{12}^{*} - \frac{i}{2}\Gamma_{12}^{*}}\}^{1/2} \quad . \tag{3.29}$$

It is also useful to introduce the quantity Z, defined by the equation

$$\frac{1+Z}{1-Z} = |(\frac{1+\varepsilon}{1-\varepsilon})|^{4} \quad , \tag{3.30}$$

because Z is very directly connected with the elements of (3.24), by the relation

$$Z = \mathrm{Im}\ (\frac{\Gamma_{12}}{M_{12}})/\{1 + \frac{1}{4}|\frac{\Gamma_{12}}{M_{12}}|^2\} \quad . \tag{3.31}$$

With CTP invariance assumed, the equality $\alpha_S = \alpha_L = \alpha$ (say) holds for the Eberhard angles in (3.9), and the angles α and θ are given in terms of the parameter ε by the relation

$$e^{i\theta}\cot\alpha = (1+\varepsilon)/(1-\varepsilon) \quad . \tag{3.32}$$

In the physical situation, the parameter ε is small, in which case it is sufficiently well given by the expression

$$\varepsilon \simeq - \frac{\mathrm{Im}\ \Gamma_{12}/2 + i\ \mathrm{Im}\ M_{12}}{\delta m - \frac{i}{2}\delta\Gamma} \quad . \tag{3.33}$$

A direct empirical determination of $\mathrm{Re}(\varepsilon)$ has been obtained from observations on the $\pi^-\ell^+\nu_\ell$ and $\pi^+\ell^-\bar{\nu}_\ell$ decay events along a K_L beam. SU(2) × U(1) electroweak theories allow only $\Delta s = \Delta Q$ transitions in the $K_{\pi\ell\nu}$ decay modes for neutral K-mesons. This rule is indeed rather well obeyed by the data, which give [16]

$$A(\Delta s = -\Delta Q)/A(\Delta s = +\Delta Q) = \{0.01\pm0.02 - (0.004\pm0.03)i\}, \tag{3.34}$$

well consistent with zero. With the $\Delta s = \Delta Q$ rule, the allowed transitions are

$$\text{(a)} \quad K^0 \to \pi^-\ell^+\nu_\ell \ , \qquad \text{(b)} \quad \bar{K}^0 \to \pi^+\ell^-\bar{\nu}_\ell \ , \tag{3.35}$$

so that the relative rates $(\ell^+)/(\ell^-)$ along a K_L^0 beam give a direct measure of the relative strengths of the K^0 and $\bar{K}^0$ components in the K_L state (3.28), namely

$$P(K^o)/P(\bar{K}^o) = |(1+\varepsilon)/(1-\varepsilon)|^2 \simeq 1 + 4\ \mathrm{Re}(\varepsilon) \quad . \tag{3.36}$$

The validity of the approximate expression depends on the inequality $|\mathrm{Im}(\varepsilon)|^2 << |\mathrm{Re}(\varepsilon)|$ and on the smallness of $\mathrm{Re}(\varepsilon)$. In the expression (3.33), we know that $|\delta m| \simeq \delta\Gamma/2$ and we expect that $\mathrm{Im}\ \Gamma_{12} << \mathrm{Im}\ M_{12}$ - the latter because Γ_{12} results from a sum over a very limited number of states, those with energy m_K, whereas M_{12} results from a sum over all energies, most of which lie well above m_K - and these two relations imply $\mathrm{Im}(\varepsilon) \simeq \mathrm{Re}(\varepsilon)$, which justifies the approximation (3.36). Assuming the $\Delta s = \Delta Q$ rule to hold precisely, so that the ratio (3.34) is zero, the ℓ^+/ℓ^- ratio measured for K_L decay leads to the value

$$\mathrm{Re}(\varepsilon) = (1.62 \pm 0.09).10^{-3} \quad . \tag{3.37}$$

In concluding this Section, it is useful to emphasize the macroscopic character of experiments on neutral K-mesons today. For momentum $p_K \simeq 100$ GeV/c, the mean distance travelled by a K_S meson in this beam, until its decay, is $\simeq 6$ metres, whereas that travelled (in vacuum) by a K_L meson in this beam is $\simeq 4$ km. It is this character which gives the study of neutral K-meson phenomena a different quality from that for the heavier neutral mesons to be discussed in Sec.5 below.

3.2 The Decay Modes $(K_S, K_L) \rightarrow \pi\pi$

In these decay modes, the final $\pi\pi$ state necessarily has orbital angular momentum $\ell = 0$, from angular momentum conservation, and is generally a linear superposition of isospin states T = 0 and 2. Since the $(\pi^+\pi^-)/(\pi^o\pi^o)$ branching ratio is observed to be 2.19 ± 0.02 for K_S decay and 2.16 ± 0.4 for K_L decay [16], the simplest interpretation is that T = 0 dominates in the final state for both K_S and K_L, the expected branching ratio (neglecting deviations due to the $\pi^\pm - \pi^o$ mass difference, the Coulomb effect, and other such effects) being 2 for a pure T = 0 state. This T = 0 dominance is an example of the general dominance of $\Delta I = 1/2$ transitions in $\Delta s = \pm 1$ nonleptonic weak processes, which is best illustrated by the smallness of the ratio $\Gamma(K^+ \rightarrow \pi^+\pi^o)/\Gamma(K_S \rightarrow \pi\pi) \simeq 1.53\times 10^{-3}$, the decay $K^+ \rightarrow \pi^+\pi^o$ being forbidden by a strict $\Delta I = 1/2$ rule, since its final $\pi\pi$ state has charge +1 and is therefore necessarily limited to T = 2. The SU(3) × U(1) model allows both $\Delta T = 1/2$ and $\Delta T = 3/2$ interactions to contribute to $\Delta s = \pm 1$ nonleptonic weak transitions. The dominance of the former is not a general feature of all such models, but presumably has some fairly specific dynamical origin lying in the particular model appropriate to Nature, which is not yet understood, despite much effort by theoreticians.

The amplitude for K^o decay to the $\pi\pi$ system with isospin T has two factors, as we discuss more generally in Appendix B. The first factor is the reaction matrix element for kaon decay to a $\pi\pi$ standing wave state, and the second factor represents the effect of the final state $\pi\pi$ scattering interaction. Thus, we have

$$\text{(a)}\ A(K^o \rightarrow (\pi\pi)_T) = A_T e^{i\delta_T}, \qquad \text{(b)}\ A(\bar{K}^o \rightarrow (\pi\pi)_T) = \bar{A}_T e^{i\delta_T} = A_T^* e^{i\delta_T}, \tag{3.38}$$

where δ_T denotes the $\pi\pi$ scattering phase for the s-wave with isospin T. That $\bar{A}_T$ is the Hermitean conjugate of A_T follows from the Hermiticity of the Hamiltonian. The phase of the K^o field may be chosen by convention, and so can be chosen to make A_o real; there is then no freedom in the phase of A_2, and A_2 is complex, in general. As a consequence of $\Delta I = 1/2$ dominance, we believe that $|A_2| << |A_o|$.

With the isospin expressions for the $\pi^+\pi^-$ and $\pi^o\pi^o$ states,

$$|\pi^+\pi^-\rangle = (\sqrt{2}|2\pi,T=0\rangle + |2\pi,T=2\rangle)/\sqrt{3}\ , \tag{3.39a}$$

$$|\pi^o\pi^o\rangle = (|2\pi,T=0) - \sqrt{2}|2\pi,T=2\rangle)/\sqrt{3}\ , \tag{3.39b}$$

we can now write down expressions for the amplitudes $A(K\to\pi\pi)$ entering into (3.2).

$$\langle\pi^+\pi^-|T|K_S^o\rangle = \{\sqrt{2}A_o + (ReA_2 + i\varepsilon ImA_2)E\}/\sqrt{\{3(1+|\varepsilon|^2)/2\}}\ , \tag{3.40a}$$

$$\langle\pi^+\pi^-|T|K_L^o\rangle = \{\sqrt{2}\varepsilon A_o + (\varepsilon ReA_2 + iImA_2)E\}/\sqrt{\{3(1+|\varepsilon|^2)/2\}}\ , \tag{3.40b}$$

where $E = \exp i(\delta_2-\delta_o)$, and

$$\langle\pi^o\pi^o|T|K_S^o\rangle = \{A_o - \sqrt{2}(ReA_2 + i\varepsilon ImA_2)E\}/\sqrt{\{3(1+|\varepsilon|^2)/2\}}\ , \tag{3.41a}$$

$$\langle\pi^o\pi^o|T|K_L^o\rangle = \{\varepsilon A_o - \sqrt{2}(\varepsilon ReA_2 + iImA_2)E\}/\sqrt{\{3(1+|\varepsilon|^2)/2\}}\ , \tag{3.41b}$$

from which we can write down full expressions for η_{+-} and η_{oo}, as follows:

$$\eta_{+-} = \frac{\varepsilon + (\varepsilon ReA_2/A_o + iImA_2/A_o)E/\sqrt{2}}{1 + (ReA_2/A_o + i\varepsilon ImA_2/A_o)E/\sqrt{2}}\ , \tag{3.42a}$$

$$\eta_{oo} = \frac{\varepsilon - 2(\varepsilon ReA_2/A_o + iImA_2/A_o)E/\sqrt{2}}{1 - 2(ReA_2/A_o + i\varepsilon ImA_2/A_o)E/\sqrt{2}}\ . \tag{3.42b}$$

Since we know from the discussion above that $|\varepsilon| \ll 1$, and $|A_2/A_o| \ll 1$, we expand these expressions, retaining only first order terms, with the result given by Wu and Yang,

$$\eta_{+-} = \varepsilon + i(E/\sqrt{2})\,ImA_2/A_o + O(|\varepsilon|^2, |\varepsilon||A_2/A_o|, |A_2/A_o|^2) = \varepsilon + \varepsilon'\ , \tag{3.43a}$$

$$\eta_{oo} = \varepsilon - 2i(E/\sqrt{2})\,\mathrm{Im}A_2/A_o + O(|\varepsilon|^2, |\varepsilon||A_2/A_o|, |A_2/A_o|^2) = \varepsilon - 2\varepsilon' , \tag{3.43b}$$

where

$$\varepsilon' = \frac{i}{\sqrt{2}} \cdot \frac{\mathrm{Im}A_2}{A_o} \cdot e^{i(\delta_2 - \delta_o)} \quad . \tag{3.44}$$

From (3.33), with $|\mathrm{Im}\Gamma_{12}| \leq |\mathrm{Im}M_{12}|$, we conclude that the phase of ε is given by

$$\phi_\varepsilon = \mathrm{artan}\ (2|\delta m|/\delta\Gamma) = 43.7^o \ . \tag{3.45}$$

From the phase shift analyses of $\pi\pi$ scattering [26], we have for the phase of ε'

$$\phi_{\varepsilon'} = (\frac{\pi}{2} + \delta_2 - \delta_o) = (37 \pm 5)^o \quad . \tag{3.46}$$

These phases are quite comparable with those determined empirically for η_{+-} and η_{oo}, as given in (3.2). On the Argand plane, η_{+-}, η_{oo}, ε and ε' all have roughly the same direction, and it is a good approximation to write:

$$(\eta_{oo}/\eta_{+-})^2 = |(\frac{\varepsilon - 2\varepsilon'}{\varepsilon + \varepsilon'})|^2 \simeq 1 \pm 6|\varepsilon'/\varepsilon| \ , \tag{3.47}$$

since we know empirically that $|\varepsilon'/\varepsilon| \ll 1$. There is an uncertainty of sign since we do not know the sign of $\mathrm{Im}A_2$. This ratio is clearly rather sensitive to ε', and its accurate determination offers our best hope for the measurement of this parameter. The existing data on this ratio corresponds to

$$|\varepsilon'/\varepsilon| = 0.004 \pm 0.006 \ . \tag{3.48}$$

We shall discuss theoretical estimates for this ratio in Sec. 4.

There are new experiments being planned, or under way,

at FNAL [27], BNL [28] and CERN [29], with the aim of measuring the ratio $(\eta_{oo}/\eta_{+-})^2$ to significantly greater accuracy. In order to achieve high precision, these experiments all aim to measure the double rate ratio

$$(\Gamma^S_{oo}/\Gamma^S_{+-})/(\Gamma^L_{oo}/\Gamma^L_{+-}) \qquad (3.49)$$

with the same apparatus, in order to cancel out systematic errors as far as possible. The FNAL experiment uses simultaneous K_L and regenerated K_S beams, side by side, alternately measuring the neutral and charged decay modes. The BNL experiment measures the two decay modes simultaneously, alternately in K_L and regenerated K_S beams. The CERN experiment proposes to measure the two decay modes simultaneously in K_L and K_S beams, the K_S beam being produced on a close up target, in order to avoid background due to neutron interactions in the regenerator and the diffraction scattered and diffraction regenerated neutral kaons. The CERN experiment is aimed at determining $|\eta_{oo}/\eta_{+-}|^2$ to an accuracy of order 10^{-3}, and therefore endeavours to reach larger acceptances and larger data-taking rates, by using a longer decay region and by making calorimetric measurements of $\pi^{\pm}$ energies rather than using analysing magnets. If successful in their aims, these experiments will provide an important new parameter with which to confront theory.

3.3 The Minimal Model: a $\Delta s = \pm 2$ Superweak Interaction

When CP conservation holds, we have $\varepsilon = 0$ in the expressions (3.28), which become

$$\text{(a)}\ K_+^0 = (K^0 + \bar{K}^0)/\sqrt{2}\ , \qquad \text{(b)}\ K_-^0 = (K^0 - \bar{K}^0)/\sqrt{2}, \qquad (3.50)$$

the eigenstates for CP = + and -, respectively. Note that we have chosen the phases of K^0 and $\bar{K}^0$ to be such that CP $K^0 = \bar{K}^0$. With $J = 0$ for the K-mesons, the final $\pi\pi$ states are s-wave and necessarily have CP = +. Consequently, with CP conservation, the decay $K^0 \to \pi\pi$ is forbidden for the state K_-^0, but allowed for K_+^0. This is, of course, the reason why the K_-^0 state is long-lived relative to the K_+^0 state, for the other decay modes $K \to 3\pi$, $\pi\ell\nu_\ell$, $\gamma\gamma$, etc. have amplitudes intrinsically weaker [30] than that for $K^0 \to \pi\pi$. These weaker decay modes are all included in the phenomenological discussion in Sec. 3.2 above, of course. The $K \to \pi\pi$ interaction also contributes to the mass of the K_+^0 state, separating it in mass from the K_-^0 state. In this $\pi\pi$ approximation, the K_+^0 state has complex mass $(\delta m_{2\pi} - \frac{i}{2}\gamma_{2\pi})$ relative to the K_-^0 state, which is a large part of the full mass difference $(\delta m - \frac{i}{2}\delta\Gamma)$.

The superweak model [31] seeks to account for the (K_+^0, K_-^0) mixing observed, by the inclusion of a single CP-nonconserving interaction H_{SW}. To achieve this, H_{SW} must have a non-zero off-diagonal matrix element,

$$\langle K_+^0 | H_{SW} | K_-^0 \rangle = \tfrac{1}{2}\langle K^0 + \bar{K}^0 | H_{SW} | K^0 - \bar{K}^0 \rangle \neq 0\ . \qquad (3.51)$$

We note first that

$$\langle K^0 | H_{SW} | K^0 \rangle = \langle \bar{K}^0 | H_{SW} | \bar{K}^0 \rangle \qquad (3.52)$$

from the (unquestioned) validity of CPT. Next, the Hermiticity of H_{SW} leads to the statement

$$<\bar{K}^o|H_{SW}|K^o>^* = <K^o|H_{SW}^\dagger|\bar{K}^o> = <K^o|H_{SW}|\bar{K}^o> \quad . \tag{3.53}$$

Taking the equations (3.52) and (3.53) into account, the matrix-element (3.51) reduces to

$$<K_+^o|H_{SW}|K_-^o> = i \, \mathrm{Im}<\bar{K}^o|H_{SW}|K^o> = i\alpha \quad , \tag{3.54}$$

say, where α is real, giving a pure imaginary off-diagonal matrix-element, a single number to characterize the CP violation. We note from (3.54) that this CP-violating interaction H_{SW} must contain elements with $\Delta s = \pm 2$, and that it is sufficient for our purpose to confine H_{SW} to these components alone. In this case, H_{SW} will not generate any other effects detectable in the foreseeable future. For example, we shall find below that even the $\Delta s = 2$ transitions $\Xi^- \to n\pi^-$ and $\Xi^o \to p\pi^-$ generated by H_{SW} have amplitudes with magnitude 10^{-3} relative to the amplitudes generated in second order from the weak interactions already known in the standard electroweak theory.

For small ε, expression (3.28) takes the form

$$|K_{S(L)}> \approx |K_{(\pm)}> + \varepsilon|K_{(\mp)}> \quad , \tag{3.55}$$

and we can calculate by perturbation theory the value of ε induced by H_{SW}, with the result

$$\varepsilon = -i\alpha/(\delta m - \tfrac{i}{2}\delta\Gamma) \quad . \tag{3.56}$$

This expression has the phenomenological form (3.33) and therefore has essentially the phase (3.45), comparable with the observed phase (3.2). The empirical value of $\mathrm{Re}(\varepsilon)$ gives us a direct measure of α,

$$(\alpha/\delta m) \approx 4 \times 10^{-3} \quad , \tag{3.57}$$

from which we can conclude that the coupling parameter for

H_{SW} is of order $10^{-3}G_F^2M^4$, where scale mass M will be taken to be the proton mass. Since $G_FM^2 \approx 10^{-6}$, the result (3.57) indicates clearly the 'superweakness' of H_{SW}. This model has one free parameter α and fits essentially one experimental fact (assuming that $\eta_{+-} = \eta_{oo}$ holds and that $\mathrm{Im}\,\Gamma_{12} \ll \mathrm{Im}M_{12}$). It predicts $\varepsilon' = 0$, so that it would be disproved if the measurements of ε' now under way give a value for ε' inconsistent with zero.

4. CALCULATIONS FOR THE $(K^o,\bar{K}^o)$ COMPLEX

A successful account of the mass difference δm_K was first given by Gaillard and Lee [32] in terms of the box diagram (b) of Fig. 2 when α and β represent the d and s quark, respectively, and the quarks i and j include both u and c quark. Indeed, they used this calculation to make a successful prediction for the c quark mass, at a time when the c quark was not yet known empirically. Later, Ellis et al. [33] extended this calculation, using the full KM matrix for the six-quark model, in order to discuss CP-nonconservation quantitatively for the $(K^o,\bar{K}^o)$ system. Their calculation included the unknown heavier t-quark, besides the c-quark now known, but their result was not sensitive to the t-quark mass m_t and did not allow any useful prediction for it, beyond giving a rough upper limit $m_t \lesssim 40$ GeV. They noted explicitly, as we would now expect generally from the discussion of the KM model given in Sec. 2, that the $\Delta s = \pm 2$ CP-violating element of the mass matrix (3.6) is non-zero only if the box diagrams for both c and t intermediate quarks are taken into account. For our purpose, it is sufficient to give their result in numerical form (see refs.[33,34] for the complete formulae),

$$(\mathrm{Im}M_{12})/\delta m \approx 18 s_2 s_3 \sin\delta \,, \qquad (4.1)$$

for the mass values $m_c = 1.5$ GeV and $m_t = 30$ GeV, the mixing angle θ_2 being assumed small. Since $\delta m \approx \Gamma/2$ empirically and

$\mathrm{Im}\,\Gamma_{12} << \mathrm{Im}M_{12}$ holds generally, the value of ε is generally given in the simplified form

$$\varepsilon = (i/\sqrt{2})(\mathrm{Im}M_{12}/\delta m)\exp(i\phi_\varepsilon) \qquad (4.2)$$

where ϕ_ε is given in (3.45). The empirical result (3.37) then tells us that

$$s_2 s_3 \sin\delta \; \widetilde{\approx} \; 1.8 \times 10^{-4} \; . \qquad (4.3)$$

CP-violation resulting from the phase angle δ in the KM matrix also occurs in the amplitudes for $K \to \pi\pi$ decay. The simplest graphs illustrating all of the possibilities are shown in Fig. 3. It is useful to trace their various contributions to the $K^o \to \pi\pi$ amplitudes:

$$K^o \to \pi^+\pi^-: \quad U_{us}U^*_{ud}(b+c+d+2e) + U_{cs}U^*_{cd}(d+2e) \qquad (4.4a)$$

$$\bar{K}^o \to \pi^o\pi^o: \quad \tfrac{1}{2}U_{us}U^*_{ud}(a+c+d) \quad + U_{cs}U^*_{cd}d \; , \qquad (4.4b)$$

where the contribution e sums all the four graphs shown, which must be taken together for a colour gauge-invariant result. With the KM matrix of form (2.11), the coefficient $U_{us}U^*_{ud}$ is real, whereas the coefficient $U_{cs}U^*_{cd}$ is not. Since they multiply different graphs in the forms (4.4), we can conclude there are non-zero $K \to \pi\pi$ transition amplitudes to both $I = 0$ and $I = 2$ final states. In particular, a non-zero phase δ generates a phase ξ for the $I = 0$ amplitude. Since A_o has been defined in Sec. 3.2 to be real, by convention, it is now necessary to re-define the phase of the K^o field, in order to maintain that convention, thus $K^o \to e^{-i\xi}K^o$. Since this implies the change $\bar{K}^o \to e^{-i\xi}\bar{K}^o$, it follows that M_{12} and Γ_{12} must be corrected by a phase factor $e^{+2i\xi}$. From (3.27), we see that δm and $\delta\Gamma$ are unaffected by this change. Since ξ is necessarily small, we have the correction

$$\mathrm{Im}M_{12} \to \mathrm{Im}(M_{12}e^{+2i\xi}) \; \widetilde{\approx} \; \mathrm{Im}M_{12} + 2\xi\,\mathrm{Re}M_{12} \; . \qquad (4.5)$$

Since $\mathrm{ReM}_{12} = \delta m$ in the present approximation, the value of ε is corrected to

$$\varepsilon = (i/\sqrt{2})(2\xi + \mathrm{ImM}_{12}/\delta m)\exp(i\phi_\varepsilon) \tag{4.6}$$

where

$$\xi = f s_2 s_3 \sin\delta \; , \tag{4.7}$$

f being a numerical coefficient determined from the graphs specified in the expressions (4.4) to be at most of the order of unity. The graphs d and e are of the class known as 'penguins' and we note that they are the only contributors to the $U_{cs}U^*_{cd}$ terms from which the CP-violation in the K^o amplitudes stems, at this level of calculation.

The graphs of Fig. 3 generate only a real I = 2 amplitude A_2. Although the factor $U_{cs}U^*_{cd}$ depends on the phase δ, it allows only $\Delta I = \frac{1}{2}$, since $\Delta I = 0$ for $c \to s$ and $\Delta I = \frac{1}{2}$ for $c \to d$. Hence the contributions (d) and (e) are zero for the final I = 2 $\pi\pi$ state. The factor $U_{us}U^*_{ud}$ does include $\Delta I = 3/2$, but has no CP-violating part. However, the phase change $K^o \to e^{-i\xi}K^o$ just made above does give A_2 a phase $(-\xi)$ and therefore a non-zero value for ε', since $\mathrm{Im}(A_2) \approx -\xi|A_2|$ then holds, to a sufficient approximation. The ratio $|A_2|/|A_o|$ is known to be of order 1/20, from the branching ratio $(\pi^+\pi^-)/(\pi^o\pi^o)$ known for K_s decay and from the $(K^+ \to \pi^+\pi^o)$ decay rate. Hence the estimate obtained for ε' is

$$\varepsilon' = (i/\sqrt{2})\xi|A_2/A_o|\exp(i(\delta_2-\delta_o)) \; , \tag{4.8}$$

leading to the estimate

$$|\varepsilon'/\varepsilon| \approx (f/18)|A_2/A_o| \approx f/400 \; . \tag{4.9}$$

Gilman and Wise [34] have emphasized the role of the Penguin graphs for the $K \to \pi\pi$ amplitudes. Their calculation does involve much uncertainty, of course, since these graphs invoke gluon exchanges. The result they obtained was

$$\xi = fs_2 s_3 \sin\delta \; c_2/\{(\ln(m_c^2/\mu^2)/\ln(m_t^2/m_c^2))-s_2^2\} \; , \qquad (4.10)$$

where f is the fraction of the amplitude due to the Penguin graphs and μ denotes a typical hadronic mass, say $\mu \approx 1$ GeV, rather than the W-meson mass, as was previously assumed. Thus, if one can trust the gluon calculations, the Penguin graphs may contribute more strongly than earlier work had suggested. This would be an attractive conclusion in that the dominance of the Penguin graphs would then account for the success of the $\Delta I = \frac{1}{2}$ rule for mesonic weak decay processes, although it would still leave open the question of why the $\Delta I = \frac{1}{2}$ rule works so well for baryonic decays. The prediction made for $(\varepsilon'/\varepsilon)$ depends on the values adopted for m_t and f. For example, the assumptions $m_t = 30$ GeV and $f = 0.75$ lead to the estimate [34]:

$$|\varepsilon'/\varepsilon| \approx 1/30. \qquad (4.11)$$

With ε and ε' almost parallel in the complex plane, the present data correspond to $\varepsilon'/\varepsilon = -0.003 \pm 0.014$, which is barely compatible with (4.11), unless f is considerably less than 0.75. Opinions differ [2] about the magnitude of f but, unless f is significantly less than this, the $(K^0, \bar{K}^0)$ experiments now underway should lead to a non-zero value for ε' and allow a definitive test as to whether or not we do have more to deal with than a CP-violating $\Delta s = \pm 2$ superweak interaction, as current theories require.

Hagelin [35] followed up the work of Gilman and Wise, making an estimate for the amplitude of the Penguin graph using the MIT bag model, but reaching essentially the same conclusions. Estimates for the $(K^0, \bar{K}^0)$ parameters using the MIT bag model were also made by Barger et al.[36], Shrock et al.[37], and McWilliams and Shanker [38], but we shall not discuss them in detail here, refering the reader instead to the review papers by Wise [39] and by Pakvasa [40] at the 1980 Madison Conference. We also refer the reader to an early paper by Wolfenstein [41] for a critical discussion of the uncertainties of semi-phenomenological analysis of the $(K^o,\bar{K}^o)$ system based on the KM model.

Carter and Sanda [42] have pointed out in a rather clear way why it is that we have no possibility of finding larger CP-violating effects in experiments concerning the $(K^o,\bar{K}^o)$ complex. It has already been pointed out in Sec.2.1 that any CP-violating contributions to an amplitude must have a factor $(s_1s_2s_3 \sin\delta)$. However, what matters in a given experiment is the ratio of the CP-violating terms to the CP-conserving terms. For example, the $K \to \pi\pi$ decay amplitude is necessarily proportional to s_1, the Cabibbo factor. The intensity of the CP-violating term relative to the total intensity for any observable quantity is given by the ratio of these two amplitudes and it will therefore have $(s_1s_2s_3 \sin\delta)/s_1 = (s_2s_3\sin\delta)$ as a necessary factor, as had always been found in any particular calculations made using the KM model.

5. NEUTRAL HEAVY QUARK SYSTEMS

5.1 Introductory Remarks

The neutral mesons in question here, the $(c\bar{u})$, $(b\bar{d})$, $(b\bar{s})$, $(t\bar{u})$ and $(t\bar{c})$ systems all decay rapidly, with lifetimes $\lesssim 10^{-12}$ sec. This makes their mean path length before decay at most about 0.01 cm, so that experimental observations on them will necessarily involve integration over their full path length, i.e., over all time. Apart from this, their discussion follows the lines laid out for the $(s\bar{d})$ system in Sec. 3. A system M^o, prepared by a specific strong interaction process, will change into a linear superposition of M^o

and $\bar{M}^o$ with increase in time, as a result of the M_S^o and M_L^o decay interactions, as follows (CPT-invariance assumed)

$$M^o(t = 0) \rightarrow F_+(t)M^o + \left(\frac{1-\varepsilon}{1+\varepsilon}\right)F_-(t)\bar{M}^o \ , \tag{5.1}$$

where

$$F_\pm(t) = \exp(-i(m_S-\tfrac{i}{2}\Gamma_S)t) \pm \exp(-i(m_L-\tfrac{i}{2}\Gamma_L)t) \ . \tag{5.2}$$

The M^o or $\bar{M}^o$ content of the beam at time t is signalled by the observation of the charge sign for the leptons emitted, the processes being $M^o \rightarrow \ell\nu_\ell$ + (hadrons) and $\bar{M}^o \rightarrow \bar{\ell}\bar{\nu}_\ell$ + + (hadrons), in the decays occurring at time t. A convenient measure of the mixing [42] is given by integrating the modulus square of these functions (5.2) and taking their ratio [38]:

$$\rho = \frac{\int_0^\infty |F_-(t)|^2 dt}{\int_0^\infty |F_+(t)|^2 dt} \tag{5.3}$$

$$= \frac{4(\delta m/\Gamma)^2 + (\delta\Gamma/\Gamma)^2}{2+4(\delta m/\Gamma)^2-(\delta\Gamma/\Gamma)^2} \ . \tag{5.4}$$

Here we have used the notation of (3.25) and (3.26). We note that

(i) leptonic decay modes have the same strength for M^o and $\bar{M}^o$, so that they do not contribute to $\delta\Gamma$ in the limit where CP invariance holds;

(ii) since the states under consideration are all pseudoscalar, decay to $(q\bar{q})$ for light quarks q will be very small, because of the helicity mismatch due to the V and A currents effective in the electroweak interactions;

(iii) the decay modes corresponding to the graphs of Fig. 4 do contribute to both Γ and $\delta\Gamma$.

For the $(D^0,\bar{D}^0)$ complex, the estimates made [45] for $\delta\Gamma/\Gamma$ and $\delta m/\Gamma$ are of order 10^{-3} and there appears little possibility for the measurement of CP-violation for the neutral D mesons. CP-violation could be sought for these systems from observations on the charge sign for leptonic decay modes, as for the $(K^0,\bar{K}^0)$ mesons, or from the observation of K^+ and K^- in their final decay products. The direct decays arise from the transitions $c \to s$ for D^0 and $\bar{c} \to \bar{s}$ for $\bar{D}^0$, so giving, for example, the modes $D^0 \to K^- +$ (hadrons) and $\bar{D}^0 \to K^+ +$ (hadrons), respectively. The observation of $D^0 \to K^+\pi^-$ would indicate that $(D^0,\bar{D}^0)$ mixing has occurred, through the sequence $D^0 \to \bar{D}^0 \to K^+\pi^-$. The searches to date give the limit [16]

$$\rho_D < 0.16, \tag{5.5}$$

a result which is far from the theoretical expectation of a value of order 10^{-6}.

The $(B^0,\bar{B}^0)$ complex has been discussed by many authors [37,42,44-55], in various aspects. Here, we consider the box diagrams of Fig. 2 again. For given quark types i and j, the evaluation of both box amplitudes leads to a U-matrix factor

$$V_{ij} = U_{\alpha i}\, U^*_{\beta i}\, U^*_{\alpha j}\, U_{\beta j} \quad . \tag{5.6}$$

From the unitarity of the matrix $U_3^c(q)$, it follows that

$$\sum_i V_{ij} = \sum_j V_{ij} = 0 \quad . \tag{5.7}$$

The calculated transition mass M_{12} has the general structure

$$M_{12} = \sum_{ij} V_{ij}\,(A\,\delta_{ij} + B_{ij})\,, \tag{5.8}$$

where A is independent of the quark masses, while B_{ij} depends on the masses (m_i,m_j), quadratically if the $\log(m_a^2/m_b^2)$ factors are taken to be constant. With (5.7), the term A in (5.8) drops out, and the largest term of M_{12} then comes when both i and j are t quarks, giving

$$M_{12} \approx B_{tt} \cdot V_{tt} \quad . \tag{5.9}$$

For the B_d meson, this leading term (5.9) is given explicitly by the expression

$$M_{12}(B) = \frac{G_F^2 f_B^2 m_B^2}{12\pi}\{(m_t^2+\frac{1}{3}m_b^2+\frac{3}{4}m_b^2 \ln(\frac{m_t^2}{m_b^2}))[s_1 s_2(c_1 s_2 s_3 - c_2 c_3 e^{i\delta})]^2 + O(m_c^2, m_b^2)\} \ , \tag{5.10}$$

B_{tt} being the expression to the left of the square brackets, and V_{tt} being the square of the expression within the square brackets. For Γ_{12}, only intermediate states which are on the energy shell can contribute and these are indicated on Fig.2 by the dotted lines. Intermediate quarks which are so heavy that their states all lie above the mass of the meson $(\alpha\bar{\beta})$ under consideration do not occur in the sum

$$\Gamma_{12} = \sum_{i,j \,<\, \text{threshold}} V_{ij}\, S_{ij} \ , \tag{5.11}$$

where S_{ij} denotes the absorptive part of the (ij) box amplitude. In the present case, this give us

$$\Gamma_{12}(B) = S_{uu}\, V_{uu} + S_{uc}\, V_{uc} + S_{cu}\, V_{cu} + S_{cc}\, V_{cc} \quad . \tag{5.12}$$

If we consider the hypothetical case $m_u = m_c << m_b$, then all these S_{ij} in expression (5.12) are equal, being given by $\bar{S}$, say. We then note that, using (5.7) for j = u,c, and t successively, this sum becomes

$$\Gamma_{12}(B) \approx \bar{S}(V_{uu} + V_{cu} + V_{cu} + V_{cc}) \tag{5.13a}$$

$$= \bar{S}(-V_{tu} - V_{ct}) \tag{5.13b}$$

$$= \bar{S}V_{tt} \quad . \tag{5.13c}$$

Hence, the phases of Γ_{12} and M_{12} both come from V_{tt}, in the reasonable approximation that m_u and m_c are different but both small relative to the yet unknown mass m_t. When we recall that the mass-mixing parameter ε_B is related by Eq.(3.30) to the quantity Z_B defined by (3.31) to have the form

$$Z_B = \mathrm{Im}(\Gamma_{12}/M_{12})/(\mathrm{real}) \ , \qquad (5.14)$$

it becomes clear that mixing due to virtual transitions $B^o \leftrightarrow \bar{B}^o$ is necessarily small, being zero in the hypothetical case $m_u = m_c << m_t$.

The general situation has also been discussed perceptively by Carter and Sanda [42].We have to do with doublets of the type

$$\begin{bmatrix} H \\ L \end{bmatrix} = \begin{bmatrix} c \\ s \end{bmatrix} \quad \text{and} \quad \begin{bmatrix} t \\ b \end{bmatrix} \ , \qquad (5.15)$$

where the upper member H is heavier than the lower member L. The structure of the KM mixing $U_3^c(q)$ is such that the element U_{HL} is dominant relative to other U_{Hq} or U_{qL}. In this situation, the largest box diagram (see Fig. 2) for H is that involving intermediate quark L, while the largest for L is that involving intermediate quark H. From the discussion at the end of Sec.4, we then conclude that

$$\text{for } (H\bar{q}): \qquad (\delta m)_H \propto m_L^2 |U_{HL}|^2 \qquad (5.16a)$$

$$\text{for } (L\bar{q}): \qquad (\delta m)_L \propto m_H^2 |U_{HL}|^2 \ . \qquad (5.16b)$$

Next, we consider the decay processes of $(H\bar{q})$ and $(L\bar{q})$ to lighter quark systems. The L-decay must involve smaller mixing amplitudes since L is energetically able to decay only out of its doublet, to lighter quarks. On the other hand, H-decay can occur to L and this involves the large matrix element U_{HL}. Also, the way the quark mass values go, the phase space for $H \to L$ is much greater than for L → q. These remarks are well illustrated by the cases of charm and strangeness. Taking these considerations together, we conclude that

$$(\delta m/\Gamma)_{H\bar{q}} << (\delta m/\Gamma)_{L\bar{q}} \quad . \qquad (5.17)$$

This conclusion (5.17) is well illustrated by the K^o and D^o meson cases, where we have

$$K^o: \quad \delta m_K \propto m_c^2, \ \Gamma_K \sim \sin^2\theta_c \gtrsim 0.05$$

$$D^o: \quad \delta m_D \propto m_s^2, \ \Gamma_D \sim \cos^2\theta_c \gtrsim 0.95 \ .$$

Emp lifetime for D-decay gives $\Gamma_D/\Gamma_K \sim 10^2$, comparable with $\cot^2\theta_c \gtrsim 19$ and a large phase-space ratio. To summarize, we have

$$(\delta m/\Gamma)_D \gtrsim (m_s/m_c)^2 \tan^2\theta_c (\delta m/\Gamma)_K \gtrsim 10^{-3} (\delta m/\Gamma)_K \ , \qquad (5.18)$$

indicating how unfavourable the situation is for observing CP-violation effects in the $(D^o, \bar{D}^o)$ complex.

5.2 Leptonic Ratios for $(M^o, \bar{M}^o)$ Systems

It appears that the most fruitful source for the study of neutral heavy mesons is likely to be the e^+e^- annihilation reaction

$$e^+e^- \to M^o + \bar{M}^o + (\text{other mesons})^o \quad . \qquad (5.19)$$

However, one can also consider reactions in which only one M^o (or $\bar{M}^o$) meson is produced, such as

$$e^+e^- \to M^o + M^- + (\text{other mesons})^+ \quad . \qquad (5.20)$$

We consider both cases, in the opposite order,

(i) where the initial state is pure M^o. The distinction between M^o and $\bar{M}^o$ is in the lepton charge sign in their decay with

$$\text{(a)}\quad M^o \to \ell^- , \qquad \text{(b)}\quad \bar{M}^o \to \ell^+ . \tag{5.21}$$

When we use the expressions (5.1) and (5.2) and integrate over all time, we obtain the ratio

$$r = \frac{N_M(\ell^-)}{N_M(\ell^+)} = \left|\frac{1-\varepsilon}{1+\varepsilon}\right|^2 \cdot \frac{\int |F_-(t)|^2 dt}{\int |F_+(t)|^2 dt} \tag{5.22}$$

$$= \left|\frac{1-\varepsilon}{1+\varepsilon}\right|^2 \cdot \rho_M . \tag{5.23}$$

A value $r \neq 1$ is, of course, an indicator that there are CP impurities in the state for times $t > 0$. When the initial state is pure $\bar{M}^o$, the corresponding ratio is given by

$$\bar{r} = \frac{N_{\bar{M}}(\ell^+)}{N_{\bar{M}}(\ell^-)} = \left|\frac{1+\varepsilon}{1-\varepsilon}\right|^2 \cdot \rho_M . \tag{5.24}$$

We emphasize here that the leptons considered here are those which result from the primary decay of M^o (or $\bar{M}^o$), not the secondary electrons coming from the decay of products resulting from M^o (or $\bar{M}^o$) decay, leptonic or otherwise. For example, in the cascades

$$b \to c + (\text{hadrons})^-, \; c \to \ell^+ \nu_\ell s , \tag{5.25a}$$

$$b \to \ell^- \bar{\nu}_\ell c, \; c \to \ell^+ \nu_\ell s , \tag{5.25b}$$

the secondary ℓ^+ is not to be included in the ratio r. This requires a low momentum cut-off in the lepton spectra measured, to exclude the leptons which could energetically have resulted from secondary decays.

(ii) Where the initial state is $(M^o\bar{M}^o)$. As the beams develop in time, they both oscillate in $(M^o,\bar{M}^o)$ content, and there will be events in which the leptons emitted are $\ell^+\ell^+$, $\ell^+\ell^-$ or $\ell^-\ell^-$; let us denote their numbers by N^{++}, N^{+-} and N^{--}, respectively. There are two quantities measured here,

$$\frac{N^{++} + N^{--}}{N^{+-}} = \frac{2\rho_B}{1 + \rho_B^2} \tag{5.26}$$

which depends only on the mixing parameter (5.4), and

$$\frac{N^{++} - N^{--}}{N^{+-}} = Z_B \tag{5.27}$$

which is the basic measure of CP impurity, as mentioned in Eq.(5.14) above, and which has been estimated from box calculations to be rather small, of order 10^{-3}. We may mention here that the ratio $(\bar{r}/r)$ defined from observations on separate beams which are initially pure M^o or pure $\bar{M}^o$ is equal to $(1 + Z_B)/(1 - Z_B)$, so that cases (i) and (ii) both determine the same parameter Z_B.

5.3 Larger CP Violation Effects?

At the present stage, we are still seeking to determine $U_3^c(q)$ empirically, as well as possible. Of course, the real question is what its structure stems from, at some deeper level, but it is of immediate importance for us to know $U_3^c(q)$ with more certainty, since its form will provide the most direct clues towards any answer to that question.

As mentioned in Sec. 2.1, to exhibit CP-nonconservation, the matrix-element must involve four distinct elements of $U_3^c(q)$. The CP-violating phases will be most apparent when they occur directly in on-shell effects. In the $(K^o,\bar{K}^o)$ complex, only U_{su}

and U_{du} can be on-shell but they are real in our convention and cannot generate any CP-violating effects. In determining $(M_{12} - \frac{i}{2}\Gamma_{12})$ the elements U^*_{sc} and U^*_{dc} are used off-shell, with energy denominators large because an intermediate c quark is involved, and the effects predicted are small, of order 10^{-3} or less.

Above the charm threshold but below "bottom" threshold, four elements can be on-shell and the factor involved is

$$X_4 = U_{du} U^*_{dc} U^*_{su} U_{sc} \quad . \tag{5.28}$$

The CP-violating effect is then necessarily measured by the ratio

$$\mathrm{Im}\, X_4/\mathrm{Re}\, X_4 \sim s_2 s_3 \sin\delta \lesssim 10^{-3} \ . \tag{5.29}$$

So we must go to higher energies, above the b quark threshold $2m_b \gtrsim 10$ GeV. Now we have more elements of $U_3^c(q)$ available on-shell. As an example, Carter and Sanda give

$$Y_4 = U_{bc} U_{bu} U_{sc} U_{su} \quad , \tag{5.30}$$

for which the CP-violating effect is measured by

$$\mathrm{Im}\, Y_4/\mathrm{Re}\, Y_4 \sim (s_2/(s_3 + s_2 \cos\delta))\sin\delta \gtrsim 0.1 \ (?), \tag{5.31}$$

which can be quite large for $U_3^c(q)$ consistent with (5.25). Off-shell effects involving virtual b quarks are still expected to be limited to relative magnitudes of order 10^{-3}.

Finally, from the discussion of Sec. 5.2, it is apparent that the new information needed will not come from the study of leptonic decay processes, since these depend on the parameter ε_B or Z_B again. Hence it is necessary to consider hadronic decay modes, for further progress. However, one must select circumstances sensitive to CP-nonconservation. For example, if we seek to find asymmetries between

$$\text{(a) } M^{o} \rightarrow \text{(final hadrons) and (b) } \bar{M}^{o} \rightarrow \text{(final antihadrons)} \tag{5.32}$$

we must pick a comparison where zero asymmetry is not already required by CPT invariance. Hence, there is no value, for our present purpose, in considering total decay rates, even total rates for hadronic decays, since

$$\Gamma_{tot}(M \rightarrow \text{all hadronic states}) = \Gamma_{tot}(\bar{M} \rightarrow \text{all hadronic states}) \tag{5.33}$$

is imposed by CPT invariance. More generally, the amplitude for '$M \rightarrow$ hadrons' has the structure shown in Fig. 5, as shown in Appendix C. The possible final states divide into groups of states which have the same strong interaction quantum numbers, I, s, P, C, etc. If we denote by f any such group of states, and by $\bar{f}$ the corresponding group of charge-conjugate states, then it follows from CPT invariance that

$$\Gamma_{tot}(M \rightarrow f) = \Gamma_{tot}(\bar{M} \rightarrow \bar{f}) \quad . \tag{5.34}$$

Let us illustrate the situation by reference to $K^{\pm}$ decays to pions. There are two groups of final states from K^{+} decay:

(i) $\pi^{+}\pi^{o}$. This group has parity (+), and only one member.
(ii) $\pi^{+}\pi^{+}\pi^{-}$ and $\pi^{+}\pi^{o}\pi^{o}$. This group has parity (-). These states can be divided further according to final I-spin, the possibilities being I = 1, 2 and 3.

CPT invariance requires that $\Gamma(K^{-} \rightarrow \pi^{-}\pi^{o}) = \Gamma(K^{+} \rightarrow \pi^{+}\pi^{o})$. CPT invariance also requires that

$$\Gamma_{tot}(K^{-} \rightarrow \pi^{-}\pi^{-}\pi^{+} \text{ and } \pi^{-}\pi^{o}\pi^{o}) = \Gamma_{tot}(K^{+} \rightarrow \pi^{+}\pi^{+}\pi^{-} \text{ and } \pi^{+}\pi^{o}\pi^{o}) \; . \tag{5.35}$$

However, it is in general quite consistent with CPT invariance to have

$$\Gamma_{tot}(K^- \rightarrow \pi^-\pi^-\pi^+) \neq \Gamma_{tot}(K^+ \rightarrow \pi^+\pi^+\pi^-) \ . \tag{5.36}$$

Unfortunately, there is little reason to expect much difference between these rates. The data show that the 3π state in K^+ decay is dominated by s-waves and is close to symmetric for pion permutations, while the branching ratio $(2\pi^+\pi^-)/(\pi^+2\pi^0)$ is very close to 1/4, the value typical of the symmetric I = 1 state. In other words, the $(2\pi^+\pi^-)$ and $(\pi^+2\pi^0)$ final states in K^+ decay are dominated by a single final state (with definite parity, isospin, internal orbital angular momenta and permutation symmetry), and CPT invariance then implies the equality for (5.36).

Let us go higher in energy to the case of D decay. We will list some of the D^+ decay modes and group them:

(i) $(\pi^+\pi^0)$, $(\rho^+\rho^0)$, $(K^+\bar{K}^0)$, etc. This group has s = 0, P = +, and several I spin values, with G = +1.

(ii) $(\pi^+\pi^+\pi^-)$, $(\pi^+\pi^0\pi^0)$, etc. This group has s = 0, P = −, and several I spins, with G = −1.

(iii) $(\bar{K}^0\pi^+)$, $(K^-\pi^+\pi^+\pi^0)$, $(K^{*-}\pi^+\pi^+)$, etc. This group has s = −1, P = +, and several I spin values.

(iv) $(K^-\pi^+\pi^+)$, etc. This group has s = −1, P = −, and several I spin values.

The CPT constraints are now quite weak. For example, CPT invariance now allows

$$\Gamma(D^+ \rightarrow \pi^+\pi^0) \neq \Gamma(D^- \rightarrow \pi^-\pi^0) \ , \tag{5.37a}$$

$$\Gamma(D^+ \rightarrow \bar{K}^0\pi^+) \neq \Gamma(D^- \rightarrow K^0\pi^-) \ . \tag{5.37b}$$

At these higher energies, there are many processes in which CP-nonconservation may show its effects. Carter and Sanda [42] and Sanda [43,44] have explored the question of where it might be fruitful for experiment to look. Since we need four U_{ij} on-shell, we are directed to processes involving a cascade of decays. For example, consider the cascade sequences:

$$b \to c\,\bar{u}d \qquad\qquad \bar{b} \to \bar{c}\,ud$$
$$\quad\hookrightarrow s\,u\bar{d} \qquad\qquad \quad\hookrightarrow \bar{s}\,\bar{u}d$$

$$V = U_{bc}\,U^*_{du}\,U^*_{sc}\,U_{du} \qquad V^* = U^*_{bc}\,U_{du}\,U_{sc}\,U^*_{du} \quad . \tag{5.38}$$

The spin and momentum factors of the amplitude remain the same under the interchange of quarks by antiquarks, whereas the KM factor V becomes V^* under CPT. When V and V^* differ in phase, there is a CP violation, measured in this case by

$$\mathrm{Im}(V/V^*) = \{2s_2(s_3+s_2\cos\delta)/(s_3^2+s_2^2+2s_2s_3\cos\delta)\}\sin\delta \ , \tag{5.39}$$

a factor which can be of order 1 over the range of parameters consistent with all CP-nonconservation data available to date.

In order to illustrate how this idea might be exploited in the search for CP-nonconserving effects, Carter and Sanda have discussed two specific transitions. These both involve B-meson decays to a particular final state f, which can proceed by two (or more) paths involving different products of the U_{ij} elements. If this is the case, the amplitude for $B \to f$ can be written in the form

$$A(B \to f) = V_1A_1 + V_2A_2 \ , \tag{5.40}$$

where, as before, the V_α denote products of U_{ij} elements and the corresponding factor A_α denotes the spin-space part of the matrix. In this case, CTP invariance tells us that, for the charge-conjugate process, we have

$$A(\bar{B} \to \bar{f}) = V_1^*\,A_1 + V_2^*\,A_2 \ , \tag{5.41}$$

and the aim is to find CP-violating effects arising from the $(V_1V_2^*)$ interference term. However, if the A_1 and A_2 have the same phase, then no asymmetry will result, since $Re(V_1^* V_2) = Re(V_1 V_2^*)$. As discussed in Appendix C, the phase of A_i arises from the final state interactions (FSI). These FSI are essential for the existence of non-zero CP-violating effects.

The example we now discuss involves the two cascade sequences

$$(1) \qquad B^- \rightarrow K_S^o D^o X_\alpha^- \rightarrow K_S^o K_S^o Y_\beta^o X_\alpha^- \tag{5.42a}$$

$$(2) \qquad B^- \rightarrow K_S^o \bar{D}^o X_\alpha^- \rightarrow K_S^o K_S^o Y_\beta^o X_\alpha^- \quad , \tag{5.42b}$$

for which there are many final states $X_\alpha^- Y_\beta^o$ which may be considered. The quark transitions involved are depicted on Fig.6. The dotted lines there show the on-shell state possible, the case where either D^o or $\bar{D}^o$ travel freely. If the D^o-$\bar{D}^o$ mixing effects were not so small, it would be necessary for the calculation of the overall amplitudes to take these effects into account; we neglect them here, for they are an inessential complication. For the two graphs, we have

$$(1) \quad F_1 = U_{bc} U_{su}^* U_{sc}^* U_{du} \qquad (2) \quad F_2 = U_{bu} U_{sc}^* U_{sc} U_{du}^* \quad . \tag{5.43}$$

The phase of F_1 is denoted by ϕ_1 and stems from the product $U_{bc}\, U_{sc}^*$, for which we have

$$\tan \phi_1 = Im(U_{bc} U_{sc}^*)/Re(U_{bc} U_{sc}^*) \approx (s_2/(s_3 + s_2 \cos\delta)) \sin\delta, \tag{5.44}$$

whereas F_2 has zero phase. We now write

$$A_{n\alpha\beta} = R_{n\alpha\beta} \exp(i\delta_{n\alpha\beta}) \tag{5.45}$$

where this amplitude refers to a particular configuration for the particles of the systems X_α^- and Y_β^o. The net amplitude (5.40) may now be written

$$A(B^- \rightarrow K_S^o K_S^o X_\alpha^- Y_\beta^o) = R_{1\alpha\beta} e^{i(\delta_{1\alpha\beta} + \phi_1)} + R_{2\alpha\beta}\, e^{i\delta_{2\alpha\beta}} \tag{5.46}$$

from which we can calculate the partial rate, denoting it by $\Gamma^-_{\alpha\beta}$. The amplitude for the charge-conjugate systems is then at once

$$A(B^+ \rightarrow K^o_s K^o_s \bar{X}^+_\alpha \bar{Y}^o_\beta) = R_{1\alpha\beta}\, e^{i(\delta_{1\alpha\beta}-\phi_1)} + R_{2\alpha\beta} e^{i\delta_{2\alpha\beta}} , \qquad (5.47)$$

from which we can now calculate the partial rate for the charge conjugate process, $\Gamma^+_{\alpha\beta}$. The quantity of interest is then the asymmetry a, given by

$$a_{\alpha\beta} = (\Gamma^+_{\alpha\beta} - \Gamma^-_{\alpha\beta})/(\Gamma^+_{\alpha\beta} + \Gamma^-_{\alpha\beta}) , \qquad (5.48)$$

where the numerator is, explicitly,

$$(\Gamma^+_{\alpha\beta} - \Gamma^-_{\alpha\beta}) = 4\, R_{1\alpha\beta} R_{2\alpha\beta} \sin(\delta_{1\alpha\beta} - \delta_{2\alpha\beta}) |F_1||F_2| \sin \phi_1 \qquad (5.49)$$

proportional to the CP-violating pahse ϕ_1. In practice, it would be necessary to sum numerator and denominator over groups of states $X^-_\alpha Y^o_\beta$. The phase difference $(\delta_{1\alpha\beta} - \delta_{2\alpha\beta})$ could vary appreciably from configuration to configuration and the asymmetry sought may tend to become washed out. The experiment is clearly very difficult to carry out, since it is limited to $K^o_s K^o_s XY$ configurations which correspond to the intermediate states specified in eq.(5.42). To make things worse, the decay modes $B^- \rightarrow K^o_s D^o X^-_\alpha$ and $B^+ \rightarrow K^o_s \bar{D}^o X^-_\alpha$ are 'Cabibbo suppressed' decays, the first involving U_{us} and the second U_{ub}, so that they will have small branching rate relative to other B^- decay modes.

The second possibility discussed by Carter and Sanda concerns the time development of B^o and $\bar{B}^o$ beams and their decay. The B^o decay cascade considered is

$$B^o \rightarrow D^o X^o_\alpha \rightarrow K_s\ Y^o_\beta\ X^o_\alpha \qquad (5.50a)$$

and its charge conjugate

$$\bar{B}^o \rightarrow \bar{D}^o X^o_\alpha \rightarrow K_s \; Y^o_\beta \; X^o_\alpha \; , \tag{5.50b}$$

and the interferences discussed arise from the fact that D^o and $\bar{D}^o$ have some decay modes in common, which can therefore interfere instructively. The beam initially B^o at t = 0 transforms to (5.1) at time t. If we denote that B^o amplitude for a specific configuration (5.50a) by VA, the $\bar{B}^o$ amplitude for the charge-conjugate configuration is V^*A. It is then possible to calculate the rates $\Gamma_{\alpha\beta}$ and $\bar{\Gamma}_{\alpha\beta}$ for specific final states, a typical asymmetry being given by

$$\frac{\Gamma-\bar{\Gamma}}{\Gamma+\bar{\Gamma}} = - \frac{2\delta m_B}{\Gamma_B} \cdot \frac{\sin 2\phi}{(1+(\delta m_B/\Gamma_B)^2)} \tag{5.51}$$

where

$$\sin\phi \approx \frac{s_3 \sin\delta}{\sqrt{(s_2^2+s_3^2+2s_2s_3\cos\delta)}} \; . \tag{5.52}$$

The complexities due to CP-nonconservation in the $(K^o,\bar{K}^o)$ system have still to be included in this discussion.

A more convenient possibility might well be its application to $B^o\bar{B}^o$ pair production

$$e^+e^- \rightarrow B^o_d\bar{B}^o_d + X^o \; , \tag{5.53}$$

where X^o is a hadronic system. As the $(B^o,\bar{B}^o)$ content of each beam develops, their leptonic decay can lead to the observation of $\ell^+\ell^-$, $\ell^+\ell^+$ or $\ell^-\ell^-$, the rates being denoted by N_{+-}, N_{++} and N_{--}, respectively. The formulae for these rates depend on the B^o-$\bar{B}^o$ internal orbital angular momentum L and may be written

$$N_{+-} = 2(G_+(x) + G_-(y)) \; , \tag{5.54a}$$

$$N_{++ \atop (--)} = \left|\frac{1\pm\varepsilon_B}{1\mp\varepsilon_B}\right|^2 (-G_+(x) + G_-(y)) \; , \tag{5.54b}$$

where all upper or all lower signs are to be taken in (5.54b); where there is a choice, the functions $G_{\pm}$ are given by

$$G_{\omega}(z) = (1-\omega(-1)^2 z^2)/(1+\omega z^2) \quad , \tag{5.55}$$

and the L arguments are $x = (2\delta m/\Gamma)$ and $y = (\delta\Gamma/\Gamma)$. Following Sanda [53], the ratio $(N_{++} + N_{--})/N_{+-}$ is plotted vs. x, for y = 0, and separately for even and odd L, on Fig. 7. The decay of the $B^o_d \bar{B}^o_d$ system to $K_s K^{\pm} Y^{\mp}$ is also of interest in this regard, since the K^- comes from B^o_d, whereas the K^+ comes from $\bar{B}^o_d$. In this case, what is to be measured is the K^+/K^- asymmetry for a given spatial configuration of the final particles $K_s K^{\pm} Y^{\mp} X^o$, the asymmetry $(N^+ - N^-)/(N^+ + N^-)$ being proportional to the asymmetry expression (5.51). The asymmetry calculated by Sanda[53] is shown on Fig. 8 against s_2, for various values of the theoretical parameter η. The parameters s_3 and δ have been constrained to fit the data on the $(K^o, \bar{K}^o)$ complex for each s_2, and the parameter η pertains to the calculation of the matrix element $\langle K^o | H_{Wk} | \bar{K}^o \rangle$, where $\eta = 1$ corresponds to the use of the vacuum saturation approximation, and $\eta = 0.4$ corresponds to the MIT bag model calculation.

6. THE ELECTRIC DIPOLE MOMENT OF THE NEUTRON

The electric dipole moment of the neutron is highly forbidden. Not only does it violate parity conservation, it also violates time reversal invariance, or equivalently, CP-invariance. Further, with the KM model, it can occur only in second order for the weak coupling constant G_F, i.e., it requires the emission and absorption of two $W^{\pm}$ bosons. This follows from the discussion in Sec. 2.1, emphasizing that CP-violation effects will be zero in the KM model unless four different elements of the matrix (U_{ij}) and its complex conjugate (U^*_{ij}). Two typical graphs are shown in Fig. 7. Even so, according to Shabalin [55], the graph of Fig. 7(a) gives result zero, and a non-zero result is obtained only when the quark is dressed with $n_g \geq 2$ gluons, as in Fig. 7(b). The upper limit obtained [56] in this way is

$$(D_N/e)_q \leq m_{u,d}\left(\frac{\alpha_s}{\pi}\right)^2\left(\frac{\alpha}{\pi}\right)s_1^2 s_2 s_3 \sin\delta\left(\frac{m_{s,c}^2}{m_W}\right)^2 F(m_t^2/m_b^2) \;, \qquad (6.1)$$

where (α_s/π) is the coupling constant of QCD, and the factor $m_{u,d}$ is present because of the helicity-flip character of V and A interactions. Numerically, this expression leads to the bound

$$(D/e)_q \lesssim 10^{-30(\pm 1)} \text{ cm} \;, \qquad (6.2)$$

whereas the best upper limit available empirically [57] is for the neutron, giving

$$D_N/e \leq 6 \times 10^{-25} \text{ cm} \;. \qquad (6.3)$$

However, as the technology of ultra-cold neutrons has been improving in the last few years, it appears that the improvement of this limit by as much as two orders of magnitude might be achievable [58] in the current round of experiments.

Here, we shall confine attention to the rather detailed calculations recently reported by Gavela et al.[59], based on the graphs shown in Fig. 8 of the type proposed by Morel [60] and by Nanopoulos et al.[61]. These depict the perturbation theory being carried out in two steps: (a) the generation of CP violating phases through heavy quark loops, and (b) summation over the low-lying intermediate baryon states. The low-lying states of most importance are the $1/2^+$ baryons (Σ^0 and Λ) and the $1/2^-$ first-excited baryons ($\Sigma^*(1620)$ and $\Lambda(1405)$), since they are excited directly by the A and V parts of the weak Hamiltonian, respectively. Graph 8(a) provides a contribution of at most 10^{-34} cm. Graph 8(b) provides a penguin diagram for the excitation of Σ^o and Λ with CP-violation, followed by weak radiative decay of the hyperon. For the $N \to \Sigma^o$ transition, this penguin diagram generates an effective interaction of the form

$$\bar{\Sigma}^o(\alpha_{\Sigma N} + \beta_{\Sigma N}\gamma_5)N \qquad (6.4a)$$

where the coefficients $\alpha_{\Sigma N}$ and $\beta_{\Sigma N}$ are complex, in general, and include the factor

$$G_F(\alpha_s/q\pi)\ln(m_t^2/m_c^2)s_1s_2c_2s_3\sin\delta \quad . \qquad (6.4b)$$

The authors note that, in this calculation, the GIM mechanism does not lead to a factor $(m_t^2-m_c^2)/(\pi M_W)^2$, as expected, but to the factor $(\alpha_s/\pi)\ln(m_t^2/m_s^2)$, which is of order unit, so increasing the amplitude from naive expectation by at least two orders of magnitude. This is the most important element in their work. The second transition, $\Sigma^o \to N\gamma$, is a process of the general class $Y \to N\gamma$, for which there exists a theoretical discussion by these authors [62] successful in its application to the well-known example, $\Sigma^+ \to P\gamma$. The transition amplitude $\Sigma^o \to N\gamma$ has the form

$$\bar{N}(p')\{\sigma_{\mu\nu}(p'-p)^\nu(C_{\Sigma N}+D_{\Sigma N}\gamma_5)\}\Sigma(p)A^\mu \ , \qquad (6.5)$$

the terms $C_{N\Sigma}$ and $D_{N\Sigma}$ representing the magnetic and electric dipole transitions, respectively. CP conservation is assumed to hold for the transitions $Y \to N\gamma$, since these involve only three types of quark. After taking into account hermiticity, the penguin diagrams giving (6.4) go together with the $\Sigma \to N\gamma$ amplitude, as is illustrated on Fig.10, to contribute

$$D_N(\Sigma) = \frac{2\mathrm{Im}(\alpha_{N\Sigma})D_{\Sigma N}}{(m_N-m_\Sigma)} + \frac{2\mathrm{Im}(\beta_{N\Sigma})C_{\Sigma N}}{(m_N+m_\Sigma)} \qquad (6.6)$$

to the neutron E1 moment D_N. Here and below, the notations Σ and Σ^* are used to specify the neutral members, Σ^o and Σ^{*o} respectively, of their charge multiplet. The second term of

(6.6) is dropped on the grounds that (i) $\beta_{N\Sigma} = 0$ in the SU(3) limit, and (ii) it has a large energy denominator. There is a companion term $D_N(\Lambda)$ for an intermediate Λ hyperon; since the discussion for Λ runs completely parallel with that for Σ, we shall henceforth mention explicitly only the Σ case. The $\Sigma \to N\gamma$ amplitude $D_{\Sigma N}$ is illustrated in Fig. 11. It is a parity-reversing operator and is therefore dominated by the low-lying Σ^* and N^* states with spin-parity $(1/2^-)$, as shown in Figs. 11(a) and 11(b), respectively. The $\Sigma \to \Sigma^*\gamma$ and $N \to N^*\gamma$ couplings are both written in the general form

$$\bar{B}(p')\sigma_{\mu\nu}(p'-p)^{\nu}\gamma_5 F_2^{BB^*} B^*(p)A^{\mu} , \tag{6.7}$$

and the weak couplings $\Sigma^* \to N$ and $\Sigma \to N^*$ due to W-exchange, as depicted in Fig. 11, are denoted by

$$\bar{N}(a_{N\Sigma^*} + b_{N\Sigma^*}\gamma_5)\Sigma^* \tag{6.8a}$$

$$\overline{N^*}(a_{N^*\Sigma} + b_{N^*\Sigma}\gamma_5)\Sigma , \tag{6.8b}$$

where $b_{N\Sigma^*}(b_{\Sigma N^*})$ and $a_{N\Sigma^*}(a_{\Sigma N^*})$ are the parity-conserving (p.c.) and parity-reserving (p.r.) couplings for the case of $N \to \Sigma^*(\Sigma \to N^*)$ transitions for spin-parity $(1/2^-)$ $\Sigma^*(N^*)$. The net contribution (6.6) then takes the form

$$D_N(\Sigma) = 2\left\{\frac{\mathrm{Im}(\alpha_{N\Sigma})F_2^{\Sigma\Sigma^*} a_{\Sigma^* N}}{(M_N-M_\Sigma)(M_N-M_{\Sigma^*})} + \frac{\mathrm{Im}(\alpha_{N\Sigma})a_{\Sigma N^*}F_2^{N^*N}}{(M_N-M_\Sigma)(M_N-M_{N^*})}\right\} . \tag{6.9}$$

This result motivated the authors to generalize this expression to all possible orderings of the penguin interaction (pen.), then electromagnetic interaction (γ) and the weak interaction (Wk). In a notation which will by this time be obvious, we shall list all of these terms,

$$(\alpha)\quad (N\Sigma)_{pen}(\Sigma\Sigma^*)_\gamma(\Sigma^*N)_{Wk}\ ,\qquad (\beta)\quad (N\Sigma)_{pen}(\Sigma N^*)_{Wk}(N^*N)_\gamma\ ,$$

$$(\gamma)\quad (N\Sigma^*)_{pen}(\Sigma^*\Sigma)_\gamma(\Sigma N)_{Wk}\ ,\qquad (\delta)\quad (NN^*)_\gamma(N^*\Sigma)_{pen}(\Sigma N)_{Wk}\ ,$$

$$(\varepsilon)\quad (NN^*)_\gamma(N^*\Sigma^*)_{pen}(\Sigma^*N)_{Wk}\ ,\qquad (\phi)\quad (NN^*)_\gamma(N^*\Sigma)_{pen}(\Sigma N)_{Wk}\ , \tag{6.10}$$

each of which is to be taken together with its hermitian conjugate, whose sequence of steps is in the reverse order.

The authors argue that, contrary to the impression given by the form (6.9), the terms (α) and (β) are finite in the SU(3) limit, and that when the flavours c, b and t are added, the GIM cancellations are very strong, so that these terms can be neglected. The terms (ε) and (ϕ) are also finite in the SU(3) limit, the denominators then being given by ω, the spacing between the $(1/2^+)$ and $(1/2^-)$ multiplets, and again the GIM cancellations are strong. The dominant contributions correspond to the terms (γ) and (δ) of (6.10), because these have denominators proportional to (m_q-m_d). The contributions from the heavier flavours are thereby suppressed and do not cancel the s-quark terms (γ) and (δ). The final expression given by Gavela et al. is

$$D_N/e = \tfrac{4}{9}s_1^2 s_2 s_3 c_1 c_2 c_3\ \sin\delta\ \ln(m_t^2/m_c^2)\left(\frac{\alpha_s}{\pi}\right)G_F^2|\psi(0)|^2/\omega m_N\delta m, \tag{6.11}$$

where δm denotes the quark mass difference (m_s-m_d), which may be estimated by $(\bar{m}_Y-m_N) \approx 200$ MeV. The result (6.11) has the value

$$D_N/e \approx 10^{-30}\ \text{cm}\quad . \tag{6.12}$$

The main question is whether other terms might somehow arise which could cancel these penguin graphs, for there are certainly many additional terms which can contribute to such a high-order amplitude but which are not considered here. None have yet been visualized which are likely to have the same degree of enhancement as is claimed for the penguin contributions discussed, so it is at least plausible that these are the dominant terms. There is still, of course, a rather large gap between this enhanced estimate (6.12) and the present upper limit (6.3).

7. CONCLUDING REMARKS

There has been considerable work on CP-violation through the Higgs mechanism, outlined here in Section 2.3, following up the work of Branco [19] who was concerned with establishing the mechanism rather than with making any fit to the data. Sanda [63] used this mechanism with the Lagrangian of Vainshtein et al.[64] which fits the data on CP-conserving $K \to \pi\pi$ decays, to calculate the parameters ε_K and ε'_K for the $(K^0, \bar{K}^0)$ complex. $\mathrm{Re}M_{12}$ is still dominated by the WW box graphs (cf. Fig.2), but $\mathrm{Im}M_{12}$ is now dominated by the additional HW and WH box graphs where one or other of the W bosons is replaced by the charged Higgs boson H. The $K \to \pi\pi$ amplitude is dominated again by penguin graphs, which are considered by many authors to be the source of the $\Delta I = 1/2$ rule, the real part by the W graph of Fig.2(d) and the imaginary part by the corresponding H graph, where W^- is replaced by H^-, so that the phase ξ defined just before Eq.(4.5) is here due dominantly to the Higgs boson. Indeed, for a reasonable choice of parameter values, Sanda finds the bound

$$|(2\xi/\varepsilon_m)| \geq 30 \,, \tag{7.1}$$

where $\varepsilon_m = (\mathrm{Im}M_{12})/(\mathrm{Re}M_{12})$ is essentially the parameter defined by Eq.(3.29). The ratio of interest is $(\varepsilon'/\varepsilon)$, which, from (4.6) and (4.8), is given by

$$\varepsilon'/\varepsilon = -\{2\xi/(\varepsilon_m+2\xi)\}|(A_2/A_0)|\exp(i(\delta_2-\delta_0-\phi_\varepsilon)) \,, \tag{7.2}$$

and this is essentially independent of ξ and ε_m for values satisfying (6.1). Sanda then concluded that

$$|\varepsilon'/\varepsilon| \approx |(A_2/A_0)| \approx 0.05 \;, \qquad (7.3)$$

the ratio $|(A_2/A_0)|$ being known empirically to have value approximately 0.05. Such a large value is really ruled out by the present data on $K_L^0 \to \pi\pi$. Essentially the same conclusion was reached by Deshpande [65] at about the same time. Korner and McKay [66] have criticized Sanda's and Deshpande's application of the matrix element evaluations, given by Vainsthein et al., to the calculation of the CP-violating terms, and carried out some additional evaluations. However, they reached the same unacceptable prediction (6.3) in the end. On the other hand, Shizuya and Tye [68] reported a calculated limit ≤ 0.03 for (6.3), but they stated the HW box graphs do not contribute to the CP-violating terms, in contradiction to the other authors. Chang [67] has given a careful discussion of the theoretical situation, pointing out that some of the disagreements between different authors have quite subtle origins which have caused unrecognized ambiguities. He finds that the HH box graphs do contribute significantly but that the HW box graphs are generally the larger part of the total. The magnitude of the penguin graphs is complicated by the contribution from soft gluons and is therefore uncertain. There is a dispersive part to M_{12} which may not be unimportant, and Chang's conclusion is that the Higgs-sector interpretation of the CP-nonconservation effects observed to data is not yet completely ruled out. As emphasized by Branco [19], the model which has been considered is "minimal", in that it has only one CP-violating phase; more complicated situations are possible for the Higgs sector and these have not yet been investigated. Also, Natural Flavour Conservation could be given up, in which case a CP-violating phase δ in $U_3^C(q)$ might also arise from this spontaneous symmetry-breaking. This is a rather unattractive possibility since the H^{0} boson would then be required to be really very heavy ($\gtrsim$ 250 GeV) in order that the model be consistent with the observed K_S-K_L mass difference.

The KM model is still quite acceptable and is not in disagreement with any known facts at present. The next step is for the experimenters, and this is under way - the measurement of the parameter ε'. If $\varepsilon' \neq 0$ is established and is comparable with the constraints we know on the KM matrix, the next question will be to understand how this small CP-nonconserving element of this matrix comes about. If CP is violated only there, why should this violation only be "milliweak" ?

To conclude, I wish to mention that the preparation of these lectures for publication was finally completed during a visit to the Stanford Linear Accelerator Center, and to acknowledge especially the warm hospitality of Professor S. Drell and the Theory Division of SLAC during that period.

REFERENCES[+]

1. G. Ecker, Schladming Lectures, 1982.
2. J. Ellis, Phenomenology of Unified Gauge Theories, Preprint TH-3174-CERN (29 September, 1981).
3. S. Weinberg, Phys. Rev. Letters 19 (1967) 1264.
4. M. Kobayashi and T. Maskawa, Progr. Theor. Phys. 49 (1973) 652.
5. E.P. Wigner, Group Theory and its Application to the Quantum Mechanics of Atomic Spectra (Academic Press, New York, 1959).
6. V. Lyubimov et al., Phys. Lett. 94B (1980) 266.

[+] The abbreviation HEP-1980 will be used for "High Energy Physics 1980, Proc. XX. Int.Conf., Madison, Wisconsin (eds. L.Durand and L.G.Pondrom, Am.Inst.Phys., New York, 1981)".

7. J. Simpson, Phys. Rev. D23 (1981) 649.
8. A. De Rujula, CERN preprint TH-3045 (1 April , 1981).
9. M. Rees, Proc. 1981 Intl. Symp. on Lepton and Photon Interactions at High Energies (ed. W. Pfeil, Phys. Inst., Univ. Bonn, 1981) p. 993; see also J. Gao and R. Ruffini, Phys. Lett. 97B (1980) 388.
10. G. Steigman, Cosmology and Neutrino Physics , in Proc. Neutrino '81 (Univ. Hawaii Press, Honolulu, 1982), in press.
11. B. Pontecorvo, Soviet. Phys. JETP 53 (1967) 1717.
12. H. Fritzsch, Phys. Lett. 67B (1977) 451.
13. N. Cabibbo, Phys. Lett. 72B (1978) 333.
14. V. Barger, K. Whisnant and R.J.N. Phillips, Phys. Rev. D22 (1980) 1636.
15. T.D. Lee, Phys. Reports 9C (1974) 143.
16. Particle Data Group, Revs. Modern Phys. 52 (1980) 51 .
17. S. Weinberg, Phys. Rev. Letters 37 (1976) 657.
18. G.C. Branco, Phys. Rev. Letters 44 (1980) 504.
19. G.C. Branco, Phys. Rev. D22 (1980) 2901.
20. R.G. Sachs, Ann. Phys. (N. Y.) 22 (1963) 239.
21. P.H. Eberhard, Phys. Rev. Letters 16 (1966) 150.
22. J.S. Bell and J. Steinberger, in Proc. Oxford Int. Conf. on Elementary Particles, 19/25 September 1965 (Rutherford High Energy Lab., Chilton, Didcot, 1966), p. 195.
23. F.S. Crawford, Phys. Rev. Letters 15 (1965) 1045.
24. P.K. Kabir, The CP Puzzle (Academic Press, New York, 1968).
25. T.T. Wu and C.N. Yang, Phys. Rev. Letters 13 (1964) 180.
26. T.J. Devlin and J.O. Dickey, Revs. Modern Phys. 51 (1979) 241.
27. J. Cronin et al., FNAL proposal No. 617.
28. R.K. Adair et al., BNL-ACS proposal.
29. D. Cundy et al., Proposal CERN/SPSC/81-110 (22 December 1981).

30. R. Dalitz, in Weak Interactions and High-Energy Neutrino Physics (Proc. Int. School of Physics 'Enrico Fermi', Course XXXII, Academic Press, New York, 1966), p. 206.
31. L. Wolfenstein, Phys. Rev. Lett. 13 (1964) 562.
32. M. Gaillard and B.W. Lee, Phys. Rev. 10D (1974) 897.
33. J. Ellis, M. Gaillard and D. Nanopoulos, Nucl. Phys. B109 (1976) 213.
34. F. Gilman and M. Wise, Phys. Lett. 83B (1979) 83.
35. J. Hagelin, ε' in the six quark model ,Harvard preprint (1979).
36. V. Barger, W. Long and S. Pakvasa, Phys. Rev. Lett. 42 (1979) 1585.
37. R. Shrock, S. Treiman and L. Wang, Phys. Rev. Lett. 42 (1979) 1589.
38. B. McWilliams and O. Shanker, Phys. Rev. D22 (1980) 2853.
39. M. Wise, HEP-1980, p. 398.
40. S. Pakvasa, HEP-1980, p. 1164.
41. L. Wolfenstein, Nucl. Phys. B160 (1979) 501.
42. A. Carter and A. Sanda, Phys. Rev. D23 (1981) 1567.
43. A. Pais and S. Treiman, Phys. Rev. D12 (1975) 2744.
44. E. Ma, W. Simmons and S.F. Tuan, Phys. Rev. D20 (1979) 2888.
45. S. Barhay and J. Seris, Phys. Lett. 84B (1979) 319.
46. A. Ali and Z. Aydin, Nucl. Phys. B148 (1979) 165.
47. J. Hagelin, Phys. Rev. D20 (1979) 2893.
48. J. Bernabeu and C. Tarlskog, Z. Phys. C8 (1980) 233.
49. M. Bander, D. Silverman and A. Soni, Phys. Rev. Lett. 43 (1979) 242.
50. B. Gaisser, T. Tsao and M. Wise, Ann. Phys. 132 (1981) 66.
51. V. Barger, W. Long and S. Pakvasa, Phys. Rev. D21 (1980) 174.
52. A. Sanda, in Physics in Collision I (eds. P. Trower and C. Bellini, Plenum Press, New York, 1981) p. 79.
53. A. Sanda, in Proc. Cornell Z^o Theory Workshop (eds. M. Peskin and S. Tye, Newman Lab., Cornell Univ.,1981), publ. CLNS 81-485, p. 190.

54. E. Franco, M. Lusignoli and A. Pugliese, Strong Interaction Corrections to CP Violation in $B^o - \overline{B^o}$ Mixing , Univ. Rome preprint No. 262 (September, 1981).
55. E. Shabalin, Sov. J. Nucl. Phys. 32 (1980) 228.
56. I.S. Altarev et al., Phys. Lett. 102B (1981) 13.
57. J. Ellis and M. Gaillard, Nucl. Phys. B150 (1979) 141.
58. N.F. Ramsey, private communication (1981).
59. M. Gavela et al., Phys. Lett. 109B (1982) 215.
60. B. Morel, Nucl. Phys. B157 (1979) 23.
61. D. Nanopoulos, A. Yildiz and P. Cox, Phys. Lett. 87B (1979) 53.
62. M. Gavela et al., Phys. Lett. 101B (1981) 417.
63. A. Sanda, Phys. Rev. D23 (1981) 2647.
64. A. Vainshtein, V. Zakharov and M. Shifman, Sov. Phys.-JETP 45 (1977) 670.
65. N. Deshparde, Phys. Rev. D23 (1981) 2654.
66. T. Korner and D. McKay, DESY preprint 81/034 (June,1981).
67. D. Chang, Phys. Rev. D25 (1982) 1318.
68. K. Shizuya and S. Tye, Phys. Rev. D23 (1981) 1613.
69. A. Sirlin, Revs. Mod. Phys. 50 (1978) 573; Phys. Rev. D23 (1981) 2899.
70. S. Pakvasa, S. Tuan and J. Sakurai, Phys. Rev. D23 (1981) 2799.
71. J. Knobloch, in Neutrino '81, Proc. Int. Conf. on Neutrino Physics and Astrophysics (eds. High Energy Group, Univ. Hawaii, Honolulu, 1981), p. 421.
72. M. Tonker et al., Phys. Lett. 105B (1981) 242.
73. V. Barger, T. Leveille and P. Stevenson, HEP-1980, p. 390.
74. J. Ellis et al., Nucl. Phys. B131 (1977) 285.
75. JADE Collaboration: W. Bartels et al., DESY preprint 82-014 (March, 1982).
74'. A. Silverman, Proc. 1981: Int. Symp. on Lepton and Photon Interactions at High Energies (ed. W. Pfeil, Phys. Inst., Univ. Bonn, 1981), p. 138.

75'. L. Chan Wang, Proc. Cornell Z^{O} Theory Workshop (eds. M. Peskin and S. Tye, Newman Lab. Nuclear Studies, Cornell Univ., 1981), preprint CLNS 81-485, p. 25.
76. L. Wolfenstein, Phys. Rev. D17 (1978) 2369.
77. J. Blietschau et al., Nucl. Phys. B133 (1978) 205.
78. V. Barger et al., Phys. Rev. D22 (1980) 2717.
79. G.C. Wick, Ann. Revs. Nucl. Sci. 9 (1959) 1.
80. R.H. Dalitz, Strange Particles and Strong Interactions (O.U.P., Bombay, 1961).
81. R.H. Dalitz, Revs. Mod. Phys. 33 (1961) 471.

APPENDIX A

Empirical Estimates of the Matrix Elements of $U_3^C(q)$

The matrix element best known is the (ud) element C_1, whose value

$$C_1 = 0.9737 \pm 0.0025 \tag{A1}$$

has been deduced from the ratio of the rate for $0^+ \to 0^+$ superallowed nuclear beta-decay and the rate for free μ^+ decay, after allowing for the ratiative corrections, as calculated for both transition rates by Sirlin [69].

The analysis of the decay rates for the leptonic decay modes known for the various $\Sigma \to N$ and $\Lambda \to N$ transitions relative to nucleonic beta-decay $N \to N$ has led to the estimate

$$|U_{us}| = |s_1 c_3| = 0.219 \pm 0.003 \ , \tag{A2}$$

which is related to the Cabibbo angle θ_c by the relation $\tan\theta_c = c_3 \tan\theta_1$.

Since the matrix $U_s^c(q)$ is unitary, the (ub) element is constrained by the relationship

$$|U_{ud}|^2 + |U_{us}|^2 + |U_{ub}|^2 = 1 \; . \qquad \text{(A3)}$$

With the estimates (A1) and (A2), this leads us to the estimate $|U_{ub}| = 0.06 \pm 0.06$, from which we deduce

$$|s_3| = 0.28 \begin{array}{l} + 0.21 \\ - 0.28 \end{array} \; . \qquad \text{(A4)}$$

Direct estimates for the matrix-elements involving heavy quarks may be obtained from the data on opposite-sign di-leptons observed in ν and $\bar{\nu}$ interactions with nucleons, which arise from charmed particle production and decay. Thus we have the possibilities:

$$U_{cd}: \quad \text{(a)} \; \nu_\mu + d \to \mu^- + c \;(\hookrightarrow e^+, \mu^+, \ldots) \qquad \text{(b)} \; \bar{\nu}_\mu + \bar{d} \to \mu^+ + \bar{c} \;(\hookrightarrow e^-, \mu^-, \ldots) \qquad \text{(A5)}$$

$$U_{cs}: \quad \text{(a)} \; \nu_\mu + s \to \mu^- + c \;(\hookrightarrow e^+, \mu^+, \ldots) \qquad \text{(b)} \; \bar{\nu}_\mu + \bar{s} \to \mu^+ + \bar{c} \;(\hookrightarrow e^-, \mu^-, \ldots) \qquad \text{(A6)}$$

The ν_μ reaction (A5a) can occur on valence d quarks, whereas the other reactions (A5b) and (A6) occur only for quarks or antiquarks from the quark-antiquark sea in the nucleon. Pakvasa et al.[70] considered the data on $\mu^- e^+$ events in ν_μ interactions on nucleons, separating out the d and s contributions on the basis of the x-distribution of these events and the known x-distributions for the valence and sea quarks in the nucleon. This discussion requires an estimate of $B(e^+)$, the charm branching ratio for e^+, for which they argued $B(e^+) = 9 \pm 1\%$, and then gives the limits

$$0.19 < |U_{cd}| < 0.34 \; . \qquad \text{(A7)}$$

A similar analysis by Ellis [2] on the basis of the $\mu^+\mu^-$ events from the CDHS collaboration leads to the limits

$$0.17 < |U_{cd}| < 0.23 \ . \qquad (A8)$$

Ellis [2] also made an estimate for U_{cs} from the CDHS dimuon data with the result

$$|U_{cs}| = 0.85 \begin{smallmatrix} +\ 0.15 \\ -\ 0.11 \end{smallmatrix} \ . \qquad (A9)$$

This requires an absolute normalization for the number of $\bar{s}$ and s quarks in the nucleon, which he obtained from a neutral current analysis , assuming the d and s quarks to have the same neutral current coupling strength.

Pakvasa et al.[70] obtained an independent estimate for U_{cs} from the rate observed for the charm decay $D \to Ke\nu$, assuming F^* pole dominance for its matrix-element. Adopting the estimate [16]

$$\text{(a)}\ \Gamma(D^+ \to K^0 e^- \nu)/\Gamma(D^+ \to all) \approx 10\%, \quad \text{(b)}\ \Gamma(D^+ \to all) \approx 10^{12}\,sec^{-1} \ , \qquad (A10)$$

they obtained the value

$$|U_{cs}| = 0.8 \pm 0.2 \qquad (A11)$$

quite comparable with the estimate (A9).

For B-hadron decay, the matrix-elements U_{bu} and U_{bc} come into play. As given by Barger et al.[73], the total decay rate based on the spectator model is

$$\Gamma(b \to all) = \{7.69|U_{bu}|^2 + 3.07|U_{bc}|^2\}\ G_F^2 m_b^5/(192\pi^3) \ . \qquad (A12)$$

The upper limit of 1.4×10^{-12} sec recently obtained for the lifetime for b-flavored hadrons places the following lower limit on U_{bu} and U_{bc},

$$\sqrt{(|U_{bc}|^2 + 2.50|U_{bu}|^2)} > 0.04 \ . \qquad (A13)$$

Further, the electron spectrum observed for the $B \to e\nu X$

modes indicates that the $B \to$ (D and D^*)$e\nu$ decay rate dominates over the rate for $B \to$ (π and ρ)$e\nu$, implying that $|U_{bu}|/|U_{bc}| < 0.3$, so that b couples more strongly to c quarks than to u quarks. This conclusion appears to be confirmed by the number of final kaons from $b\bar{b}$ events at CLEO , these resulting from the decays $c \to u$ and $\bar{c} \to \bar{u}$. These two results lead to the limit $|U_{bc}| > 0.04$.

The constraints imposed by the above considerations have been discussed in greatest detail by Pakvasa et al.[70]. If some evidence from theoretical calculations on the $(K^o, \bar{K}^o)$ complex is also included, the constraints on the KM angles θ_2, θ_3 and δ are now quite strong, but they are also rather complicated. For example, Pakvasa et al. have given the curve shown in Fig. 12 here, where acceptable values for s_2 and s_3 are plotted as function of $|U_{cs}|^2$, with two branches, one for δ near π and the other for δ near 0. This curve has taken into account the $(K^o, \bar{K}^o)$ calculations of Barger et al.[73] which used the bag model and adopted mass 30 GeV for the t quark. However, the fact is that it is not really possible to determine all of the KM angles, nor of the elements of $U_3^c(q)$, today. We will simply quote a "central value" matrix which was specified recently by L. Chau Wang[75'] as an illustration to guide the experimenters, namely

$$U = \begin{bmatrix} 0.97 & -0.22 & -0.046 \\ 0.22 & 0.85-0.66\times10^{-3}i & 0.48+3.2\times10^{-3}i \\ 0.068 & -0.48+2.1\times10^{-3}i & -0.88-1.0\times10^{-3}i \end{bmatrix} \quad \text{(A14)}$$

in the format U_{ij}, where $i = \bar{u}, \bar{c}$ and $\bar{t}$ specify the rows and $j = d, s$ and b specify the columns, in the order given. The magnitude $|U_{ij}|$ is largest along the diagonal in (A14) and falls as we move away from the diagonal, the smallest elements being $U_{\bar{u}b}$ and $U_{\bar{t}d}$. Its structure corresponds to cascade sequences for the decay of heavy quarks; for example $b \to c$ is stronger than $b \to u$, then $c \to s$ is stronger than $c \to u$, and finally only $s \to u$ is possible. However, the final form for U can well be greatly different, and even qualitatively different, from (A14). For example, the CP violation effects in the $(K^0-\bar{K}^0)$ complex depend on the product $(s_2 s_3 \sin\delta)$, which has the value 0.66×10^{-3} in (A.14), but it is not yet known for sure whether its smallness results from the smallness of s_3 or the smallness of δ, or both.

APPENDIX B

CP-Nonconservation and Neutrino Oscillations

The neutrino states ν_i which couple with e, μ and τ leptons to form the charged leptonic weak currents are related with the mass eigenstates n_α by a unitary mixing matrix U:

$$|\nu_i\rangle = \sum_\alpha U_{i\alpha} |n_\alpha\rangle \quad , \tag{B1a}$$

$$|n_\alpha\rangle = \sum_i U^*_{i\alpha} |\nu_i\rangle \quad . \tag{B1b}$$

We shall restrict ourselves to the case of three neutrinos. The matrix U can then be written in the KM form

$$\begin{bmatrix} \nu_e \\ \nu_\mu \\ \nu_\tau \end{bmatrix} = \begin{bmatrix} c_1 & s_1c_3 & s_1s_3 \\ -s_1c_2 & c_1c_2c_3+s_2s_3e^{i\lambda} & c_1c_2s_3-s_2c_3e^{i\lambda} \\ -s_1s_2 & c_1s_2c_3-c_2s_3e^{i\lambda} & c_1s_2s_3+c_2c_3e^{i\lambda} \end{bmatrix} \begin{bmatrix} n_1 \\ n_1 \\ n_3 \end{bmatrix} \tag{B2}$$

where (c_i, s_i) denote $(\cos\phi_i, \sin\phi_i)$, the ϕ_i being the mixing angles, and λ the phase which gives rise to the CP non-conservation.

For a beam of definite momentum, the mass eigenstates propagate in time in accord with their energy $E_i \approx (p+m_i^2/2p)$. A neutrino beam which is of the type ℓ at $t = 0$ will become a linear superposition of neutrino types at time t, with amplitudes

$$T(\ell \rightarrow \ell';t) = \sum_\alpha U^*_{\ell'\alpha} U_{\ell\alpha} \exp(-im^2_\alpha t/2p) , \qquad (B3)$$

and corresponding intensities

$$P(\ell \rightarrow \ell';t) = |T(\ell \rightarrow \ell';t)|^2 . \qquad (B4)$$

CPT invariance requires that the mixing for antineutrinos be related with that for neutrinos. Explicitly, we have

$$|\bar{\nu}_i\rangle = \sum_\alpha U^*_{i\alpha}|\bar{n}_\alpha\rangle \qquad (B5a)$$

$$|\bar{n}_\alpha\rangle = \sum_i U_{i\alpha}|\bar{\nu}_i\rangle \qquad (B5b)$$

so that

$$T(\bar{\ell} \rightarrow \bar{\ell}';t) = \sum_\alpha U^*_{\ell\alpha} U_{\ell'\alpha} \exp(-im^2_\alpha t/2p)$$

$$= T(\ell' \rightarrow \ell;t) . \qquad (B6)$$

The requirement of CP invariance is that

$$T(\bar{\ell} \rightarrow \bar{\ell}';t) = T(\ell \rightarrow \ell';t)^* \qquad (B7)$$

and this is not generally the case, unless the phase λ is zero, in which case the transformation matrix (B2) is real. The symmetry which measures the degree of CP nonconservation is

$$Q(\ell,\ell';t) = \{P(\bar{\ell} \rightarrow \bar{\ell}';t) - P(\ell \rightarrow \ell';t)\}. \qquad (B8)$$

With the parametrization (B2), this is given by the following expression:

$$4s_1^2 c_1 s_2 c_2 s_3 c_3 \sin\lambda \ \{\sin(\Delta_{12}t) + \sin(\Delta_{23}t) + \sin(\Delta_{31}t)\} \qquad (B9)$$

for the transitions $\nu_\mu \rightarrow \nu_\tau$, $\nu_\tau \rightarrow \nu_e$ and $\nu_e \rightarrow \nu_\mu$, where

$$\Delta_{\alpha\beta} = (m_\alpha^2 - m_\beta^2)/2p \quad . \tag{B10}$$

We note that (B9) reverses in sign for a neutrino transition in the opposite sense; also that $(\Delta_{12} + \Delta_{23} + \Delta_{31}) \equiv 0$.

We now list a series of remarks [13,14,40]:

(i) For the diagonal element, CPT invariance (cf. Eq.(B6)) implies that $T(\bar{\ell} \to \bar{\ell};t) = T(\ell \to \ell;t)$ and the CP-asymmetry $Q(\ell,\ell;t)$ is necessarily zero.

(ii) The CP-asymmetry is zero unless all three masses m_α are different. This follows from the general considerations in Sec. 2.2, as well as from (B9). If two masses are equal, say $m_1 = m_2$, then $\Delta_{12} = 0$ and $\Delta_{23} = -\Delta_{31}$, and the curly bracket of (B9) vanishes.

(iii) For small t, the CP-asymmetry is proportional to t^3. If the curly bracket of (B9) is expanded in powers of t, the term of order t has coefficient $(\Delta_{12} + \Delta_{23} + \Delta_{31})$, which is identically zero.

(iv) The CP asymmetry is bounded in magnitude by unity.The bound is achieved for $|c_1| = 1/\sqrt{3}$, $|c_2| = |c_3| = 1/\sqrt{2}$, $\sin\delta = \pm 1$, and $\sin(\Delta_{12}t) = \sin(\Delta_{23}t) = \sin(\Delta_{31}t)$ are all $+\sqrt{3}/2$ or all $-\sqrt{3}/2$.

(v) For large t, such that all $|\Delta_{\alpha\beta}t| >> 1$, the CP-asymmetry averages to zero.

(vi) The CP-asymmetry vanishes if any one element $U_{i\alpha}$ is zero.

(vii) While any one of the $(\Delta_{\alpha\beta}t)$ is small, the CP asymmetry is necessarily small. This remark is closely related with remark (ii) above. The curly bracket of (B9) may be written, with $\bar{\Delta}_3 = (m_3^2 - \frac{1}{2}(m_1^2 + m_2^2))/2p$,

$$\{\sin(\Delta_{12}t) + 2\cos(\bar{\Delta}_3 t)\sin(\tfrac{1}{2}\Delta_{12}t)\} \tag{B11}$$

which is small when $(\Delta_{12}t)$ is small, vanishing when it vanishes. Thus, the CP asymmetry can be large only when all three $(\Delta_{\alpha\beta}t)$ have developed. When $|\Delta_{12}| << |\Delta_3|$, the oscillations due to Δ_3

have very high frequency relative to those due to Δ_{12} and will essentially average out in the CP-asymmetry. After this averaging, this curly bracket is bounded by ± 1, so that the mean CP asymmetry is bounded by $\pm 2/(3\sqrt{3})$, still a subtantial value.

(viii) CP-violation has effects on neutrino oscillations even the observations are limited to configurations consistent with CP conservation. To give one example [76], if the parameters are such that $P(\nu_e \to \nu_e; t = \infty)$ takes its minimum value of 1/3, an interesting possibility because of the solar neutrino problem, then $P(\nu_\mu \to \nu_\mu; t = \infty) = (\frac{1}{2} - \frac{2}{3} c_2^2 s_2^2 \sin^2\delta)$ and $P(\nu_\tau \to \nu_\tau; t = \infty) = (\frac{1}{2} - \frac{2}{3} c_3^2 s_3^2 \sin^2\delta)$, so that we have the sum

$$\sum_{\alpha=e,\mu,\tau} P(\nu_\alpha \to \nu_\alpha; t=\infty) = \frac{4}{3} - \frac{2}{3}(s_2^2 c_2^2 + s_3^2 c_3^2)\sin^2\delta \; . \tag{B12}$$

If CP conservation holds, $\sin\delta = 0$ and this sum is constrained to the magnitude 4/3. With CP-nonconservation, however, the sum can lie between 1 and 4/3, and the observation of a value in this range would imply either that CP is violated or that there are more than three types of neutrino. Barger et al.[14] have made a systematic study of how the existence of CP-violation can extend the domain of values over which two expectation values of this kind can be consistent.

(ix) The only experimental limit on CP-violation at present comes from work of Blietschau et al.[77],

$$P(\bar{\nu}_\mu \to \bar{\nu}_e) - P(\nu_\mu \to \nu_e) = (0.5 \pm 1.2) \times 10^{-3} \; . \tag{B13}$$

Barger et al.[14] have also surveyed the parameter ranges over which various classes of experiment are sensitive. The empirical factor which is open is L/E, where L is the distance from the neutrino source to the observing equipment and E is the neutrino energy. For experiments involving solar neutrinos, $L/E \sim 10^{12}$ cm $(\text{MeV})^{-1}$ and the $\delta m^2_{\alpha\beta}$ must be $\gtrsim 10^{-10}$ $(\text{eV})^2$ for the effects to be observable. Deep mine experiments involve $L/E \sim 10^4$ cm $(\text{MeV})^{-1}$ and need $|\delta m^2_{\alpha\beta}| \gtrsim 10^{-3} (\text{eV})^2$ for the effects to be visible. Reactor and meson factor experiments at present involve L/E $\sim$ 30 to 300 cm $(\text{MeV})^{-1}$. Finally, high energy accelerator data involve typically L/E $\sim$ 4 cm $(\text{MeV})^{-1}$ and exclude ν_μ oscillations with $|\delta m^2_{\alpha\beta}| \geq 10\ (\text{eV})^2$. There are no positive indications from any of the experiments to date, but these experiments were not designed for this purpose and do not yet cover the range of possibilities in a systematic way. The low flux observed by Davis for $\bar{\nu}_e$ neutrinos from the Sun has been the constant factor directing attention towards more general investigations of the possibilities of neutrino oscillation phenomena.

Wolfenstein [76] has pointed out that, since the neutrinos observed sometimes travel large distances through matter, it is necessary to pay attention to the possibility that their interaction with matter, although very tiny, may affect their phase development in time and so need to be taken into account in the analysis of data. Barger et al.[78] have given a broad discussion of these effects and their importance but it is not appropriate for us to go into further detail on this here.

APPENDIX C

Time Reversal Invariance and its Failure

We begin with a simple statement about the consequences of time-reversal invariance. We shall speak in terms of a set of two-particle channels, with potential interactions which act both within channels and between channels.

1. The T-matrix may be defined as the sum of the series

$$T = V + V \frac{1}{E-H_o-i\varepsilon} V + V \frac{1}{E-H_o-i\varepsilon} V \frac{1}{E-H_o-i\varepsilon} V + .. \tag{C1}$$

to be taken with the limit $\varepsilon \to 0$. This series may be partially summed, leading to the Lippmann-Schwinger equation for T, but we are not concerned here with the practical calculation of T. We recognize the role of the $i\varepsilon$ in the denominators of (C1), that they guarantee that there are only outgoing waves in all channels except the incident channel.

2. The K-matrix may be defined as the sum of the series

$$K = V + V \frac{P}{E-H_o} V + V \frac{P}{E-H_o} V \frac{P}{E-H_o} V + \dots \qquad (C2)$$

where the symbol P denotes that a principal value integral is to be taken at the singularities $E-H_o = 0$. This series may be partially summed to give an integral equation for K, but we shall not be concerned with this here. The physical meaning of the principal value integration is that there are only standing waves in all channels except the incident channel.

The difference between $(E-H_o-i\varepsilon)^{-1}$ and $P/(E-H_o)$ is $i\pi\delta(E-H_o)$, and its contribution concerns only those intermediate states "on the energy shell", i.e. with free particle energy (H_o) equal to the initial energy E. In channel space, i.e. using basis states which are diagonal in the free energy H_o, the contribution $\pi\delta(E-H_o)$ may be normalized to be k, the c.m. momentum for each channel. In this remark, we have assumed each channel to have just two particles: multiparticle channels can be considered as the limiting case of an infinite set of two-particle channels, which means that the general properties we shall discuss will carry over to the case of multiparticle channels, although this limit does not provide a practical procdure for calculations in this case.

A general relationship between K and T may be deduced from Eq.(C1) and (C2). This is given by

$$T = K/(1-ikK) \; . \qquad (C3)$$

For a finite number of two-particle channels, this is an algebraic relationship. We note that it has a more symmetrical form when we expand the second term,

$$T = K + iKkK + i^2 \, KkKkK + i^3 \, KkKkKkK + \ldots , \qquad (C4)$$

and this can be resummed to give the alternative form

$$T = (1-iKk)^{-1}K . \qquad (C5)$$

The great advantage in the use of the K-matrix in parametrizing multichannel systems is that, for any K-matrix which is Hermitean, the T-matrix given by (C3) or (C4) is necessarily unitary. To discuss unitarity, it is more convenient to use the S-matrix, which is related with T as follows:

$$S = 1 + 2i\sqrt{k}\ T\sqrt{k} \quad , \qquad (C6)$$

which can be written in the form,

$$S = \frac{1 + i\sqrt{k}K\sqrt{k}}{1 - i\sqrt{k}K\sqrt{k}} \quad , \qquad (C7)$$

as can be seen using (C4) to evaluate $\sqrt{k}\ T\sqrt{k}$. This form (C7) has the advantage that it is a function of only one matrix $K' = \sqrt{k}K\sqrt{k}$. For an energy above all thresholds considered, we have from (C7)

$$S^{\dagger} = \frac{1 - i\sqrt{k}K\sqrt{k}}{1 + i\sqrt{k}K\sqrt{k}} = S^{-1} \qquad (C8)$$

or equivalently

$$S^{\dagger}S = SS^{\dagger} = I \ , \qquad (C9)$$

which is the statement of unitarity (i.e. probability conservation) for the S-matrix.

For a one-channel situation, the scattering may be characterized by a real phase δ, related with K by the equation

$$k \cot \delta = 1/K \quad . \tag{C10}$$

We then have, from (C3),

$$T = (e^{i\delta} \sin\delta)/\mathbf{k} \ , \tag{C11}$$

and from (C6)

$$S = e^{2i\delta} \ . \tag{C12}$$

3. <u>Time reversal in the Wigner sense</u> [5] has been discussed for elementary particle physics by Wick [79] and others [80,81]. All we need to say here is that, if the potential V is invariant under time reversal, its matrix element for a definite partial wave is not only Hermitean (which guarantees unitarity for T) but real. Since the operator $P/(E-H_o)$ is real (and also symmetric, being diagonal in channel space), it follows from the form of (C2) that time-reversal invariance of the interaction energy implies that <u>K is real and symmetric.</u>

We see then from the form of (C4) that time-reversal invariance implies that T is symmetric (but not real). Equivalently, we may see this for S, using the form (C7).

4. <u>A weak process $i \to f$,</u> to lowest order in the weak interaction, is described by a T-matrix of the following form

$$T = \left(\frac{1}{1-iK_s k}\right) K_W \left(\frac{1}{1-ikK_s}\right) \ . \tag{C13}$$

Here, the interaction energy has been separated into V_s, the strong interaction, and V_W, the weak interaction. K_s is then defined by (C2) using V_s alone, and K_W is the first order term in V_W when the complete K-matrix is defined in terms of (V_s+V_W).

Our present interest is in the case where the initial state is a particle M. From (C10), the T-matrix describing its decay is simply

$$T = K_W/(1 - ikK_s) \tag{C14}$$

since there is no scattering in the initial state. K_W is of the lowest nonvanishing order in the weak interaction, and the second factor of (C13) expresses the effect of the strong final state interactions (FSI) on the decay. The FSI factor has all the symmetries which are appropriate for the strong interactions, e.g. isospin, parity, charge conjugation (or G-parity), strangeness, etc.. K_W contains the effects of all the symmetry violations appropriate to the weak interactions.

Let us consider a particle which can decay into two final channels, labelled 1 and 2, unconnected by the strong interactions. The FSI factor is then simply

$$\begin{bmatrix} (1 - ik_1K_{11})^{-1} & 0 \\ 0 & (1 - ik_2K_{22})^{-1} \end{bmatrix} \tag{C15}$$

where K_{ij} is the K-matrix describing the strong interactions in the final state. K_W may be written (w_1,w_2), say, and the T-matrix is diagonal,

$$\begin{bmatrix} \frac{w_1}{(1-ik_1K_{11})} & 0 \\ 0 & \frac{w_2}{(1-ik_2K_{22})} \end{bmatrix} . \tag{C16}$$

The decay amplitude for one channel is just the same as would hold if the other channel did not exist. To fix ideas, we may think of the case of a "D-meson" which could decay only to $\pi\pi$ and to $K\pi$. Each one-channel decay amplitude has the form

$$w/(1-ikK) = (w \cos \delta)e^{i\delta} , \tag{C17}$$

so that the decay amplitude picks up a phase δ due to the FSI.

5. The $M \to f$ and $\bar{M} \to \bar{f}$ decay amplitudes are related. We consider two two-particle channels, since this is sufficient to demonstrate all the properties of interest to us.

The amplitudes for $M \to f$ are given by

$$(w_1,w_2)\begin{bmatrix}1-ik_1K_{11} & -ik_1K_{12}\\ -ik_2K_{21} & 1-ik_2K_{22}\end{bmatrix}^{-1} = \frac{(w_1,w_2)}{D}\begin{bmatrix}1-ik_2K_{22} & ik_1K_{12}\\ ik_2K_{21} & 1-ik_1K_{11}\end{bmatrix} \tag{C18}$$

where D is the determinant det(1-ikK),

$$D = (1-ik_1K_{11})(1-ik_2K_{22}) + k_1k_2K_{12}^2 , \tag{C19}$$

recalling that $K_{12} = K_{21}$.

The amplitudes for $\bar{M} \to \bar{f}$ are governed by the same K-matrix, since C holds for the strong interactions. However, since all the interactions are Hermitean, the elements of K_W for the antiparticle $\bar{M}$ are (w_1^*,w_2^*). If the weak interactions were T-invariant, then K_W would be real, and we would have $K_W = (w_1,w_2)$ also for $\bar{M}$. In the general case, we have for $T(M \to f_i)$ and $T(\bar{M} \to \bar{f}_i)$:

i	$T(M \to f_i)$	$T(\bar{M} \to \bar{f}_i)$	
1.	$\{w_1(1-ik_2K_{22})+iw_2k_2K_{21}\}/D$,	$\{w_1^*(1-ik_2K_{22})+iw_2^*k_2K_{21}\}/D$,	(C20a)
2.	$\{iw_1k_1K_{12}+w_2(1-ik_1K_{11})\}/D$,	$\{iw_1^*k_1K_{12}+w_2^*(1-ik_1K_{11})\}/D$.	(C20b)

We may now consider the intensities for $M \to f_1$ and $\bar{M} \to \bar{f}_1$,

$$|T(M \to f_1)|^2 = |w_1|^2(1+k_2^2K_{22}^2) + |w_2|^2k_2^2K_{12}^2 +$$

$$+ 2k_2K_{12}\mathrm{Im}(w_1^*w_2) - 2k_1k_2K_{12}K_{22}\mathrm{Re}(w_1^*w_2) , \tag{C21a}$$

$$|T(\bar{M} \to \bar{f}_1)|^2 = |w_1|^2(1+k_2^2K_{22}^2) + |w_2|^2k_2^2K_{12}^2 -$$

$$- 2k_2K_{12}\mathrm{Im}(w_1^*w_2) - 2k_1k_2K_{12}K_{22}\mathrm{Re}(w_1^*w_2) \quad , \qquad (C21b)$$

and we note that they are different, in general, so that

$$\Gamma(M \to f_1) \neq \Gamma(\bar{M} \to \bar{f}_1) \ . \qquad (C22)$$

If T holds, then w_1 and w_2 are real, and $\mathrm{Im}(w_1^*w_2) = 0$; it then follows that $\Gamma(M \to f_1)$ and $\Gamma(\bar{M} \to \bar{f}_1)$ are equal. However, irrespective of T-invariance, it follows from direct calculation that (note that the final phase space k_i must be included in the rate)

$$|T(M\to f_1)|^2k_1 + |T(M\to f_2)|^2k_2 \equiv |T(\bar{M}\to\bar{f}_1)|^2k_1 + |T(\bar{M}\to\bar{f}_2)|^2k_2 \ , \qquad (C23)$$

in other words, that

$$\Gamma_{tot}(M \to f) = \Gamma_{tot}(\bar{M} \to \bar{f}) \ , \qquad (C24)$$

where Γ_{tot} refers to the full set of final states, f_1 and f_2 in the present case.

These remarks generalize to the case of n connected channels. We have

$$T(M \to f_i) = (w)(1 - ikK)^{-1}|_{f_i} \qquad (C25)$$

and for the total $M \to f$ decay rate,

$$\Gamma_{tot}(M \to f) = \sum_i |T(M \to f_i)|^2k_i \qquad (C26)$$

$$= (w(1-ikK)^{-1}k(1+iKk)^{-1}w^\dagger) = \sum_{i,j} w_iO_{ij}w_j^* \ . \qquad (C27)$$

The total $\bar{M} \to \bar{f}$ decay rate is obtained by replacing w by w^*,

$$\Gamma_{tot}(\bar{M} \rightarrow \bar{f}) = (w^*(1 - ikK)^{-1}k(1 + iKk)^{-1}\tilde{w}) \qquad \text{(C28)}$$

$$= \sum_{i,j} w_i^* O_{ij} w_j \quad . \qquad \text{(C29)}$$

These two expressions are equal if $O_{ij} = O_{ji}$, and this is at once apparent if we express O as a power series in K, namely

$$O = (1+ikK+i^2kKkK+\ldots)k(1-iKk+(-i)^2KkKk+(-i)^2KkKkKk+\ldots)$$

$$= k - kKkKk + kKkKkKkKk - \ldots \qquad \text{(C30)}$$

It is easiest to demonstrate this by using $K' = \sqrt{k}\, K\, \sqrt{k}$ in (C28), since O then takes the form

$$O = \sqrt{k}(1-iK')^{-1}(1+iK')^{-1}\sqrt{k}$$

$$= \sqrt{k}(1+(K')^2)^{-1}\sqrt{k} \quad , \qquad \text{(C31)}$$

whose expansion is in accord with (C30).

When the final n channels divide into blocks of $(n_1, n_2, n_3, \ldots)$ channels which have different strong-interaction quantum numbers, then each group n_i of channels may be discussed as we have just discussed the case of n connected channels above, and the equality (C24) holds for each block n_i separately.

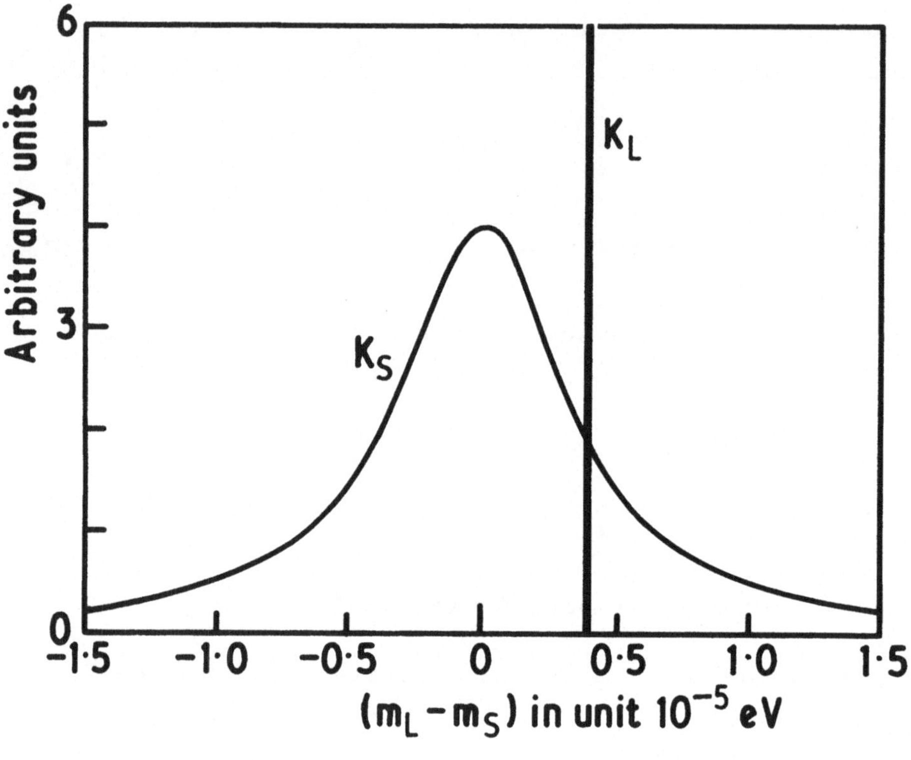

Fig. 1. Displays the mass relationship between the K_S and K_L mesons.

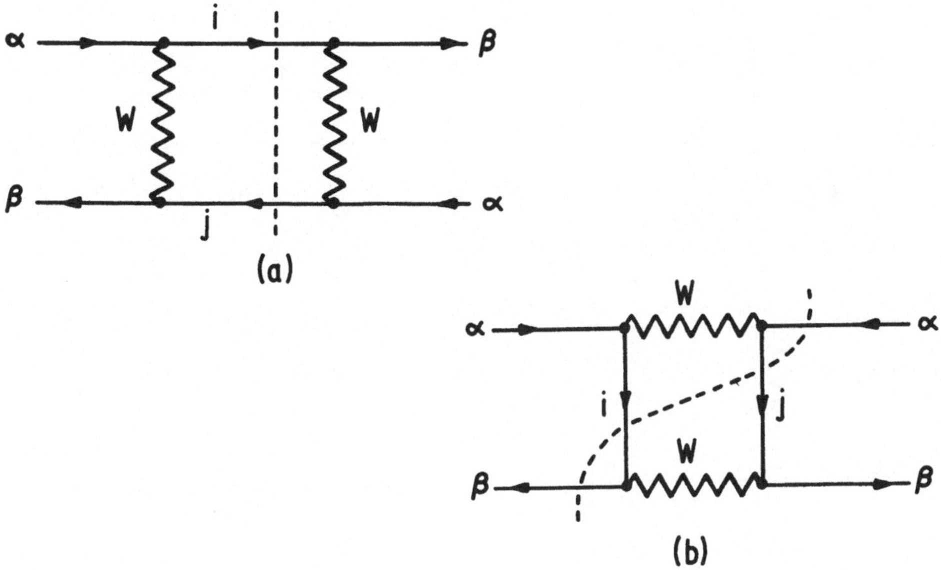

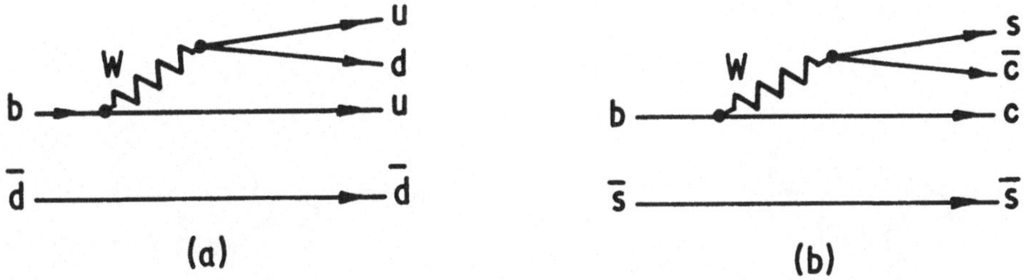

Fig.2. The box diagrams for the calculation of $(M_{12}-\frac{i}{2}\Gamma_{12})$ for neutral meson state $(\alpha\ \bar{\beta})$. The dotted line indicates the intermediate states which can be on-shell and so contribute to Γ_{12}.

Fig.4. Two graphs for B^{o} decay to hadrons, analogous to Fig.3(a) for K^{o} decay, to show the increased possibilities for heavy quark systems.

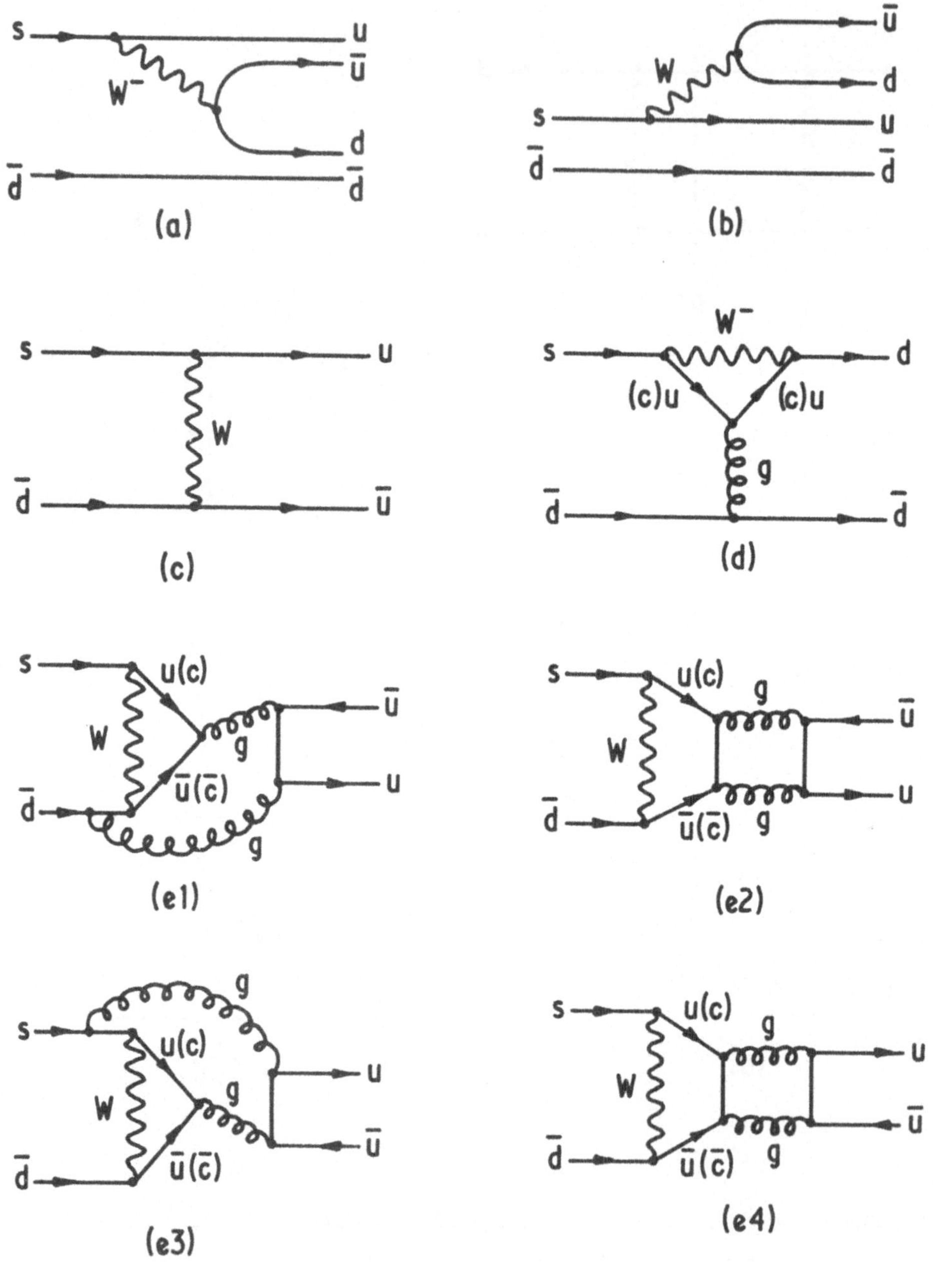

Fig.3. The classes of graphs contributing to the matrix-element for the weak decay K → ππ. Graphs (d) and (e) are the penguin graphs.

$$M(i \to f) = (K_{i\alpha}, K_{i\beta}, K_{i\gamma}, K_{i\delta}, K_{i\varepsilon}, K_{i\phi} \ldots) \left[\begin{array}{c|c|c} & 0 & 0 \\ \hline 0 & & 0 \\ \hline 0 & 0 & \end{array} \right] \begin{array}{c} \text{Final State} \\ \alpha \\ \beta \\ \gamma \\ \delta \\ \varepsilon \\ \phi \end{array}$$

Fig.5. shows the structure of the transition amplitude from initial meson to a series of final states $\alpha, \beta, \gamma, \delta, \varepsilon$, etc. related hadronically within the groups (α), (β, γ, δ), $(\varepsilon \ldots)$ etc..The FSI matrix is nonzero only in diagonal blocks, each one corresponding to one group of hadronically connected final states.

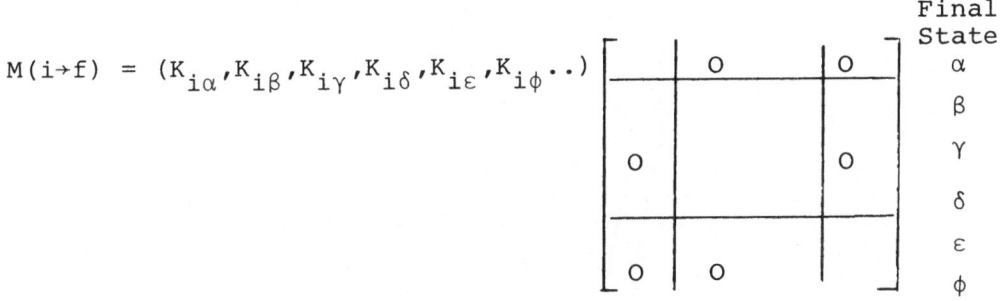

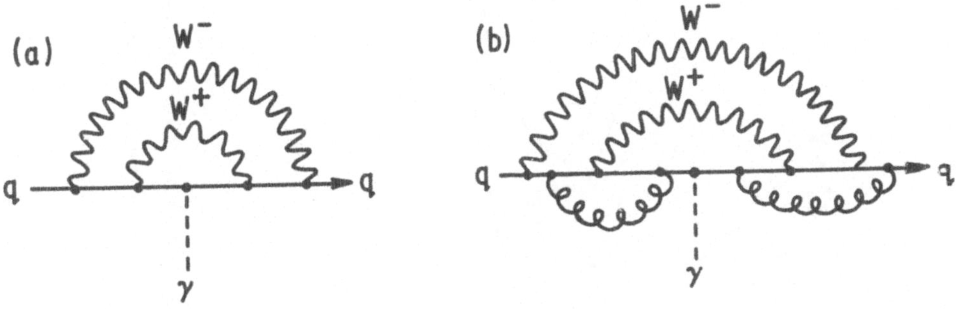

Fig.9. Graphs for calculation of the neutron electric dipole (E1) moment (a) for the simplest case, which gives zero, and (b) for the simplest non-vanishing graph obtained from it by adding QCD gluon exchanges.

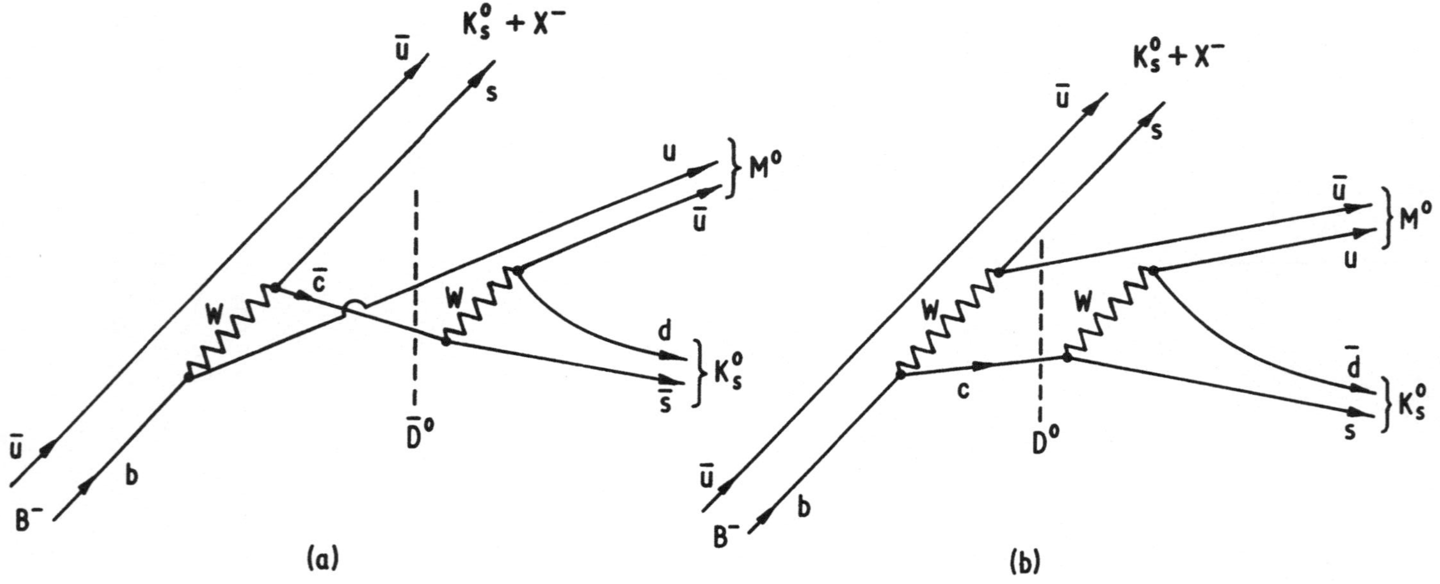

Fig.6. Graphs for the two sequences of Eqs. (5.42), leading from a B^- meson to the same final state $K_s^o K_s^o X_\alpha^- Y_\beta^o$.

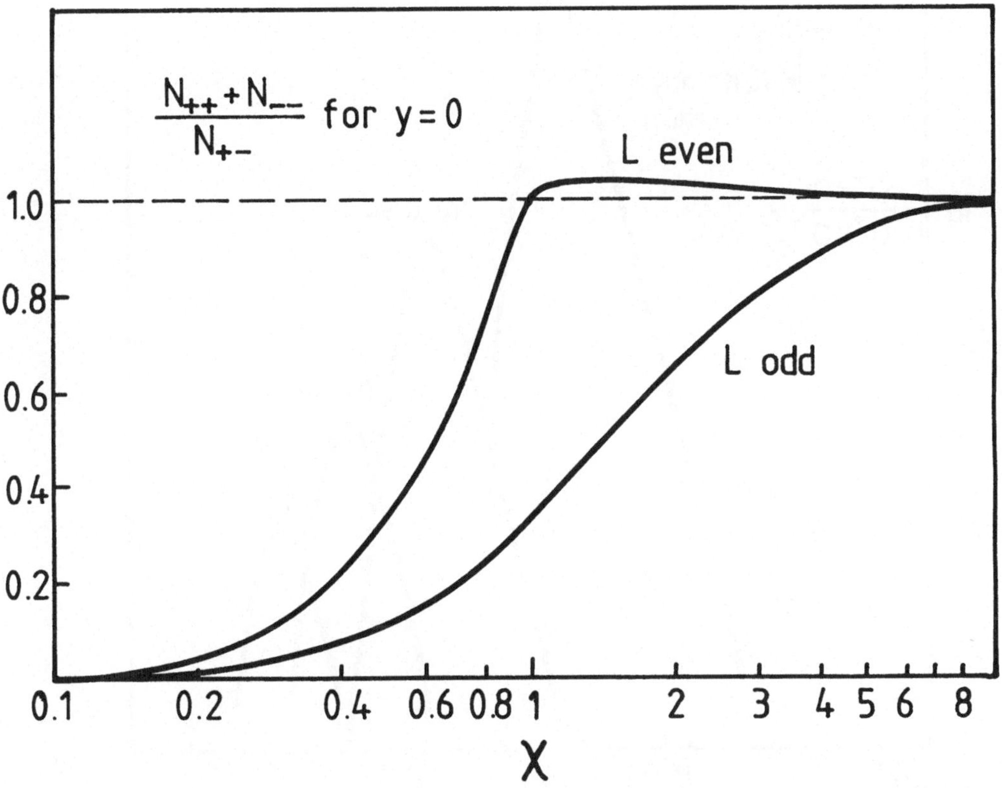

Fig.7. The di-lepton ratio$\{(\ell^+\ell^+)+(\ell^-\ell^-)\}/(\ell^+\ell^-)$ for a system initially $B^O\bar{B}^O$ is plotted, separately for the cases L = even and L = odd for its internal orbital angular momentum, as a function of $x = 2\delta m_B/\Gamma_B$, for the case $x = \delta\Gamma_B/\Gamma_B = 0$.

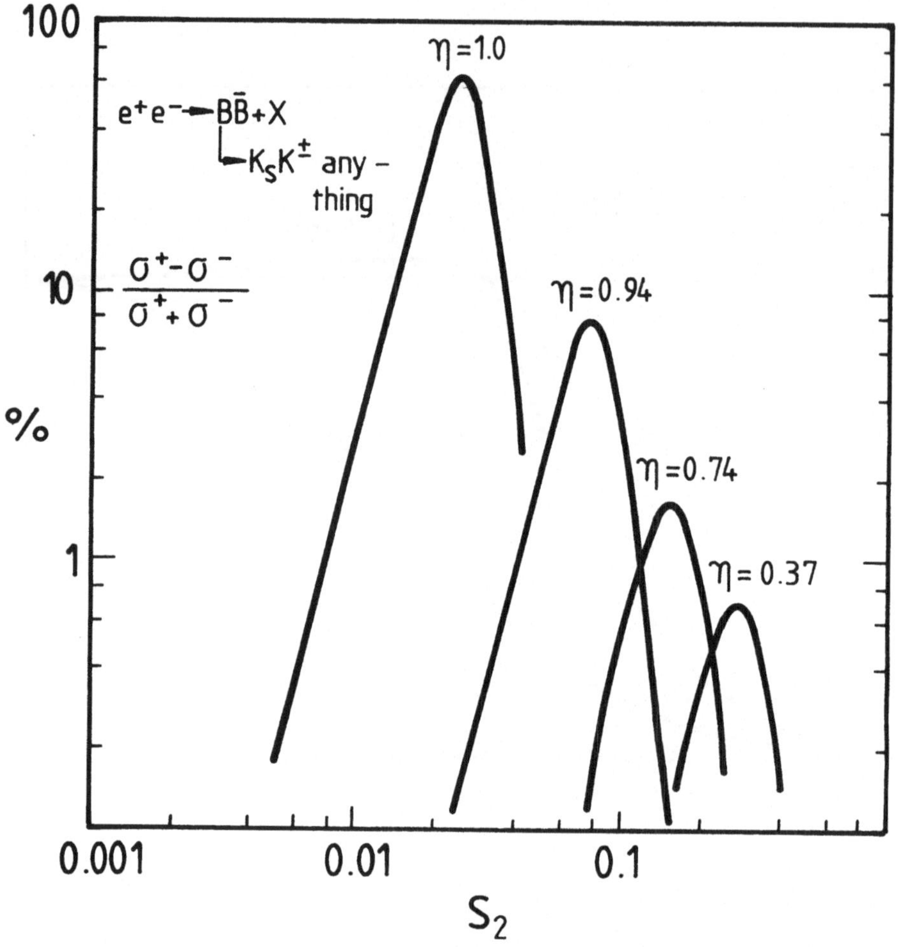

Fig.8. The K^+/K^- asymmetry calculated for a system initially $B_d^o B_d^o$ is plotted as a percentage for the ratio $(\sigma^+-\sigma^-)/(\sigma^++\sigma^-)$, as function of the KM parameter s_2, the other KM parameters being constrained to fit the data on the $(K^o,\bar{K}^o)$ complex, for a series of values for $\eta = \langle K^o|H_{Wk}|\bar{K}^o\rangle$.

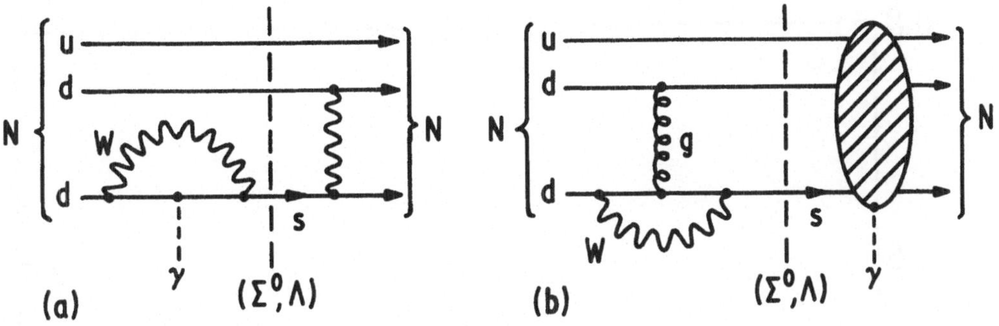

Fig.10. Graphs for the neutron E1 moment, discussed by Gavela et al. [59], for Σ^{o} and Λ intermediate states.

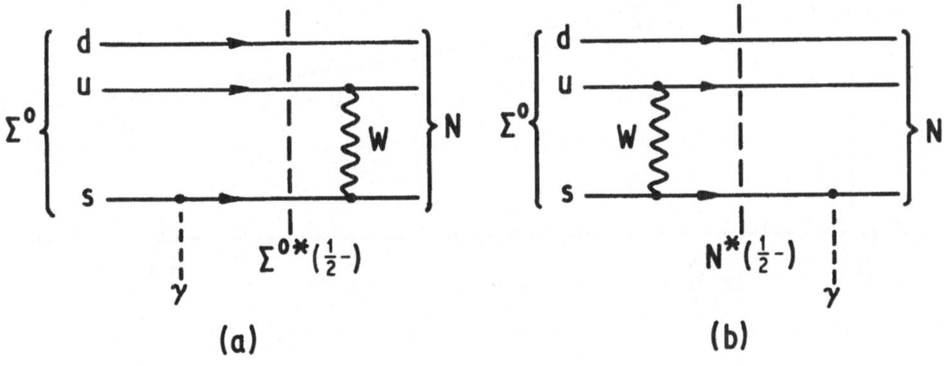

Fig.11. Graphs for the radiative transition $\Sigma^{o} \to N\gamma$, discussed by Gavela et al. [59], for intermediate states Σ^{*o} and Λ^{*o} with spin-parity $(1/2^{-})$.

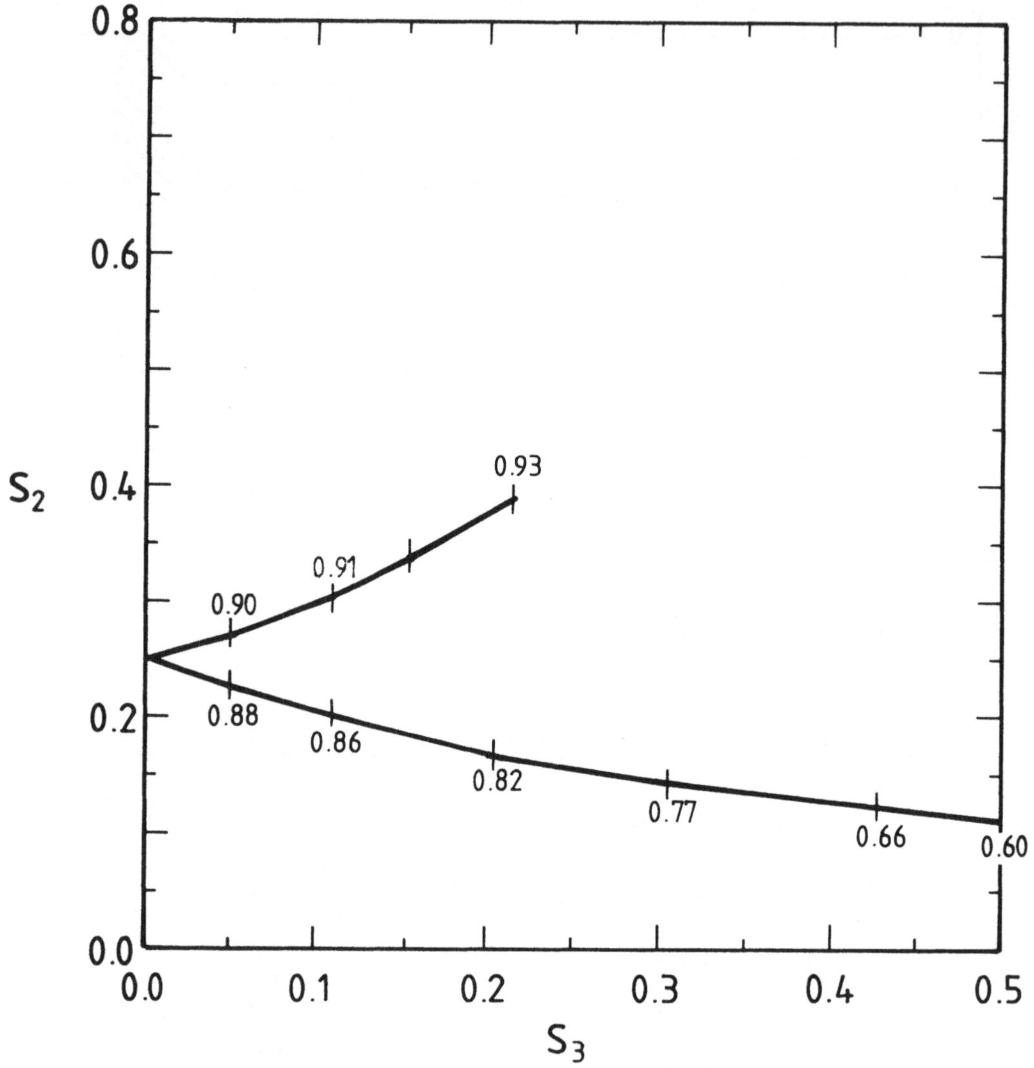

Fig.12. Shows the result of combining the constraints obtained by Pakvasa et al. [55] with the calculations of Barger et al. [52] on the $(K^0-\bar{K}^0)$ complex. The upper branch is for δ near π, the lower branch for δ near 0, and the number specified is the corresponding value for $|U_{cs}|^2$.

Acta Physica Austriaca

Supplementum XXIII

New Developments in Mathematical Physics

Edited by **H. Mitter** and **L. Pittner**

1981. 54 figures. VII, 701 pages.
Cloth DM 158,–, S 1130,–. ISBN 3-211-81676-3

Supplementum XXII

Field Theory and Strong Interactions

Edited by **P. Urban**

1980. 245 figures. V, 815 pages.
Cloth DM 166,–, S 1190,–. ISBN 3-211-81615-1

Supplementum XXI

Quarks and Leptons as Fundamental Particles

Edited by **P. Urban**

1979. 184 figures. V, 716 pages.
Cloth DM 149,–, S 1070,–. ISBN 3-211-81564-3

Springer-Verlag Wien New York